INFRARED AND MILLIMETER WAVES

VOLUME 1   SOURCES OF RADIATION

CONTRIBUTORS

E. AFFOLTER

THOMAS A. DETEMPLE

V. L. GRANATSTEIN

J. L. HIRSHFIELD

G. KANTOROWICZ

F. K. KNEUBÜHL

H. J. KUNO

K. MIZUNO

S. ONO

P. PALLUEL

P. SPRANGLE

ROBERT A. SMITH

INFRARED AND MILLIMETER WAVES

VOLUME 1 SOURCES OF RADIATION

Edited by **KENNETH J. BUTTON**

NATIONAL MAGNET LABORATORY
MASSACHUSETTS INSTITUTE OF TECHNOLOGY
CAMBRIDGE, MASSACHUSETTS

1979

ACADEMIC PRESS *New York San Francisco London*

A Subsidiary of Harcourt Brace Jovanovich, Publishers

ACADEMIC PRESS, INC.
111 Fifth Avenue, New York, New York 10003

United Kingdom Edition published by
ACADEMIC PRESS, INC. (LONDON) LTD.
24/28 Oval Road, London NW1 7DX

Library of Congress Cataloging in Publication Data
Main entry under title:

Infrared and millimeter waves.

Includes bibliographies.
  1. Infra−red apparatus and appliances.
2. Millimeter wave devices. I. Button, Kenneth J.
TA1570.I52   621.36'2   79−6949
ISBN 0−12−147701−0 (v. 1)

PRINTED IN THE UNITED STATES OF AMERICA

79 80 81 82   9 8 7 6 5 4 3 2 1

CONTENTS

Chapter 1 Gyrotrons

J. L. Hirshfield

Chapter 2 IMPATT Devices for Generation of Millimeter Waves

H. J. Kuno

Chapter 3 Pulsed Optically Pumped Far Infrared Lasers

Thomas A. DeTemple

Chapter 4   Backward Wave Oscillators

G. Kantorowicz and P. Palluel

Chapter 5   The Ledatron

K. Mizuno and S. Ono

Chapter 6   Infrared and Submillimeter-Wave Waveguides

F. K. Kneubühl and E. Affolter

Chapter 7   Free Electron Lasers and Stimulated Scattering from Relativistic Electron Beams

P. Sprangle, Robert A. Smith, and V. L. Granatstein

LIST OF CONTRIBUTORS

Numbers in parentheses indicate the pages on which the authors' contributions begin.

E. AFFOLTER (235), *Infrared Physics Group, Solid State Physics Laboratory, ETH, Zurich, Switzerland*

THOMAS A. DeTEMPLE (129), *Electro-Physics Laboratory, Department of Electrical Engineering, University of Illinois, Urbana, Illinois 61801*

V. L. GRANATSTEIN (279), *Plasma Physics Division, Naval Research Laboratory, Washington, D.C. 20375*

J. L. HIRSHFIELD (1), *Department of Engineering and Applied Science, Yale University, New Haven, Connecticut 06520*

G. KANTOROWICZ (185), *Electron Tube Division, Thomson-CSF, Velizy-Villacoublay, France*

F. K. KNEUBÜHL (235), *Infrared Physics Group, Solid State Physics Laboratory, ETH, Zurich, Switzerland*

H. J. KUNO (55), *Hughes Aircraft Company, Torrance, California 90509*

K. MIZUNO (213), *Research Institute of Electrical Communication, Tohoku University, Sendai, Japan*

S. ONO (213), *Research Institute of Electrical Communication, Tohoku University, Sendai, Japan*

P. PALLUEL (185), *Electron Tube Division, Thomson-CSF, Velizy-Villacoublay, France*

P. SPRANGLE (279), *Plasma Physics Division, Naval Research Laboratory, Washington, D.C. 20375*

ROBERT A. SMITH (279), *Jaycor, Alexandria, Virginia 22304*

PREFACE

In an emerging technology, textbooks and monographs may not become available for a number of years. In the meantime, a worker undertaking the study of a new subject finds his only introduction to be in the form of journal letters and short articles that would be difficult to understand even if much of this literature were not obsolete. The present volumes on infrared, far-infrared, submillimeter, and millimeter waves are intended to serve as personal textbooks for the professional engineer and scientist. The professional will wish each chapter to stand alone so that he will not be required to study the entire volume at the first encounter. The professional may also wish his book to have a theme—in this case, sources of radiation. Additional volumes will soon appear dealing with instrumentation and techniques, interaction of radiation with matter, and particular emphasis will be given to volumes on millimeter wave propagation, components, and systems. The nature of this sort of treatise is such that the emphasis will fall on new material, but an attempt will be made to avoid *perishable* new material. Somewhat older topics will also be reviewed, not for the sake of completeness, but to pursue the theme of the volume when the older material cannot easily be acquired by references to other sources.

Subsequent volumes will include new advances as well as basic material. Often new material first emerges at traditional conferences, such as the periodic International Conference on Infrared and Millimeter Waves (which will always be referred to as the Submillimeter Wave Conference) and the annual IEEE Microwave Symposium; the Electromagnetic Wave Propagation Panel of NATO will also play a role. The meetings of Commission D (Physical Electronics) of the International Union of Radio Science (URSI) accommodates topical sessions that will surely become traditional because of the well-known URSI hospitality to new technologies and areas of science. Indeed it has been the generosity and hospitality of the IEEE Microwave Theory and Techniques Society, the IEEE Quantum Electronics and Applications Society, Commission D of the International Union of Radio Science, and the Optical Society of

America that have made it possible to bring the workers in these areas together during these recent, formative years.

This first volume contains chapters on the important modern sources of radiation available for use in the far-infrared and near-millimeter wavelength range of the spectrum. Operating principles and comparative performance are emphasized. All three types of device are represented: electron tubes, solid state devices, and optically pumped lasers. The emphasis has been placed on the "here and now" devices, giving priority to the principles of operation for those who wish to get the most performance for their money. Most of these sources of radiation are available commercially, but this book should permit the reader to decide whether to "build or buy."

T. A. DeTemple's chapter on the optically pumped laser is a presentation of the principles and performance of a variety of models of high power pulsed lasers. The gyrotron is perhaps the most promising high power source and is described by a pioneer in its development, J. L. Hirshfield, although Bott's bottle and the work of the Soviet scientists are fully recognized. The best solid state device for power generation is the IMPATT and H. J. Kuno, representing the Hughes leadership in this field, has prepared a scholarly treatment. It would be difficult to argue that the carcinotron (backward wave oscillator) is not the best tunable source of radiation in the submillimeter wave range of the spectrum and it is certainly the best established commercial device. It is available only from the French firm Thomson-CSF, so our chapter has been prepared by G. Kantorowicz and P. Palluel of Thomson-CSF. As you open this book for inspection, turn first to their Fig. 1, which compares the output performance of *all* of the far infrared and submillimeter sources of radiation. It is an honest comparison: It does not show the Thomson-CSF carcinotron producing the highest power at the highest frequency, but it is very near to the top of both parameters. For the future, the very near future, please focus your attention on the development of the relativistic electron beam devices that have been described in a "state of the art" chapter by the recognized leaders in the field, P. Sprangle, R. A. Smith, V. L. Granatstein, and their colleagues at the U.S. Naval Research Laboratory. Finally, a neglected source of radiation, the ledatron, is described in a brief chapter by K. Mizuno, the organizer of the 1978 Japanese Submillimeter Wave Conference. Mizuno and D. H. Martin produced an incomparable review on tunable sources of radiation that was published only a few years ago. Then there is the obvious omission of the extended interaction oscillator and amplifier. There is no need to include a chapter on this topic. This commercial device is available from Varian Associates of

Canada,* and their standard brochure is an excellent chapter on the subject. Its performance is compared with other sources of radiation in Fig. 1 of the chapter by Kantorowicz and Palluel.

Finally, there are the Gunn oscillator and other microwave sources of radiation that will be extended into the millimeter and far-infrared region of the spectrum as solid state materials are improved. These subjects will be covered thoroughly in what is now planned as Volume 5, which is devoted to millimeter wave techniques and systems, and is scheduled to be published about 1981.

Future volumes in this edited treatise will fill some of the obvious gaps in sources of radiation. The chapter on continuous wave optically pumped lasers from 2 to 40 μm is being written by the best available authors on this subject, and will appear in the next volume.

As an editor, I am grateful to the contributors to these volumes and to their employers, who provided the atmosphere for their professional activity. I wish to express my personal appreciation (1) to Professor Benjamin Lax and the Massachusetts Institute of Technology and (2) to the United States National Science Foundation, which supported my own research for so many years, thus making it possible for me to participate in this new technology. Finally, I thank my wife Athena Fillios-Button for her patience during weekends when I do things like this.

*Address: 45 River Drive, Georgetown, Ontario, Canada.

CONTENTS OF VOLUME 2

CHAPTER 1

Gyrotrons

J. L. Hirshfield[*][†]

I. Introduction

This review is intended as an extension of the historical survey on development of the cyclotron maser which was published in 1977 (Hirshfield and Granatstein, 1977). Thus, we will not present a complete review of the early work. However, a *precis* of the original linear cyclotron maser gain mechanism is given here, so that this review may be read by scientists who have not studied the foregoing literature in detail. The main points of emphasis shall be on summarizing recent developments in the U.S.A. in the design and construction of high-power, high-efficiency millimeter sources and in the nonlinear theory necessary for this design. Furthermore, we shall attempt to present a cogent summary of the very extensive Soviet work on the subject which led, in 1974, to announcement of millimeter and submillimeter

---

[*] Consultant: Plasma Physics Division, Naval Research Laboratory, Washington, D.C.

[†] Permanent address: Department of Engineering and Applied Science, Yale University, New Haven, Connecticut.

1

gyrotrons (Zaytsev *et al.*, 1974; Kisel' *et al.*, 1974) with kilowatt continuous outputs at efficiencies between 6 and 30%. Indeed, these two dramatic Soviet papers have provided important stimulus to the very active resurgence of interest in these devices in the U.S.A. and elsewhere. As shall be seen, however, these Soviet devices were not isolated results, but represented the culmination of a massive effort[1] dating from the publication by Gaponov (1959) of his discovery of the cyclotron maser gain mechanism, only slightly later than the publication in the West of the first results by Twiss (1958) and by Schneider (1959).

It is hoped that this review will provide present and future researchers and users of cyclotron masers with material in the field sufficient for their needs. Achievements over the next few years can be expected to be rapid, so that this work may be of only transitory value; since further development is only to be welcomed, so will the obsolescence of this review. In commenting on his theoretical prediction of the cyclotron maser gain mechanism in 1959, Jurgen Schneider wrote: "It does not appear unlikely that this effect could be used for a new type of maser, which would require no microwave 'pump' and no low-temperature operation." It is probably safe to say that this prophecy has indeed materialized, and more. Applications of gyrotrons are already at hand, principally for electron cyclotron resonance heating in Tokamaks (Temkin *et al.*, 1977; Manheimer and Granatstein, 1977; Wolfe *et al.*, 1978), and for advanced radar systems (Granatstein, 1977).

II.   The Linear Cyclotron Maser Gain Mechanism

As is by now well known, free electrons gyrating in a uniform magnetic field B can provide gain, leading to amplification of electromagnetic waves. The gain mechanism originates with the relativistic energy dependence of the electron gyrofrequency $\Omega = eB/m\gamma$, as evidenced by the relativistic factor $\gamma = W/mc^2$, where W is the total electron energy. As the electrons interact with an electromagnetic disturbance oscillating at radian frequency ω a phase-focusing effect arises: Those electrons whose initial phases favor energy *loss* to the radiation field experience $\Delta\Omega > 0$ and *advance* in phase, while those electrons whose initial phases favor energy *gain* from the radiation field experience $\Delta\Omega < 0$ and *recede* in phase. The tendency is therefore for an electron bunch to develop centered at a phase value, relative to the electric field phase, which depends upon ω/Ω. When $\omega/\Omega > 1$ this phase value is such that $-e\mathbf{v} \cdot \mathbf{E}$, the work done by the field on the electrons, is negative for more than half the electrons; the field energy grows at the

---

[1] I have counted a total of sixty-seven different Soviet authors of papers published on this subject between 1959 and 1978.

expense of the electrons. For $\omega/\Omega < 1$ the reverse situation prevails and the field energy decays.

This phase-focusing phenomenon is responsible for the instability of electromagnetic waves in a plasma with a sufficiently non-Maxwellian electron energy distribution $F(W)$, and for the phenomenon of negative radiation temperatures (Bekefi *et al.*, 1961). The energy distribution must have $dF(W)/dW > 0$ over a range of energy in order that the phase bunching not cancel off after integrating the effect over a finite spread of energies. It is possible to study this instability, and its companion the so-called Weibel (1959) instability, by appeal to plasma dispersion relation theory. In this introductory section we shall take an alternative approach, which we believe better clarifies the underlying physics of the interaction.

This approach models the interaction in as straightforward a way as possible, for a conventional oscillator configuration. A two-cavity amplifier can be modeled in a similar fashion (Wachtel and Hirshfield, 1966). We consider a simple cylindrical cavity resonator immersed in a uniform axial magnetic field, with an electron beam drifting through the cavity along the magnetic field. A suitably designed electron gun is assumed to have launched the electron beam with a distribution of velocities perpendicular and parallel to the magnetic field; the optimum form for this distribution is governed by nonlinear considerations, and will be discussed below. Perturbations in the vacuum spatial distribution due to the presence of the electrons is neglected (the beam is of sufficiently low current to permit this), as are perturbations due to apertures in the cavity for access by the beam and for coupling out the radiation. We have developed a linear model for this system (Hirshfield *et al.*, 1965), so that the *weak* coupling of the radiation fields to the electrons is assumed not to alter the electron energy distribution appreciably; the achievement of high efficiency for such an oscillator implies *strong* coupling, which we shall discuss below in the context of nonlinear theory.

The fields of the cavity (TE_{0mn} in this example) are

$$\mathbf{E}_1 = \mathbf{a}_\theta E_0 J_1(k_\perp r) \sin k_\perp z \cos \omega t$$

$$\mathbf{B}_1 = \mathbf{a}_r E_0(k_\perp/\omega) J_1(k_\perp r) \cos k_\perp z \sin \omega t \qquad (1)$$

$$- \mathbf{a}_z E_0(k_\perp/\omega) J_0(k_\perp r) \sin k_\perp z \sin \omega t$$

where, in terms of cavity length L and radius R, $k_\perp = \chi_{1m}/R$ and $k_\perp = n\pi/L$ with χ_{1m} the mth zero of the Bessel function $J_1(\chi)$. The linearized kinetic equation is integrated along unperturbed trajectories for the electrons $d\mathbf{v}/dt = -\Omega(\mathbf{v} \times \mathbf{a}_z)$, subject to given initial values, to give the perturbed part of the electron distribution function, and thence the *rf* current density $\mathbf{J}$ induced on the electron beam. The power flow $P = \frac{1}{2}\text{Re} \int d^3r \, \mathbf{J} \cdot \mathbf{E}$ from the fields to the electrons may then be calculated. The result of such a calculation is (Hirshfield et al., 1965).

$$P = (\pi e^2 E_0^2/4mk_\parallel^2)\rho(k_\perp R) \int_0^\infty dw \int_{-\infty}^\infty du \, f_0(u, w)(w^2/u^3)(2k_\parallel u/\omega)G_\omega(x)$$

$$\times [(\Omega/k_\parallel u)(u^2/w^2) + Q_\omega(x)] \tag{2}$$

where

$$Q_\omega(x) = x + \tfrac{1}{2}(d/dx)(\ln G_\omega)[x^2 - \Omega\omega/k_\parallel^2(c^2 - w^2 - u^2)]$$

with

$$x = (\Omega - \omega)/k_\parallel u, \quad G_\omega(x) = \cos^2(\pi x/2)/(1 - x^2)^2,$$

$$\rho(k_\perp R) = 2\pi \int_0^R dr \, rJ_1^2(k_\perp r)n(r),$$

where $n(r)$ is the (azimuthally symmetric) electron density in the beam.

Before evaluating (2) in any detail, it is instructive to analyze its structure. First one sees that the power absorbed from the fields is proportional to the product of beam current and stored cavity energy density [i.e., to $\rho(k_\perp R)E_0^2$]. Second, we see that P is positive (energy flow from fields to particles) unless $Q_\omega(x)$ is negative, since with the exception of $Q_\omega(x)$ all quantities in the expression are positive definite. The function $G_\omega(x)$ can be viewed as a line shape function; we have plotted this somewhat unusual function in Fig. 1, and for comparison the rather more standard line shape functions e^{-x^2} and $(1 + x^2)^{-1}$. The peculiar algebraic form of $G_\omega(x)$ comes from the orbit integration along the z-varying cavity fields. It should be emphasized that Eq. (2) represents the interaction of the electrons with both the electric and magnetic fields of the cavity; both interactions are indispensible to a complete description.

From the integrand of Eq. (2) we can define a one-electron response function

$$R_\omega(x) \equiv G_\omega(x)[(\Omega u/k_\parallel w^2) + Q_\omega(x)]$$

which is seen to be a nontrivial function of $G_\omega(x)$ and the parameters of the orbit (u, w, ω, Ω); $R_\omega(x)$ reflects the overall linear interaction of one electron of velocity (u, w) with the fields of the cavity. As we shall see, $R_\omega(x)$ can become negative in the vicinity of $x = -1$; depending upon the extent of this negative part of $R_\omega(x)$ and upon the particular beam distribution $f_0(u, w)$, the sign of P may be either negative or positive. Near $x = -1$ the second term in square brackets in $Q_\omega(x)$ dominates the expression. This second term can be traced directly to the relativistic dependence of the electron gyrofrequency Ω upon energy. (Indeed as $c \to \infty$ this term vanishes.) It is seen

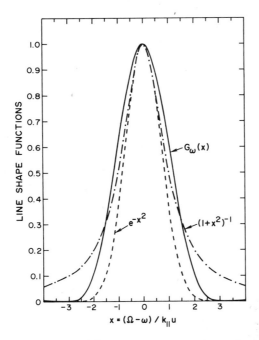

FIG. 1. Line shape function $G_\omega(x)$ for cyclotron resonance interaction of drifting electrons in a TE$_{011}$ cavity. For comparison the functions $(1 + x^2)^{-1}$ and $\exp(-x^2)$ are also shown, where $x = (\Omega - \omega)/k_\parallel u$.

then that, when this term dominates, oscillation would be expected for $x < 0$, i.e., for $\omega > \Omega$, since $dG_\omega(x)/dx > 0$ for $x < 0$.

If the relativistic energy dependence of the gyrofrequency had been over-looked in the derivation of Eq. (2), negative values of the response function $R_\omega(x)$ might still occur. In such a case the mechanism for gain originates with *axial* bunching of the electrons, brought about by combined action of the cavity electric and magnetic fields. A detailed discussion of this bunching mechanism, and the way in which it competes with the cyclotron maser *azimuthal* bunching, is provided in a recent paper (Chu and Hirshfield, 1978). It is sufficient to note that, in a fast-wave interaction (where a cavity or wave-guide determines frequency-to-wave number relationship), the azimuthal bunching mechanism dominates.

Equation (2) can be used to obtain a start-oscillation condition for the system, once the beam distribution function $f_0(u, w)$ is specified. Start-oscillation is the condition where power extracted from the electrons is just sufficient to balance the losses of the cavity, as designated by the cavity quality factor Q. (If Q is the loaded value, then this condition may also include the coupling out of a certain amount of power, so long as this does not lead

to any violation of our weak perturbation assumption taken at the outset.)
For the TE_{011} mode, this condition can be written

$$P = -0.0811(\pi R^2 L)(\omega \varepsilon_0 E_0^2/2Q) \tag{3}$$

where the numerical factor comes from integrating the field energy, from
Eq. (1), over the cavity volume.[2]

We will evaluate the start-oscillation current I_s for the idealized distribution $f_0(u, w) = (2\pi w_0)^{-1} \delta(w - w_0) \delta(u - u_0)$, i.e., where all electrons have
identical values of parallel and of perpendicular velocity. For a thin annular
beam, positioned at radius a, the beam current is $I = eu_0 J_1^{-2}(k_\perp a)\rho(k_\perp R)$.
Therefore, after evaluating Eq. (2) for this distribution, we find from Eq. (3)

$$I_s Q R_\omega(x_0) = -[0.1622\pi^4/J_1^2(k_\perp a)][(R^2/L^2)\alpha^2](u_0 c/w_0^2)(u_0^2/c^2)(mc^3\varepsilon_0/e) \tag{4}$$

where x_0 designates the value of x with $u = u_0$ and $w = w_0$ at which $R_\omega(x)$
has its minimum (i.e., negative) value, and where $\alpha^2 = 1 + k_\perp^2/k_\parallel^2 = 1 +
\chi_\parallel^2 L^2/\pi^2 R^2$. For an annular beam positioned at the maximum of the Bessel
function $J_1(k_\perp a) = 0.5819$, the starting current is minimized and the above
expression becomes

$$I_s Q R_\omega(x_0) = -5.065 \times 10^4[(R^2/L^2)\alpha^2](u_0 c/w_0^2)(u_0^2/c^2) \quad \text{A.}$$

In addition to the two parameters (R/L) or α and $(u_0 c/w_0^2)$, we find that a
third one $(u_0/c)^2$ enters in determining the start-oscillation current, once the
cavity Q is specified.

In Fig. 2 we have plotted the response function $R_\omega(x)$ versus x for three
values of cavity length-to-radius ratio L/R (1, 2, and 5) and for several values
of $\beta = u_0 c/w_0^2$, the parameter which measures the inverse of the free energy
available to drive the process $(mw_0^2/2)$ times the time available for the interaction to proceed (L/u_0). Examination of the form of $R_\omega(x)$ reveals that for
$\alpha < 3$ there is a critical value of β below which $R_\omega(x)$ becomes negative in
the interval $0 > x > -3$. For $\alpha > 3$, $R_\omega(x)$ will always be negative in this
interval. Therefore, for $\alpha < 3$, one may define a critical value β_c for which
$R_\omega(x_c) = R'_\omega(x_c) = 0$; for $\beta < \beta_c$ oscillations are possible. The cross marks on
Figs. 2a,b indicate the points x_c at which $R_\omega(x_c) = R'_\omega(x_c) = 0$. The horizontal
lines indicate the minimum possible value of $R_\omega(x)$, i.e., its minimum value
for $\beta = 0$. Calculations have recently appeared extending these numerical
results to resonators two and three half-wavelengths long (Symons and
Jory, 1977).

In Table I we have provided a few values of starting current I_s for the
parameters used in Fig. 2. From this table, for example, we see, for a cavity

---

[2] A factor-of-two numerical error appears in the expression for the start-oscillation current
in the Hirshfield and Granatstein (1977) reference.

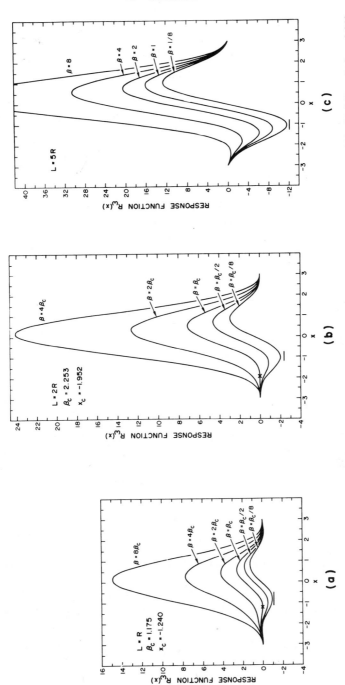

FIG. 2.   Response functions $R_\omega(x)$ for three values of cavity aspect ratio. (a) $L/R = 1$, $\beta_c = 1.175$, and $X_c = -1.240$; (b) $L/R = 2$, $\beta_c = 2.253$, and $X_c = -1.952$; and (c) $L/R = 5$. The short horizontal lines indicate the minimum values of $R_\omega(x)$; the cross marks [(a) and (b) only—see text] indicate the point at which $R_\omega(x)$ first becomes negative.

TABLE I

VALUES OF $R_{\omega\,\mathrm{min}}$ AND STARTING CURRENT FOR
SELECTED T_{011} CAVITY LENGTH-TO-RADIUS RATIOS

$L = R(\alpha = 1.577), \beta_{\mathrm{c}} = 1.1751$

β/β_{c}	$R_{\omega\,\mathrm{min}}$	$I_{\mathrm{s}}Qc^2/u_0^2$ (amps)
1.0	0	—
0.5	−0.5039	14.69 × 10⁴
0.25	−0.8009	4.62 × 10⁴
0.125	−0.9580	1.93 × 10⁴
0	−1.075	0

$L = 2R(\alpha = 2.636), \beta_{\mathrm{c}} = 2.2526$

β/β_{c}	$R_{\omega\,\mathrm{min}}$	$I_{\mathrm{s}}Qc^2/u_0^2$ (amps)
1.0	0	—
0.5	−0.8376	11.83 × 10⁴
0.25	−1.566	3.16 × 10⁴
0.125	−1.978	1.25 × 10⁴
0	−2.452	0

$L = 5R(\alpha = 6.1798)$

β	$R_{\omega\,\mathrm{min}}$	$I_{\mathrm{s}}Qc^2/u_0^2$ (amps)
4.0	−2.774	11.2 × 10⁴
2.0	−6.138	2.52 × 10⁴
1.0	−8.783	8.81 × 10³
0.5	−10.358	3.73 × 10³
0.25	−11.215	1.72 × 10³
0.125	−11.644	8.31 × 10²
0	−12.086	0

with $L = 5R$ and β about unity, that one requires a starting current of about $10^4 Q^{-1}(u_0/c)^2 = 19.6 Q^{-1} V_{\parallel}$ A, where $V_{\parallel}$ is the parallel beam energy in kilovolts.

An approximate analytic formula for the starting current can be obtained in the limit $\alpha \gg 1$, i.e., for $L \gg R$, since in this limit (provided β is not too large) $R_{\omega}(x) \simeq -(\alpha^2/2)(d/dx)G_{\omega}(x)$. Now the maximum value of $dG_{\omega}(x)/dx$ is $\pi^2/16$, at $x = -1$. Therefore, we can approximate

$$I_{\mathrm{s}} = 1.642 \times 10^5\, Q^{-1}(R/L)^2(u_0 c/w_0^2)(u_0/c)^2 \quad \mathrm{A}$$

TABLE II

PARAMETERS FOR A KILOWATT LEVEL 8 mm CYCLOTRON MASER
OSCILLATOR, BASED ON LINEAR ANALYSIS[a]

Operating frequency	37.5 GHz ($\lambda = 8.0$ mm)
Cavity Q	1000
Cavity length	2.472 cm
Cavity radius	0.494 cm
Annular electron beam radius	0.237 cm
Annular electron beam thickness	0.05 cm
Magnetic field	13.8 kG ($\Omega = 0.984\omega$)
Starting current	0.131 A
Operating current	0.33 A
Beam voltage	25 kV ($\gamma = 1.05$)
Parallel energy	5 kV
Output power[a]	2000 W
Assumed perpendicular efficiency[a]	30%

[a] The values of efficiency and output power are estimates.

As a first approximation for the design of a kilowatt level oscillator for a wavelength of 8 mm, we have used the results of the above analysis to give the parameters shown in Table II.

Table II is not intended as a final design guide by any means, but only is given to show that the orders of magnitude are not overly exotic. Clearly, one must employ the nonlinear theory to calculate the expected efficiency, and thus output power, in an actual design; but we shall see that the above estimates are not far off. Indeed, careful design of the resonator structure, including higher mode operation and higher cyclotron harmonic operation, have enabled overall efficiencies of 30% to be achieved in a Soviet cw device operating at 2.78 mm, with a 12-kW power output (Zaytsev et al., 1974). The Naval Research Laboratory pulsed amplifier operating at 35 GHz is expected to have an overall efficiency of 50%, with a power output of about 300 kW (Granatstein, 1977).

In subsequent sections of this review, we shall outline the steps necessary to move from the linear theory of this section to actual design of devices with parameters in the ranges mentioned.

III. Nonlinear Considerations

The linear analysis presented in the preceding section has been shown to be exceedingly useful in providing order-of-magnitude estimates for the parameters of an actual millimeter-wave oscillator. Parameters for a proto-type amplifier could have been determined in a similar way. But a number of deficiencies are evident: (a) The ultimate efficiency was unknown; (b) no

optimum resonator configuration or choice of mode was dictated; and
(c) no optimum beam energy distribution function was specified.

It can be easily demonstrated that our estimate for the perpendicular
efficiency was not totally out of line. Moreover, it will be evident at the same
time that optimum overall beam energies are in the range $W \ll mc^2$, or less
than about 100 keV. Thus our choice of a 25-kV beam was motivated by more
than considerations of technical simplicity.

From Fig. 2c, corresponding to the resonator aspect ratio $L/R = 5$ chosen
for the example, one sees that the curve for $\beta = 2$ has a half-width for the
negative (gain) portion of $\Delta x = x_2 - x_1 \simeq -1$, where x_2 and x_1 are the
final and initial values of x for an electron passing through the resonator.
Since ω is fixed by the narrow response width of the resonator, and since $k_\parallel u$
is constant if the wave–particle interaction only extracts energy from the
perpendicular electron motion, we have $\Delta x = \Delta \Omega / k_\parallel u = -(\Delta \gamma / \gamma)(\Omega / k_\parallel u)$.
Thus there is a limit to the relative energy loss $\Delta \gamma / \gamma$ which an electron can
suffer before the excursion in x brings it out of the gain portion of the response function $R_\omega(x)$. For the example cited above $\Omega / k_\parallel u = 30$, so that
$|\Delta \gamma| < 0.033$ in order to remain within the gain portion of the response
function. A perpendicular efficiency of 30% with $\gamma_\perp = 1.05$ corresponds to
$\Delta \gamma = 0.0128$ or less than half the maximum allowed value, according to the
simple criterion $|\Delta x| < 1$. We can also conclude that highest efficiency
would be expected for $\gamma \sim O(1)$, since $\Delta \gamma / \gamma = \Delta W / (W + mc^2)$ is minimized
for $W \ll mc^2$. It may be remarked in passing that one way of increasing the
efficiency would be to provide a slowly varying magnetic field so that Ω
could remain constant in the electron's transit through the resonator,
despite changes in γ.

It remains for a rigorous nonlinear theory to determine if the full excursion
$|\Delta \gamma| = 0.03$ could be realized in practice for our example, or if, in fact, energy
depletion causing this shift $\Delta \gamma$ is the only saturation mechanism. Moreover
optimization of the beam distribution function, the resonator profile, and
the choice of mode all require a more careful theoretical justification. In
this section we will outline a number of theoretical contributions which have
been made in recent years to bring about these optimizations.

A. Optimization of Parameters for a Cyclotron Maser Amplifier

We shall begin by summarizing a recently completed analytical/numerical
study (Chu et al., 1977) of optimization of parameters for a cyclotron maser
(gyrotron) traveling wave amplifier. By presenting this advanced result first
in our survey of the nonlinear theoretical considerations we hope to accomplish two objectives: first, to demonstrate the power of available techniques in providing solid design criteria; second, to briefly convey the flavor
of a calculation relevant to an amplifier, in contrast to the linear theory for an

oscillator presented in Section II. Following this presentation of optimization of amplifier parameters, we shall present background material fundamental to the nonlinear physics underlying saturation in cyclotron maser devices.

The study by Chu et al. (1977) was for the purpose of optimizing design parameters for the Naval Research Laboratory prototype amplifier, which is intended to produce 340-kW pulsed power at 35 GHz (Granatstein, 1977). The study models the amplifier geometry as an unterminated smooth cylindrical TE_{01} waveguide in a uniform axial magnetic field penetrated by a coaxial annular cold electron beam. If the beam density is low enough to allow the waveguide fields to retain the forms they assume in the absence of the beam, the linear dispersion relation for the system may be derived. It is

$$\omega^2 - k_{\parallel}^2 c^2 - \omega_0^2 = \frac{\omega_p^2}{\gamma} \left[\left(\frac{\omega}{\omega - \Omega} \right) N\left(\frac{r_L}{r_w}, \frac{r_0}{r_w} \right) \right.$$
$$\left. - \frac{w^2}{c^2} \frac{(\omega^2 - k_{\parallel}^2 c^2)}{(\omega - \Omega)^2} M\left(\frac{r_L}{r_w}, \frac{r_0}{r_w} \right) \right], \qquad (5)$$

where ω_0 is the waveguide cutoff frequency, and ω_p the plasma frequency. For convenience, this dispersion relation is written for waves propagating as $\exp[i(k_{\parallel} z - \omega t)]$ in a frame of reference in which the beam's axial velocity u is zero (i.e., the so-called "beam" frame). The variables have the same definitions as in Section II, except for the aforementioned specialization to the beam frame; transformation to the laboratory frame will alter the quantities, but this transformation need not be taken into account in order to follow our condensed presentation. The functions $N(r_L/r_w, r_0/r_w)$ and $M(r_L/r_w, r_0/r_w)$ are dimensionless combinations of geometrical factors determined by the waveguide radius r_w, the annular beam location r_0, and the electron gyration radius r_L.

The first term on the right-hand side of Eq. (5) is analogous to the first term of $R_\omega(x)$ from Section II, namely $(\Omega u/k_{\parallel} w^2)G_\omega(x)$, while the second term on the right-hand side of Eq. (5) is analogous to $G_\omega(x)Q_\omega(x)$; this second term leads to instability, or growth.

A first point is to find the approximate range of values of ω and $k_{\parallel}$ where highest efficiency would be expected to occur. The phenomena limiting efficiency are energy depletion, as discussed in the introduction of this section, and particle trapping, discovered by Sprangle and Drobot (1977), and discussed below. These phenomena are minimized near the so-called "point of grazing incidence," i.e., the point in the ω–$k_{\parallel}$ plane at which the dispersion curve for the empty waveguide mode

$$\omega^2 - k_{\parallel}^2 c^2 - \omega_0^2 = 0$$

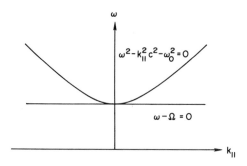

FIG 3. Dispersion curves for the empty waveguide mode and for the cold beam cyclotron mode (beam frame) shown under condition for grazing incidence ($k_\parallel = 0$). [From Chu *et al.* (1977).]

intersects the dispersion curve for the cold beam cyclotron mode ($u = 0$)

$$\omega - \Omega = 0$$

with equal phase and group velocities. Figure 3 portrays this intersection for these two modes, while Fig. 4 shows actual evaluations of a dispersion function similar to Eq. (5). It is easy to show that, at the point of grazing incidence ($k_\parallel = 0$), the sensitivity of solutions of the linear dispersion relation to small changes in Ω or $k_\parallel$ is minimized. These small changes enter as nonlinear effects arise. A further virtue of analyzing the system where phase and group velocities are equal is that it allows straightforward transformation of a temporal growth rate ω_i into a spatial growth rate $k_i = \omega_i/v_\parallel$.

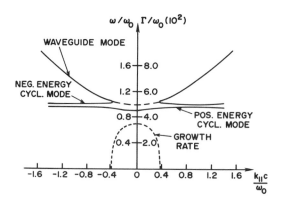

FIG. 4. Typical dispersion curve for cyclotron maser instability. This case is for $\gamma_{0\perp} = 1.2\omega_b/\sqrt{\gamma_{0\perp}} = 0.05\omega_0$, $\omega_0 = ck_n = \Omega/\gamma_{0\perp}$ and $l = n = 1$. [From Sprangle and Drobot (1977).]

Since grazing incidence occurs at $k_\parallel = 0$, we see trivially that optimum operation will be very close to the waveguide cutoff $\omega = \omega_0 = \Omega$. A parameter $y = \Omega/\omega_0$ (in the beam frame) is introduced, which is then expected to take on a value close to unity at the point of optimum operation. The second crucial parameter in the optimization is the beam density (which relates to the beam current in the lab frame) as measured by the parameter v, equal to the number of beam electrons per classical electron radius $e^2/4\pi\varepsilon_0 mc^2$ of beam length. Other adjustable parameters are the mean radius of the annular beam r_0, and the transverse (i.e., beam frame) value of γ. Further specification of the axial beam velocity is also required, but here one is at the mercy of unavoidable space-charge forces in the electron gun, which tend to produce increased velocity spread on the beam as ξ, the ratio of perpendicular-to-parallel electron velocity, is increased; this velocity spread is deleterious to achievement of high gain and efficiency. Existing numerical design codes for magnetron injection guns used in prototype gyrotron designs indicate that it is imprudent to undertake a design with ξ greater than about 1.5. Thus the beam density parameter v is determined once the design value of beam power is specified. If the beam radius r_0 is chosen to be the value for which linear gain is maximized (i.e., $r_0 = 0.48r_w$), then there remain two parameters, γ and y, which may be varied to optimize the amplifier efficiency.

To bring about this two-parameter optimization, the single-wave numerical simulation code of Sprangle and Drobot (1977) was converted to cylindrical geometry. The simulated wave–particle interaction could thus be followed until the wave amplitude saturated. At this point the ratio of wave energy to initial beam energy gives the efficiency, provided the device is designed to terminate the wave–particle interaction at this point. This simulation process was repeated as values of γ and y were scanned, so as to find the optimum set of parameters.

Results of this scan process are shown in Figs. 5 and 6. Figure 5 shows the values of beam frame efficiency η found as the value of (beam frame) γ was varied. Two curves are shown, one for $y = 1$, and one for y itself optimized at each value of γ. While a beam frame efficiency of 40% was achieved for the $y = 1$ case, it is seen that the optimized choice of y allows values of over 70% to be achieved. This optimization of y is a delicate matter however, as shown in Fig. 6. Here, for a fixed value of $\gamma = 1.10$ one sees that the efficiency increases steeply as y is reduced slightly below unity, but below $y = 0.956$ in this case the amplifying interaction ceases and the efficiency plummets to zero. This sharp break point can be predicted from linear theory; it corresponds, for an amplifier, to the point where $R_\omega(x) = 0$, i.e., where the interaction switches from gain to loss.

Based upon such numerical simulation scans for optimization of design parameters, the Naval Research Laboratory group has built the device shown

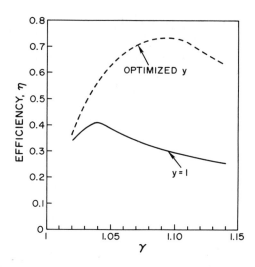

FIG. 5.   Efficiency vs. γ for optimized and nonoptimized values of y. [From Granatstein and Godlove (1978).]

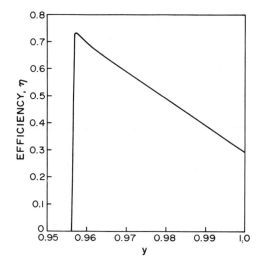

FIG. 6.   Efficiency vs. y for $\gamma = 1.10$ showing critical dependences as y approaches its optimum value (for $y < 0.956$ linear theory predicts no gain). [From Chu *et al.* (1977).]

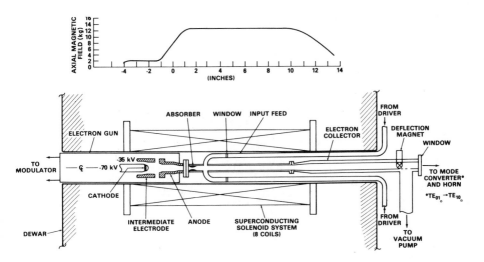

FIG. 7. Sketch of the planned NRL Gyro-Traveling Wave Amplifier (TE_{01}° mode 35 GHz operating at the fundamental of the cyclotron frequency). The solenoid is of the superconducting type. [From Granatstein and Godlove (1978).]

schematically in Fig. 7. Figure 8 shows the electrode configuration and some typical electron trajectories for the electron gun designed for use in the device. The final design parameters are listed in Table III.

B. FUNDAMENTAL STRUCTURE OF THE NONLINEAR THEORY

In order to evolve nonlinear calculations leading to design parameters by Sprangle and Drobot (1977). These authors made one crucial assumption which allowed this seemingly intractable problem to be attacked: They assumed that the electromagnetic fields could be represented as a single pure mode of the waveguide; mode–mode coupling was taken to be absent. Thus, for the model parallel-plate waveguide system analyzed by Spangle and Drobot (1977) (see Fig. 9 for the relevant coordinates) the electric field in the cutoff frame took the form

$$\mathbf{E}(x, y, t) = \mathbf{e}_y E_y(x, t) = \mathbf{e}_y E_0(t) \cos[\omega_0 t + \alpha(t)] \sin[k_n(x - a)] \quad (6)$$

where the waveguide boundaries are at $x = \pm a$, and $k_n = n\pi/2a$. In this representation ω_0 is constant, but the amplitude $E_0(t)$ and phase $\alpha(t)$ are slowly varying quantities, e.g.,

$$E_0(t) = E_0 \exp \int_0^t dt' \, \Gamma(t') \quad (7)$$

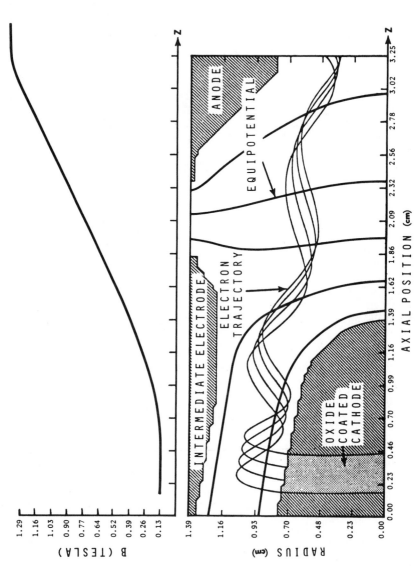

FIG. 8. Computer design of electron trajectories for 35 GHz distributed amplifier, showing outlines, cathode, intermediate electrode, and anode. [From Granatstein and Godlove (1978).]

TABLE III

DESIGN PARAMETERS FOR THE U.S. NAVAL
RESEARCH LABORATORY GYROAMPLIFIER[a]

v (density parameter)	2.06×10^{-3}
V (electron energy)	70.82 keV
I_b (beam current)	9.48 A
η (efficiency)	51.0%
P_b (beam power)	671.5 kW
P_W (wave power)	342.5 kW
B_0 (magnetic field)	12.87 kG
$k_{\parallel}$	1.96 cm^{-1}
r_w	5.37 mm
r_0	2.52 mm
r_L	0.61 mm
$v_{\perp}/c$	0.401
$v_{\parallel}/c$	0.268
G (power gain)	2.0 dB/cm
$\Delta\omega/\omega$ (3 dB bandwidth with center gain of 20 dB)	11%

[a] Chu et al. (1977); Chu and Drobot (1978).

where $\Gamma(t)$ is the time-dependent growth rate [$\alpha(t)$ is real]. For early times $\Gamma(t)$ is constant and is determined by solution of the linear dispersion relation, i.e., $\Gamma(0^+) = \Gamma_L$, while $\alpha(0^+) = (\Delta\omega_L)t$. Since k_n is fixed (and since k_z is zero in the cutoff frame), the linear dispersion yields $\omega(k_n) = \omega_0 + \Delta\omega_L + i\Gamma_L$.

The nonlinear particle–wave simulations of Sprangle and Drobot (1977) involve solutions of the particle orbit equations coupled, via the current

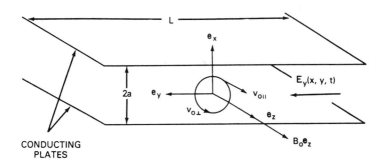

FIG. 9.   The electron cyclotron maser configuration in plane geometry. [From Sprangle and Drobot (1977).]

density, to the wave equation. For the electric field given by Eq. (6) the associated vector potential is

$$A_y(x, t) = -\omega_0^{-2}\{(\omega_0 - \dot{\alpha}) \sin[\omega_0 t + \alpha(t)]$$
$$+ \Gamma(t) \cos[\omega_0 t + \alpha(t)]\}E_0(t) \sin[k_n(x - a)] \quad (8)$$

valid to first order in the assumed small parameters $\dot{\alpha}(t)/\omega_0$ and $\Gamma(t)/\omega_0$. (The dot indicates time differentiation.) The wave equation for $A_y(x, t)$ is

$$c^{-2}[\partial^2 A_y(x, t)/\partial t^2 - \partial^2 A_y(x, t)/\partial x^2] = \mu_0 J_y(x, t). \quad (9)$$

Substituting Eq. (8) into Eq. (9) and again retaining only first-order terms in the small quantities gives

$$\{[\omega_0(\omega_0^2 - k_n^2 c^2) + \dot{\alpha}(\omega_0^2 + k_n^2 c^2)] \sin(\omega_0 t + \alpha)$$
$$- (\omega_0^2 + k_n^2 c^2)\Gamma \cos(\omega_0 t + \alpha)\}E_0 \sin[k_n(x - a)] = \omega_0^2 \varepsilon_0^{-1} J_y(x, t) \quad (10)$$

where, for a sheet charge of surface density σ_0, the current density is

$$J_y(x, t) = (\sigma_0 e/2\pi) \int_0^{2\pi} d\phi \, v_y(\phi, t) \, \delta[x - x(\phi, t)] \quad (11)$$

with ϕ the orbit phase angle (measured from the x axis) and $x(\phi, t)$ the x coordinate of the particle. This simple nonlinear form for the current density can be written since it is assumed that the electron beam has no spread in velocity initially; thereafter every particle's velocity is determined only by its orbit angle, and once the orbit is determined, the instantaneous orbit angle is also, by the initial value ϕ_0 of this angle. We can thus substitute the initial value ϕ_0 for ϕ in Eq. (11).

The wave Eq. (9) is then separated into equation for $\dot{\alpha}(t)$ and $\Gamma(t)$ alone. This is accomplished by multiplying Eq. (10) by $\sin[k_n(x - a)]$, integrating over the waveguide cross section, and performing the in-phase and quadrature rapid time scale averages. That is, one performs the two operations

$$\int_{-a}^{a} dx \sin[k_n(x - a)] \int_t^{t + 2\pi(\omega_0 + \dot{\alpha})^{-1}} dt' \begin{Bmatrix} \sin[\omega_0 t' + \alpha(t')] \\ \cos[\omega_0 t' + \alpha(t')] \end{Bmatrix} \quad \text{[Eq. (10)]}$$

to obtain the two slow time scale equations

$$\dot{\alpha}(t) = \frac{\omega_0}{2\pi a \varepsilon_0 E_0(t)} \int_t^{t + 2\pi(\omega_0 + \dot{\alpha})^{-1}} dt' \sin(\omega_0 t' + \alpha)$$
$$\times \int_{-a}^{a} dx \sin[k_n(x - a)]J_y(x, t') \quad (12)$$

and

$$\Gamma(t) = -\frac{\omega_0}{2\pi a \varepsilon_0 E_0(t)} \int_t^{t + 2\pi(\omega_0 + \dot{\alpha})^{-1}} dt' \cos(\omega_0 t' + \alpha)$$

$$\times \int_{-a}^a dx \, \sin[k_n(x - a)] J_y(x, t'). \tag{13}$$

It may be noted that introduction of a linear lossy dielectric into the waveguide, such that an in-phase current

$$J_y(x, t) = \sigma_r E_0 \cos(\omega_0 t + \alpha) \sin[k_n(x - a)]$$

was present, would give a decay of the electric field with a decay constant $\Gamma = \sigma_r \omega_0^2/\varepsilon_0(c^2 k_n^2 + \omega_0^2) \simeq \sigma_r/2\varepsilon_0$. (Here σ_r is the real part of an rf conductivity.) The equivalent of a *negative* real part of the conductivity is needed for growth.

The formal expressions Eqs. (12) and (13) can only be dealt with after the orbit equation is solved, so as to allow explicit substitution for $J_y(x, t)$. This can be done using the equation of momentum balance for the electrons, together with the definition Eq. (11). Toward this end, one writes the momentum balance equation for the normalized velocity variable $\beta = \beta_x + i\beta_y$, with $\beta_x = v_x/c$ and $\beta_y = v_y/c$, as follows:

$$d\beta/dt = (i\Omega/\gamma_\perp)\beta + (ie/\gamma_\perp m)\beta(\partial A_y/\partial x) + (ie/\gamma_\perp mc)[1 - \tfrac{1}{2}\beta(\beta - \beta^*)]\partial A_y/\partial t \tag{14}$$

and

$$dx/dt = (c/2)(\beta + \beta^*), \qquad \text{with} \qquad \gamma_\perp = (1 - \beta\beta^*)^{-1/2}.$$

These equations admit a solution of the form

$$\beta(\phi_0, t) = (\beta\beta^*)^{1/2} \exp[i \tan^{-1}(\beta_y/\beta_x)]. \tag{15}$$

Substituting Eq. (15) into Eq. (14) and equating real and imaginary parts gives

$$d\beta_\perp/dt = (e/\gamma_\perp^3 mc)(\partial A_y(x, t)/\partial t) \sin[\phi(\phi_0, t)] \tag{16}$$

and

$$d\phi/dt = (\Omega_0/\gamma_\perp) + (e/\gamma_\perp mc\beta_\perp)(\partial A_y(x, t)/\partial t) \cos[\phi(\phi_0, t)]$$
$$+ (e/\gamma_\perp m)(\partial A_y(x, t)/\partial x) \tag{17}$$

where the coordinate equation follows as $dx/dt = c\beta_\perp \cos[\phi(\phi_0, t)]$.

Relations (16) and (17), together with Eqs. (12) and (13), form a closed set for determination of the full nonlinear time evolution for the amplitude and phase of the electric field in this problem, subject to the important assumptions of a single mode excitation together with slow growth rate and slow phase shift changes. Sprangle and Drobot (1977) have shown how the linear behavior can be deduced from these relations by going to the limit of weak fields and small gyroradii. They also derive from energy conservation considerations an approximate expression for the wave electric field amplitude in terms of the (time-dependent) energy parameter $\gamma_\perp$.

$$E_0(t) \simeq \sqrt{8}(mc/e)(\sigma_0 e^2/ma\varepsilon_0)^{1/2}[\gamma_\perp(0) - \langle\gamma_\perp(\phi_0, t)\rangle]^{1/2}, \qquad (18)$$

where the brackets indicate an ensemble average over all phases.

Interpretation of the results of Sprangle and Drobot's (1977) numerical simulations of the coupled set of Eqs. (12), (13), (16), and (17) is greatly aided by reference to the slow time scale constant of the motion which may be deduced from Eqs. (16) and (17). In terms of the perpendicular momentum $p = mc\beta_\perp\gamma_\perp$ and the electric field $E_0(t) = -\partial A_y/\partial t$, the fast time scale averages of Eqs. (16) and (17) in the small excursion limit (i.e., $x \ll a$) are

$$dp/dt = -\tfrac{1}{2}eE_0 \sin \lambda \qquad (19)$$

and

$$d\lambda/dt = \omega - \Omega + \tfrac{1}{2}(eE_0/p) \cos \lambda \qquad (20)$$

where $\lambda(t) = \omega t - \phi(t)$ is the phase slip between the particle and the field. Equations (19) and (20) have been analyzed for the cyclotron maser in the context of a general theory of the laser (Borenstein and Lamb, 1970). It may be inferred by direct substitution that the quantity

$$[eE_0 p/\Omega_0(mc)^2] \cos \lambda + (\omega/\Omega_0)(\gamma_\perp - \Omega_0/\omega)^2 \equiv C \qquad (21)$$

obeys the equation $dC/dt = 0$, provided E_0 and ω are held constant. Thus, if at the initial time t_0 a particle has phase $\lambda(0)$ and momentum $p(0) = mc[\gamma_\perp(0) - 1]^{1/2}$, then throughout the interaction the subsequent values of p and λ will obey the relationship

$$[eE_0 p(t)/\Omega_0(mc)^2] \cos \lambda(t) + (\omega/\Omega_0)[\gamma_\perp(t) - \Omega_0/\omega]^2$$
$$= [eE_0 p(0)/\Omega_0(mc)^2] \cos \lambda(0) + (\omega/\Omega_0)[\gamma_\perp(0) - \Omega_0/\omega]^2.$$

The character of the family of curves in the $p\lambda$ plane given by Eq. (21) depends upon the field strength through the dimensionless parameter $eE_0/mc\Omega_0 = E_0/cB_0$, and upon the wave dispersion relation, which determines ω/Ω_0. (The real part of ω is understood in this context.) Figure 10 shows an example of one such family, in this case for

$$\gamma_\perp(0) = 1.05 \ (W_\perp = 25 \text{ keV})$$

and

$$\mu = (m\gamma_\perp(0)\sigma_0/a\varepsilon_0 B_0^2)^{1/2} = 5 \times 10^{-2}.$$

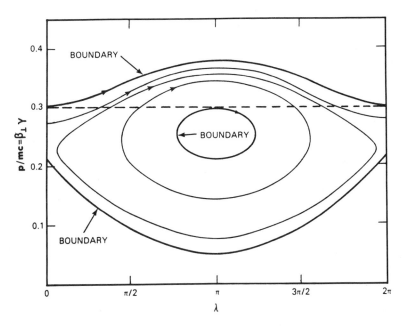

FIG. 10. Particle trajectories in velocity phase space as found from the constant of motion. The figure shows the boundaries of the regions accessible to particles initially uniformly distributed so that $0 < \lambda < 2\pi$. [From Sprangle and Drobot (1977).]

The dimensionless parameter $\mu = \omega_p \gamma_\perp^{1/2}(0)/\Omega_0$ is a measure of the electron beam concentration, through the modified plasma frequency ω_p. For these values of μ and $\gamma_\perp(0)$, the linear dispersion relation gives $\omega/\Omega_0 = 0.9695$ (i.e., $\omega/\Omega = 1.018$), and a value for E_0/cB_0 of 2.5×10^{-3} was selected. It is seen in Fig. 10 that two classes of orbit are possible, those which are "open" so that all values of λ from 0 to 2π are accessible, and those which are "closed" such that the particle oscillates in a bounded zone between two limiting values of λ; these latter orbits are termed "trapped" by Sprangle and Drobot (1977) since the entire phase circle is not accessible. Of course this simple picture for the orbits fails when the actual time variations of the phase shift and the amplitude are taken into account.

In following the orbits in the $p\lambda$ plane one assumes the distribution initially to be uniform in λ and singular in p, i.e., uniformly distributed along the horizontal dashed line in Fig. 10. The subsequent motion will carry the particles along whatever $C = \text{const}$ lines they are on initially. In Fig. 10 three bounding curves exist, beyond which no particle moves. These boundaries place maximum and minimum limits on p (at $\lambda = \pi$) and define an inner proscribed zone which is free of orbits. These features combine to reduce the overall amplifier efficiency below the value predicted from energy depletion

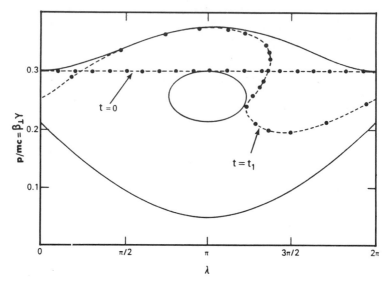

FIG. 11.   The deformation of a monoenergetic beam in the presence of a constant amplitude field showing the actual particle positions in phase space. The curve for $t = t_0$ indicates the initial positions and the one for $t = t_1$, the positions when the beam indicates trapping. [From Sprangle and Drobot (1977).]

arguments alone, as shall be demonstrated in the numerical results. This follows since energy depletion assumes that orbits exist which after a time τ carry all particles from $p(0)$ down to a value $p_0(\tau)$ where gain vanishes; if a lower bound to p is introduced by the trapping process which is greater than p_0, it follows that the efficiency is reduced. At this juncture it is worth remarking, therefore, that any process which delays or breaks the trapping process is likely to allow higher amplifier efficiencies. Examples of this will be given below.

Figure 11 shows the locations of twenty quasi-particles at $t = 0$ and at $t = t_1$, a time at which a well formed bunch is in evidence around $\lambda = 1.3\pi$. Of the twenty quasi-particles, eight have gained and eleven have lost energy: The net flow of energy has been from the particles to the fields.

At a later time $t_2 > t_1$, the phase-space locations are as shown in Fig. 12. Now only three quasi-particles out of the twenty have gained energy. The state depicted here is for a time when the maximum energy has left the particles. Thereafter the particles begin to recover lost energy from the fields. At t_2 the wave energy has reached its maximum—or *saturation*—value.

If the field parameter E_0/cB_0 is small enough, the trapping phenomenon is absent. This is evident from Eq. (12) which admits solutions for all λ when

$$(eE_0/B)(p/mc) < (\omega/\Omega_0)(\gamma - \Omega_0/\omega)^2.$$

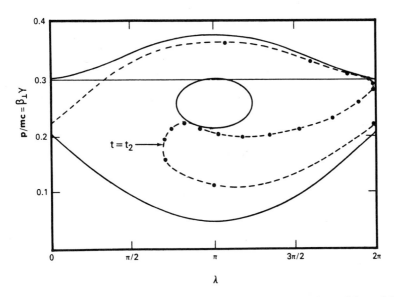

FIG. 12.   The particle positions in phase space for a monoenergetic beam deformed by a constant amplitude electric field at $t = t_2$, when the particles have lost the maximum energy and are in a state corresponding to saturation. [From Sprangle and Drobot (1977).]

Figure 13 shows a set of quasi-particles in the $p\lambda$ plane at three times for a weak field case: during the linear phase of the motion, during the nonlinear phase, and at saturation. A smooth undulation of the distribution is seen, without pathological distortions, or excluded zones of $p\lambda$. Corresponding momentum distributions are shown in Fig. 14, where a general diffusion in momentum is evident, with a gradual shift of the mean momentum value to lower energy.

Figure 15 shows a strong field case

$$(eE_0/B) > (mc/p)(\omega/\Omega_0)(\gamma - \Omega_0/\omega)^2$$

for those time phases (linear, nonlinear, and saturation) in which a strong electron bunch is seen to form in the nonlinear phase. Two strong bunches form in the saturated state, the second of which evidently acts to feed energy back from the wave to the particles. Corresponding momentum distributions are shown in Fig. 16; here one sees first a diffusion, then a coalescence in momentum as saturation sets in.

Of course in the actual full evolution of the field, the interaction progresses from the weak field to the strong field case, and the orbits in the $p\lambda$ plane do not fall on the $C = $ const curves given by Eq. (21). One example of such an

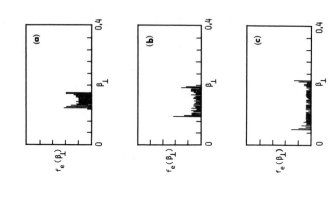

FIG. 14. The distribution function of the electrons from a case of saturation by energy depletion at various times as in Fig. 13. [From Sprangle and Drobot (1977).]

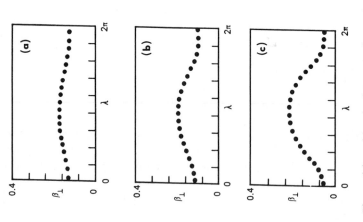

FIG. 13. The particle positions in phase space from a simulation of the cyclotron maser instability in the case of saturation by energy depletion: (a) $t = t_1$ (linear phase); (b) $t = t_2$ (nonlinear phase); and (c) $t = t_3$ (at saturation). [From Sprangle and Drobot (1977).]

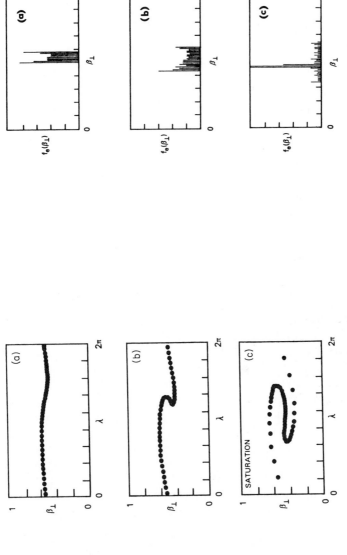

FIG. 16.   The distribution function for saturation by trapping for times given in Fig. 15. [From Sprangle and Drobot (1977).]

FIG. 15.   Particle positions in phase space from a simulation in the case of saturation by trapping: (a) linear phase; (b) nonlinear phase; ... on phase. [From Sprangle and Drobot (1977).]

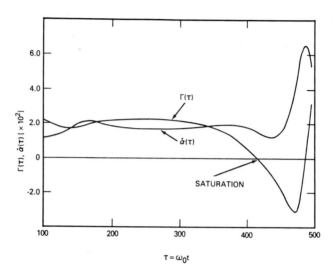

FIG. 17.   The nonlinear phase shift and growth rate as a function of time for a typical simulation. [From Sprangle and Drobot (1977).]

evolution is shown in Fig. 17 in which values of $\Gamma(t)$ and $\dot{\alpha}(t)$ are plotted as functions of ωt. Both the growth rate and phase slip are constant for about three quarters of the time prior to saturation. At saturation, $\Gamma = 0$; thereafter becomes negative and $\dot{\alpha}$ undergoes strong variations. The electric field olution and mean energy $\langle \gamma_\perp \rangle$ are shown in Fig. 18 for this same simula- . In this example one sees that the average particle kinetic energy has eased by 40 % at saturation. Figure 19 shows efficiency values from many simulations for $\mu = 0.1$. The results (solid curve) show a peak efficiency ut 40 %. The two dashed curves are efficiency estimates based on energy on and on particle trapping as the limiting mechanisms. Naturally lepletion gives an upper limit to the efficiency. At higher values of $\gamma_\perp$, lated efficiencies approach values consistent with the phase trapping

'0 shows calculated efficiency values as the beam concentration is varied. At low energy ($\gamma < 1.05$) highest efficiency is achieved t μ; at higher energy ($\gamma > 1.3$) the reverse is true. At the lower practical factors may dictate perpendicular beam energies less $\gamma_\perp = 1.1$) it is seen that attempts to achieve higher output) will probably result in lower efficiency. hich defeats or postpones phase trapping is likely to allow One such scheme is to detune the magnetic field slightly value $\Omega_0/\gamma_0 = \omega$. Examples in this instance are shown in

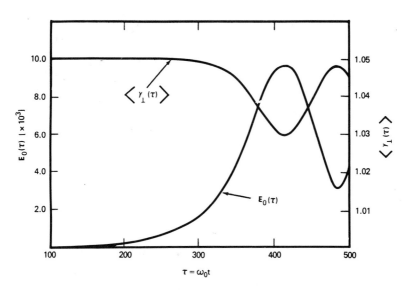

FIG. 18.   The field amplitude and average beam γ as a function of time for a typical simulation. [From Sprangle and Drobot (1977).]

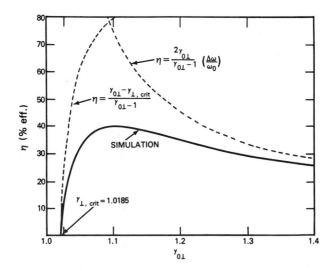

FIG. 19.   The efficiency of the cyclotron maser as a function of energy found from the two mechanisms of saturation and from simulations. [From Sprangle and Drobot (1977).]

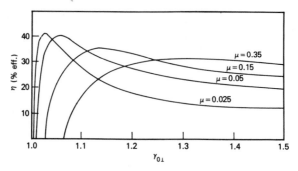

FIG. 20.   The beam frame efficiency of the electron cyclotron maser as a function of energy for various densities when $ck_n = \Omega_0/\gamma_{0\perp}$. [From Sprangle and Drobot (1977).]

Fig. 21 where efficiencies exceeding 60 % are predicted for $\mu = 0.1, \gamma_{0\perp} = 1.155$ ($W_\perp \sim 80$ keV). Another possibility for postponing phase trapping would be to provide, in a distributed interaction amplifier, a spatially decreasing magnetic field along the beam interaction path, so as to make greater use of the free energy available. It is obvious that operation at as low an impedance as possible is indicated, so as to keep the field magnitude lower at a given power level and thus to further postpone trapping.

IV.   Summary of Soviet Gyrotron Development

A.   REVIEWS

Two reviews of gyrotron development in the USSR have appeared recently (Flyagin et al., 1977; Andronov et al., 1978). It would be presumptuous therefore to attempt a definitive parallel review here. But a selective summary

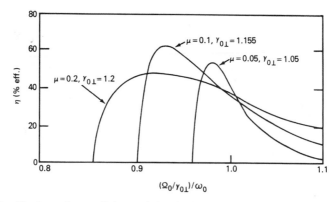

FIG. 21.   The beam frame efficiency of the electron cyclotron maser when the cyclotron frequency is varied. [From Sprangle and Drobot (1977).]

of the major contributions is altogether in order, both as a guide for the reader into the extensive Soviet literature, and so as not to withhold even a small fraction of the immense credit due the Soviet workers for their signal accomplishments in gyrotron evolution.

An important overview of the interaction of free electrons with various configurations of electric and magnetic fields is given in a tutorial article (Gaponov et al., 1967). Here a class of interactions not requiring slow-wave circuits is surveyed, with a view towards practical limitations encountered with these circuits at millimeter and submillimeter wavelengths. Both phase-bunching and space-bunching mechanisms are considered. General considerations for the linear dielectric response of an ensemble of classical non-linear oscillators led to conclusions identical to those published earlier (Bekefi et al., 1961), namely that if the elementary oscillator's eigenfrequency, or alternately its damping rate, were to be energy-dependent, then one could expect phase bunching of the oscillators, with attendant coherent stimulated emission.

Nonlinear considerations are outlined in this tutorial article leading to coupled equations for the amplitude and phase of the elementary oscillators similar to Eqs. (12) and (13), discussed above. This allowed numerical solutions of the governing equations so as to determine the maximum efficiency $\eta_\perp$ for the transfer of energy from the oscillators to the fields, together with the values of the system parameters required to achieve the maximum efficiency. For harmonics of the fundamental oscillator frequency $n = 1-5$, the corresponding maximum efficiencies were determined to be 41.5%, 29.5%, 21.5%, 17.0%, and 14.0%, respectively. High efficiency values such as these gave strong impetus to the search for practical configurations for exploitation of these interactions.

These authors outlined a number of arrangements for promoting non-isochronous electron oscillations, including helical electron beams in uniform magnetic fields, trochoidal beams in crossed electric and magnetic fields, and beams in transversely inhomogeneous fields. Specific considerations for the electron cyclotron maser case led to expressions for the start oscillation current for a single-cavity oscillator (gyromonotron) similar to Eq. (4).

B. LINEAR AND NONLINEAR SOLUTIONS OF THE GYROTRON EQUATIONS

An important early article (Gaponov and Yulpatov, 1967) established the necessity for employing relativistic equations of motion in order to correctly include the cyclotron maser azimuthal bunching mechanism, in addition to the axial bunching mechanism. An approximate cubic dispersion relation was analyzed to determine the growth rates for fast and slow modes on a helical electron beam in a waveguide. Conclusions are drawn that the cyclotron maser azimuthal bunching mechanism dominates for fast waves, and that the axial bunching mechanism dominates for slow waves. This point has been recently reiterated in the detailed comparison of the two mechanisms (Chu and Hirshfield, 1978).

An analysis of the nonlinear relativistic equations of motion for electrons of a homogeneous helical electron beam and the large amplitude electromagnetic field of a cavity has been carried out (Rapoport et al., 1967). The novel feature in this work is that the electromagnetic field was taken to be a homogeneous standing wave $E_x = E_0 \sin ky \sin \omega t$ resembling that found in a cavity whose size is much larger than a wavelength, or as in an open mirror Fabry–Perot resonator. Field variations over the electron orbit diameter were taken into account so that cyclotron harmonic interactions could be included. The phase-angle averaged equations of motion were analyzed numerically for weakly relativistic electrons. Each electron beam layer of half-wave width was simulated by eight infinitessimal charge layers, each of which was represented by a sequence of sixteen groups of electrons uniformly distributed over one period of phase in the initial reference plane. This gave a coupled set of $2 \times 8 \times 16 = 256$ difference equations for the quasi-particle amplitudes and phase. Critical parameters were the normalized electron transit time through the interaction zone $T = \alpha \omega t$, the initial frequency detuning parameter $\sigma = (\omega - n\Omega)/\alpha\omega$, and the field parameter $v^{-1} = \alpha(2\omega/n\Omega\beta_\perp^2)$, where $n = 1, 2, 3, \ldots$, is the order of the harmonic, E_0 is the peak value of the standing wave electric field, $\beta_\perp = v_\perp/c$, and $\alpha = en^nE_0\beta_\perp^{n-2}/2mc\omega n!$. In terms of these variables the equations solved were for the normalized orbit amplitude p_{ij}

$$dp_{ij}/dT = \sin \xi_j p_{ij}^{n-1} \cos \theta_{ij}, \tag{22}$$

and for the orbit angle θ_{ij}

$$d\theta_{ij}/dT = -n \sin \xi_j p_{ij}^{n-2} \sin \theta_{ij} + \sigma + v^{-1}(p_{ij}^2 - 1) \tag{23}$$

where $\xi_j = ky_j$ is the normalized location of each charged layer. The index i runs from 0 to 15 to designate the sixteen electron groups. The transverse efficiency of each layer was then

$$\eta_{\perp j} = 1 - \frac{1}{16} \sum_{i=0}^{15} p_{ij}^2$$

while the overall transverse beam efficiency was

$$\eta_\perp = 1 - \frac{1}{128} \sum_{i=0}^{15} \sum_{j=1}^{8} p_{ij}^2.$$

The principal results are displayed in Figs. 22 and 23. Figure 22 shows the maximum interaction efficiency $\eta_\perp$ as a function of the field parameter v^{-1}

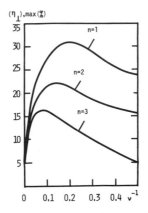

FIG. 22. Maximum interaction efficiency plotted versus the parameter v^{-1}, proportional to the electromagnetic field strength. [From Rapoport *et al.* (1967).]

for $n = 1$, 2, and 3. Figure 23 shows the optimum transit time (a) and the optimum frequency detuning (b) plotted versus v^{-1} for $n = 1$, 2, and 3. For a thin beam, maximum attainable efficiencies were found to be 42% for $n = 1$, 29% for $n = 2$, and 22% for $n = 3$. Calculations were also presented by these authors for normalized output power to be expected when the controlling parameters above were optimized, both for open and closed resonator configurations.

C. IRREGULAR WAVEGUIDES AND CAVITIES

It appeared from this work of Rapoport *et al.* (1967) that the theoretical transverse efficiency was limited to values no greater than about 40% for $n = 1$, and to successively smaller values for the higher harmonics, when regular (i.e., uniform cross section) waveguides or cavities were employed.

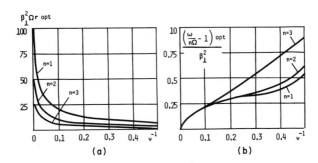

FIG. 23. Optimum transit time of electrons in interaction space (a), and optimum frequency detuning (b), plotted versus the parameter v^{-1}, proportional to the electromagnetic field strength. [From Rapoport *et al.* (1967).]

Moreover it is apparent that low-Q devices are preferable to high-Q ones from the point of view of widening the electronic tuning range (oscillators), and broadening the amplification band (amplifiers). But low Q would imply either very lossy walls (a clear disadvantage to achievement of high efficiency) or an open structure with a tightly coupled output segment. Thus the interaction field would be nonuniform, and would not correspond to that of any regular rf structure. But this nonuniformity could be used to advantage to further increase the efficiency. The original germ of the idea was stated: "...(the efficiencies) can be much higher if the electron bunching takes place in a fairly weak, and the energy take-off in a stronger, rf field..." (Gaponov et al., 1967). This idea was apparently pursued in some detail in the mid-sixties, but not then published in the archival literature (Moiseyev et al., 1968). The first available discussions of this appeared much later (Bratman et al., 1973; Bykov and Gol'denberg, 1975) where the gyrotron equations were analyzed for a system in which the field quantities (rf electric and magnetic fields, static magnetic field) have an arbitrary dependence upon distance along the symmetry axis of the device. The simplest example of this considered by Bratman et al. (1973) was a semi-infinite regular waveguide, bounded at the electron beam input end by an abrupt cross section change to beyond cutoff, and at the collector end by an abrupt drop in the dc magnetic field. The magnetic field is taken to be uniform in the intervening segment. Numerical calculations based on the governing equations were performed by the method of successive iterations. An example of the results is shown in Figs. 24 and 25. Figure 24 shows a schematic of the waveguide interaction zone and the electron beam, together with the longitudinal field amplitude

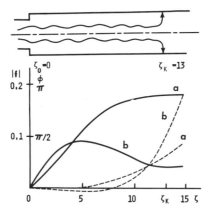

FIG. 24. Schematic of waveguide interaction zone and electron beam, together with longitudinal field amplitude and phase for (a) $\delta = 0.6$, $I = 0.015$, $\eta_{\perp max} = 75\%$; and (b) $\delta = 0$, $I = 0.015$, $\eta_{\perp max} = 10\%$. [From Bratman et al. (1973), with permission of Plenum Press, New York.]

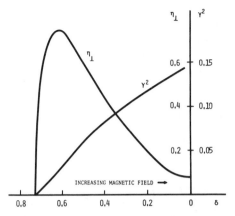

FIG. 25. Perpendicular efficiency and operating frequency of gyrotron vs. δ. (Here $\gamma^2 = 8\beta_{\parallel}^2(\omega - \omega_0)/\beta_{\perp}^4$.) [From Bratman et al. (1973), with permission of Plenum Press, New York.]

and phase for two cases: (a) $\delta = 0.6$, $I = 0.015$, $\eta_{\perp max} = 75\%$, and (b) $\delta = 0$, $I = 0.015$, $\eta_{\perp max} = 10\%$. The parameter I is proportional to the beam current and $\delta = 2\beta_{\perp}^{-2}(1 - n\Omega/\omega_0)$ where ω_0 is the waveguide cutoff frequency. In both examples $n = 1$ and $\beta_{\perp} = 2\beta_{\parallel}$. The strong dependence of efficiency on Ω/ω_0 is evident. Figure 25 shows the variation of perpendicular efficiency $\eta_{\perp}$ (up to the maximum value of 75%) with the parameter δ, together with the operating frequency given in units of $\gamma^2 = 8\beta_{\parallel}^2(\omega - \omega_0)/\omega\beta_{\perp}^4$. It is stressed by these authors that the variation of the field phase should not exceed $\pi/2$ throughout the interaction region.

A second example presented by these authors is for a so-called gyro-twistron amplifier, depicted in Fig. 26. A low level input signal is supported

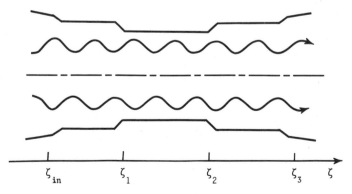

FIG. 26. Schematic of gyrotwistron amplifier. [From Bratman et al. (1973), with permission of Plenum Press, New York.]

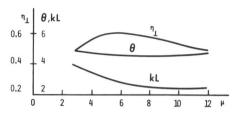

FIG. 27. Calculated transverse efficiency $\eta_\perp$, electron drift angle θ, and phase advance kL in the output segment versus length of output segment $\mu = \zeta_3 - \zeta_2$. Bunching parameter $X_c = 1.84$. [From Bratman *et al.* (1973), with permission of Plenum Press, New York.]

by the waveguide section $\zeta_{in} < \zeta < \zeta_1$, where $\zeta = (\pi\beta_\perp^2/\beta_\parallel)(z/\lambda)$ is the normalized axial coordinate. The segment $\zeta_1 < \zeta < \zeta_2$ is cut off at the frequency of interest and forms a drift section for the electron beam to bunch. The segment $\zeta_2 < \zeta < \zeta_3$ is the output waveguide section. This overall system should not support self-excited oscillations, a point of considerable practical importance. The calculated transverse efficiency $\eta_\perp$ for this system is plotted in Fig. 27, as a function of the length of the output segment $\mu = \zeta_3 - \zeta_2$. Also shown as functions of μ are the electron drift angle $\theta = (\omega - \Omega)L/v_\parallel$ and the phase advance kL in the output segment, where $kL = \gamma\mu$. These calculations are for a current just equal to the starting value, and for an optimized bunching parameter $X = X_c = 1.84$. (The boundary condition for the electron phase distribution at the entrance of the drift tube was taken to be

$$p(\theta) = \exp[i(\theta_0 - X \sin \theta_0)]$$

where the perpendicular energy of rotation of the electron is proportional to $|p|^{2/n}$.) It is noted that the maximum value of transverse efficiency predicted was 60%, but that this was a fairly weak function of μ, indicating that some variation in device length could be tolerated without undue sacrifice of efficiency.

D. HARMONIC OPERATION

A comparison was published of the maximum attainable efficiencies at the first and second cyclotron harmonics using irregular rf field distributions (Kolosov and Kurayev, 1974). These authors characterized the axial variation of the rf electric field expressed as a multiparameter polynomial. The parameters were adjusted, while numerically solving the nonlinear gyrotron equations, so as to optimize the efficiencies. Thus the azimuthal components of the electric field (TE_{0n} mode of a circular resonator) had the form

$$E_\phi = \sum_{n=1}^{\infty} C_n J_1(\chi_n r) \sin \pi n z/l$$

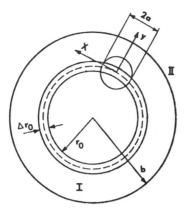

FIG. 28.   Cross section of device: I is the tubular electron beam, II the side wall of resonator, r_0 the average radius of leading center of electron orbit with diameter $2a$ at entrance to interaction region, and Δr_0 the spread in radii of leading center. $\Delta r_0 \ll a$. The magnetic field is normal to the plane shown. [From Kolosov and Kurayev (1974).]

and the optimization led to determination of the C_n's. Figure 28 shows a cross section of the device considered, where a thin annular beam of thickness Δr is positioned at $r = r_0$, and the outside radius b is a function of axial coordinate z. Results for the first and second harmonic cases are shown in Figs. 29 and 30 for both the gyromonotron oscillator and the gyroamplifier. For the first harmonic examples the function $Ag^{opt}(z/l)$ is plotted, where $A = eE_m l(\beta_\perp/kr_0)^{n-1}/2v_\perp v_\parallel$ is the field parameter, and where $E_\phi(r_0, z) = E_m g(z/l)$. For the second harmonic examples, the function $AG^{opt}(z/l)$ is plotted, where $r_0\, \partial E_\phi(r_0, z)/\partial r - E_\phi(r_0, z) = E_m G(z/l)$. The coefficients C_n are thus given by

$$C_n = \frac{2E_m}{J_1(\chi_n r_0)} \int_0^1 dT\, g(T) \sin \pi n T$$

or

$$C_n = \frac{2E_m}{\chi_n r_0 J_2(\chi_n r_0)} \int_0^1 dT\, G(T) \sin \pi n T.$$

Table IV gives the parameters and maximum efficiencies calculated by these authors. The definitions of the parameters here are: transit angle $\mu = \beta_\perp^2 \Omega l/2v_\parallel$, detuning $\Phi = (\Omega - \omega)l/nv_\parallel$, and prebunching parameter X as before. Two important points are seen from the values in the table: (a) The magnitude of the achievable efficiencies is in the range 88–97%; and (b) the efficiencies at the second harmonic are higher than those at the fundamental.

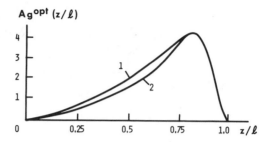

FIG. 29. Optimum high-frequency field distributions for interaction at first harmonic ($n = 1$): (1) $Ag^{opt}(T)$ for gyromonotron; (2) $Ag^{opt}(T)$ for selector of gyroamplifier. [Reproduced from Kolosov and Kurayev (1974).]

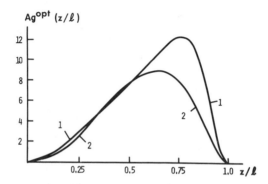

FIG. 30. Optimum high-frequency field distributions for $n = 2$: (1) $AG^{opt}(T)$ for gyromonotron; (2) $AG^{opt}(T)$ for selector of gyroamplifier. [Reproduced from Kolosov and Kurayev (1974).]

TABLE IV

CALCULATED MAXIMUM EFFICIENCIES FOR GYROMOMOTRONS AND
GYROTWISTRONS, FOR FIRST AND SECOND HARMONIC OPERATION,
TOGETHER WITH OPTIMIZING PARAMETERS[a]

	μ	Φ	A	X	$\eta_{\perp \, max}$
Gyromonotrons					
$\omega \simeq \Omega$	28.55	15.79	4.25	—	88.1%
$\omega \simeq 2\Omega$	22.35	10.77	12.49	—	90.0%
Gyrotwistrons					
$\omega \simeq \Omega$	27.7	16.0	4.3	−2.16	95.1%
$\omega \simeq 2\Omega$	16.16	7.187	9.05	2.47	97.2%

[a] Kolosov and Kurayev (1974).

For millimeter and submillimeter wave devices, the decided advantage of second harmonic operation with half the magnetic field required for fundamental operation is most dramatically brought out here. It apparently has yet to be established whether the efficiency continues to remain high for harmonic numbers higher than two, on the basis of a calculation such as this.

E. MODE STABILITY

The problem of mode stability is rather serious in these devices, since higher modes of the cavity or waveguide structure are employed. When other nearby modes are accessible competition occurs for the available free energy, as in an optical laser when the gain curve embraces more than one resonator mode (Bennett, 1973). Stability is measured by comparing the separation between neighboring modes with the frequency width of the gain curve. Numerical techniques have been developed for optimizing the rf field distribution to allow high efficiency gyroklystron operation, while maximizing the stability margin (Kurayev et al., 1974; Nusinovich, 1974; Moiseev and Nusinovich, 1974; Zarnitsyna and Nusinovich, 1975a,b). Two examples of rf field profiles for the output section of a two-stage gyroklystron which accomplish good mode stability are shown in Fig. 31. The figures show the field profile itself $Ag^{opt}(z/l)$ and the cavity boundary which produces the profile. For the first example (Fig. 31a) $\mu = 9.012$, $\Phi = 5.69$, $X = 2.929$, $\gamma = 0.541$, and $\eta_{\perp}^{opt} = 80\%$. For the second example (Fig. 31b) $\mu = 12.72$, $\Phi = 7.41$, $X = 2.757$, $\gamma = -0.9988$, and $\eta_{\perp}^{opt} = 90\%$. The annular electron beam location was $r = r_0 = 0.3\lambda$. The parameter k^{-1} is the specified modal current stability within a band $\Delta\Phi$ (which should be 3–10 times greater than the relative passband of the cavity mode chosen). For the first example, $k^{-1} = 5$; for the second example $k^{-1} = 25$, so that high stability can be achieved without significant sacrifice of device efficiency. It should be noted however, that while the resonators shown are relatively long (five to six wavelengths), their radii are of the order of, or less than, one wavelength. The design of the electron beam to properly thread this type of resonator at the lower millimeter, or at submillimeter, wavelengths, would present a difficult challenge.

F. INFLUENCE OF SPACE CHARGE

To this point in this review, the influence of the quasi-static electric field of the electron beam has been neglected in formulating and analyzing the gyrotron equations; this neglect was justified since, for the range of beam currents and magnetic fields employed, one has $\omega_p \ll \Omega$. It is important to determine a posteriori to what extent this neglect is legitimate. This determination has been undertaken by a number of authors using perturbation theory to modify the dynamical equations by including space-charge forces arising

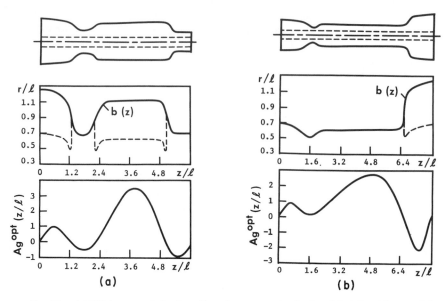

FIG. 31.   (a) Efficiency-optimized profiles of output cavity for $k = 0.2$, $\Delta\Phi = 15$, $\eta_\perp = 0.8$, and $m = 11$. Top: overall shape of one version of cavity profile; dashed lines—level of guiding center in beam $r_0 = 0.3\lambda$. Center: two versions of cavity profile—solid curve, smooth profile; dashed profile with corners. Bottom: distribution $Ag(z/l)$ achieved in cavity for $r = r_0 = 0.3\lambda$. (b) Efficiency-optimized profiles of output cavity for $k = 0.04$, $\Delta\Phi = 8$, $\eta_\perp = 0.9$, and $m = 15$ [notation is the same as in (a)]. [Reproduced from Kurayev et al. (1974).]

from orbits calculated in the zero space-charge limit (Kurayev and Slepyan, 1975; Kolosov and Kurayev, 1974; Bratman and Petelin, 1974). We will cite here the main results of Kolosov and Kurayev, who analyzed the effect of space charge on the bunching of a premodulated beam in a drift tube, a situation relevant to a two-element gyroamplifier. The system analyzed is pictured schematically in Fig. 28, subject to the constraint $\Delta r_0 \ll a$, i.e., the thin beam limit. Thick beams have been considered by Kurayev and Slepyan.

The parameters governing gyrotron operation in the presence of quasi-static space-charge perturbations are μ and Φ, as defined above in connection with Table IV, together with a velocity modulation parameter ξ (related to X) such that the initial boundary condition for the perpendicular velocities at the entrance to the drift tube is ($\xi \ll 1$)

$$u_m(0) = 1 - \xi \sin (2\pi mn/N) \qquad \text{(in-phase momentum)}$$

and

$$v_m(0) = 0 \qquad\qquad\qquad \text{(quadrature momentum)}$$

where m is the index for the phase segment in the discrete distribution ($0 < m < N$, $N = 64$ in the example), and $n = 1, 2$ is the harmonic number. The parameter governing the space-charge field magnitude was

$$S = eI_0 l/2\pi^2 m\varepsilon_0 v_\parallel^2 v_\perp r_0,$$

where I_0 is the total beam current. Two functions were computed to adjudge the influence of space-charge, the phase-bunching function $F_n(z/l)$ and the transverse electron distribution function $f_n(m)$ (assuming an initially mono-energetic distribution). The phase-bunching function is defined such that $F_n = 1$ represents a uniform distribution (all phase angles equally populated), while $F_n = 0.5$ represents a fully bunched beam (all electrons at the same phase, with maximum work done against the electric field).

Figure 32 shows results for the fundamental case ($n = 1$). The four examples are for increasing space-charge influence, through increases in both bunching strength ξ and current density S. The figure caption indicates the numerical values of the parameters, with case 1 having zero space-charge effect and case 4 the strongest. What is evident from Fig. 32a, showing $F_1(z/l)$, is the shortening of the drift distance z/l for maximum bunching from 0.92 to 0.512, while the

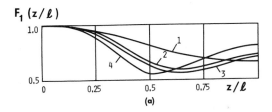

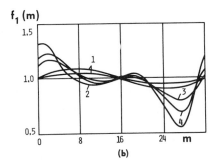

FIG. 32. The functions $F_1(T)$ and $f_1(m)$ for $n = 1$: (a) family of $F_1(T)$ for different μ, S, and ξ: (1) $\mu = 10, 20$; $\xi = 0.1, 0.05$; and $S = 0$; (2) $\mu = 10$, $\xi = 0.1$, and $S = 2$; (3) $\mu = 20$, $\xi = 0.05$, and $S = 1$; and (4) $\mu = 20$, $S = 2$, and $\xi = 0.05$; and (b) family of $f_1(m)$ for the same values of μ, S, and ξ (in the same order). [Reproduced from Kolosov and Kurayev (1974).]

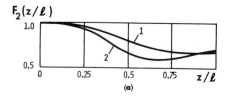

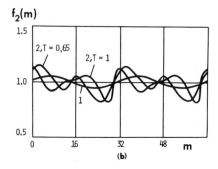

FIG. 33.   The functions $F_2(T)$ and $f_2(m)$ for $n = 2$: (a) $F_2(T)$ for the cases: (1) $\mu = 10$, $\xi = 0.05$, and $S = 0$; (2) $\mu = 10$, $\xi = 0.05$, and $S = 2$; and (b) $f_2(m)$ for the same cases. [Reproduced from Kolosov and Kurayev (1974).]

minimum value of F_1 reaches 0.55 instead of 0.664 without space charge. The associated distribution functions are shown in Fig. 32b, for the same set of parameters ($f_1 = 1$ for no bunching). The distributions also exhibit a marked difference in comparing the zero space-charge case (curve 1) where a weak sinusoidal perturbation is imposed ($0 < m < 32$), as compared with the strong space-charge cases, where a large velocity spread is induced (both as modulation index and space-charge increase).

Figure 33 shows the corresponding functions for second harmonic operation. Here, case 1 is for $S = 0$, and case 2 for $S = 2$, with $\mu = 10$ and $\xi = 0.05$ in both cases. Again, as with $n = 1$, one sees in Fig. 33a an improvement in phase bunching and a shortening of the optimum drift tube length, as space-charge forces come into play. The distribution functions in Fig. 33b show a smaller velocity spread than for $n = 1$, presumably since two bunches are formed per cycle, oscillating radially in opposite phase. The fields of these opposing bunches must partially compensate for one another to produce a smaller velocity spread than would a single bunch. As a result the bunch remains compact (shallow minimum for $F_2(z/l)$ for a longer range in z/l).

One concludes therefore that space-charge forces, at least in a drift tube, may actually aid in the bunching process, both for second-harmonic and fundamental operation, provided the physical parameters, such as μ and Φ, are carefully chosen.

G. RECENT DEVICE RESULTS

Between June and October 1973 at least four major Soviet papers were submitted for publication which described experimental device results of major significance. Publication and translation delays meant that this work did not gain much attention in the West until 1974–5, when its impact was overwhelming, to say the least. We shall summarize these results here chronologically, and briefly describe more recent work including devices announced in 1978.

The first device we shall describe was an oscillator built to operate at the second harmonic at a wavelength of 8.9 mm. It included a magnetron injection gun to obtain an annular electron beam with small velocity spread, and an irregular cavity designed to obtain high efficiency (Kisel' et al., 1974). A cross section drawing of the device is shown in Fig. 34, where the various elements are identified in the caption. A drawing of the cavity (TE_{021}) itself, together with the axial variation of the electric field, is shown in Fig. 35. The device was operated in both pulsed and continuous modes. Figures 36–39 show the results obtained, which reached an output of up to 40 kW and efficiency of up to 43 % (at 30 kW) in pulsed operation ($\tau \sim 5 \mu sec$), and up to 10 kW output at 40 % efficiency in continuous operation.

Shorter wavelength oscillators were also described (Zaytsev et al., 1974). A diagram of the apparatus used here is shown in Fig. 40, where the super-conducting coil is seen which was used to obtain the high continuous magnetic fields required (up to 65 kG in a 40-mm diam room temperature

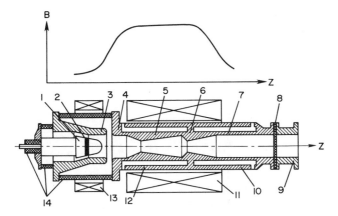

FIG. 34. Outline drawing of Soviet gyrotron prototype: (1) cathode; (2) emitting strip; (3) first anode; (4) second anode; (5) cavity; (6) output coupling aperture; (7) beam collector; (8) output window; (9) output waveguide; (10) and (12) water jackets; (11) main solenoid; (13) electron gun solenoid; (14) insulators. Overall length of this device is approximately 20 cm. [Reproduced from Kisel' et al. (1974).]

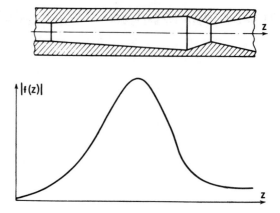

Fig. 35. Cross section of resonator cavity TE_{021} and corresponding axial electric field distribution. [Reproduced from Kisel' *et al.* (1974).]

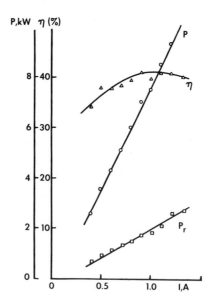

Fig. 36. Dependence of efficiency η, dissipated resonator power P_r, and output power of gyrotron P vs beam current I, operated in cw with $U_r = 19$ kV. [Reproduced from Kisel' *et al.* (1974).]

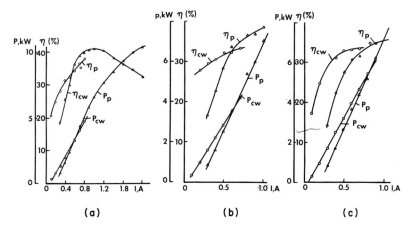

FIG. 37. Dependence of efficiency and output power vs beam current in continuous and pulsed regimes, for resonance (U_r) and anode (U_a) voltages: (a) $U_r = 16$ kV, $U_a = 11$ kV; (b) $U_r = 18$ kV, $U_a = 11$ kV; and (c) $U_r = 20$ kV, $U_a = 12$ kV. [Reproduced from Kisel' et al. (1974).]

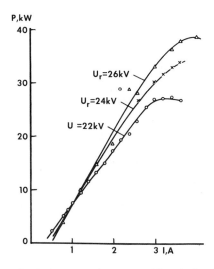

FIG. 38. Dependence of output power on beam current in pulsed regime. [Reproduced from Kisel' et al. (1974).]

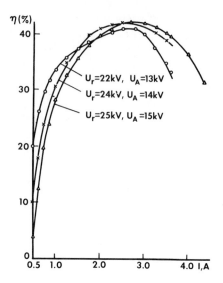

FIG. 39. Dependence of efficiency on beam current in pulsed regime. [Reproduced from Kisel' *et al.* (1974).]

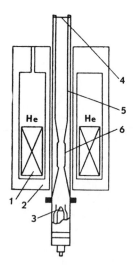

FIG. 40. Arrangement of the gyrotron in the cryostat: (1) solenoid; (2) cryostat; (3) injector; (4) cavity; (5) collector; (6) high-frequency window. [Reproduced from Zaytsev *et al.* (1974).]

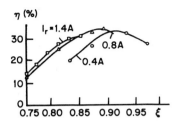

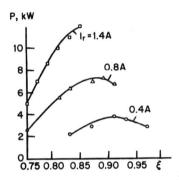

FIG. 41.   The dependence of the output power and efficiency on the parameter $\xi = U_a/U_{a\,cr}$ in the TE_{021} mode, $n = 1$, $\lambda = 2.78$ mm (ξ is the ratio of the operating anode potential to the critical potential). [Reproduced from Zaytsev *et al.* (1974).]

opening in the cryostat). These tubes were demountable, and operated with continuous vacuum pumping and forced water cooling of the gun anode, collector, and cavity. The measured variations in output power and efficiency for fundamental ($n = 1$) operation at 2.78-mm wavelength using the TE_{021} cavity mode are shown in Fig. 41. Corresponding measurements for second harmonic operation ($n = 2$) at 0.92-mm wavelength are shown in Fig. 42. A summary of optimum results is given in Table V.

Achievement of short millimeter wavelength output without use of a superconducting magnet has also been reported (Nikolayev and Ofitserov, 1974). These workers point out that reliable pulsed fields up to 450 kG can be achieved, allowing cyclotron resonance masers to operate at 0.25- ($n = 1$) and 0.12-mm wavelength ($n = 2$). In their experiments a pulsed field up to 200 kG was used ($\tau = 500$ μsec) with special attention paid to the axially symmetric electron gun (Gol'denberg *et al.*, 1972) and a high-selectivity quasi-optical cavity with diffraction output. The high voltage pulse (35 kV at 5 A) had a duration of 3.5 μsec. Oscillation was observed at discrete frequencies between 1.2- and 2.2-mm wavelength, but mode identification was complicated because of the high density of the cavity mode spectrum.

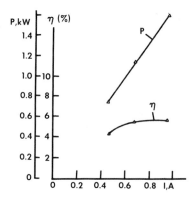

FIG. 42. The current dependence of the output power and efficiency (TE$_{231}$ mode, $n = 2$, $\lambda = 0.92$ mm). [Reproduced from Zaytsev *et al.* (1974).]

Over the entire band, output power reached 10–20 kW, corresponding to an overall efficiency of about 10%.

Gyrotrons for short centimeter wavelengths have also been built (Gaponov *et al.*, 1975). An outline drawing of this class of devices is shown in Fig. 43. One device was designed to operate with the TE$_{011}$ mode at 2.0-cm wavelength at the fundamental cyclotron resonance ($n = 1$); a second device was designed to operate with the TE$_{021}$ mode at 1.2 cm at the second harmonic ($n = 2$). The measured power output and efficiency of these devices are shown in Fig. 44.

High-power gyrotron operation at millimeter wavelengths can introduce serious cooling problems, since the cavity and gun structures tend to be only millimeters in diameter. One way to mitigate this cooling problem is to scale

TABLE V

GYROTRON OPERATING CONDITIONS AND OUTPUT PARAMETERS[a]

Model no.	Oscillation mode	λ, mm	Operation	$n = \dfrac{\omega}{\omega_{\mathrm{H}}}$	He, kOe	U, kV	I, A	P, kW	η, %	η_t, %
1	TE$_{021}$	2.78	Cont.	1	40.5	27	1.4	12	31	36
2	TE$_{031}$	1.91	Cont.	2	28.9	18	1.4	2.4	9.5	15
	TE$_{231}$	1.95	Pulsed	2	28.5	26	1.8	7	15	20
3	TE$_{231}$	0.92	Cont.	2	60.6	27	0.9	1.5	6.2	5

[a] (I, U) Electron beam current and voltage; (P, η) Gyrotron output power and efficiency; (η_t) Theoretical efficiency value for stated operating conditions; and (TE$_0$) Operating magnetic field. [From Zaytsev *et al.* (1974).]

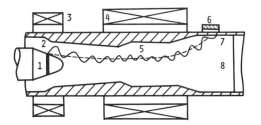

FIG. 43. Basic arrangement of gyrotron: (1) cathode; (2) anode; (3) auxiliary solenoid; (4) main solenoid; (5) resonator; (6) magnetic shield; (7) collector and output waveguide; (8) output port. The wavy line is the electron path, and the dashed line the magnetic field line of force. [From Gaponov et al. (1975), with permission of Plenum Press, New York.]

up the diameters of the cathode and the resonator, but conventional resonators are unsuitable for this because of their overcrowded mode spectrum. Successful utilization of whispering-gallery modes (i.e., modes with high electromagnetic energy density near the walls) has been achieved, with a coaxial conducting rod inserted in the resonator to aid in unwanted mode suppression (Bykov et al., 1975). Thus TE_{521} whispering-gallery mode operation at a wavelength of 2.0 cm was achieved in a pulsed (40 kV, up to 40 A) cyclotron resonance maser. Figure 45 shows output power and efficiency for this device. These power levels (hundreds of kilowatts) at efficiencies of between 30 and 50% represented (at the time) new levels of achievement for gyrotrons.

The most recent Soviet results were announced in 1978 (Andronov et al., 1978). The summary in Table VI was presented by these authors to indicate the present level of achievement. In particular, megawatt level pulsed devices at 3.0- and 6.7-mm wavelength, which were previously unpublished were announced.

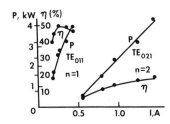

FIG. 44. Power and efficiency of gyrotrons as functions of electron current. The dots correspond to the TE_{011} mode ($n = 1$), the circles to the TE_{021} mode ($n = 2$). [From Gaponov et al. (1975), with permission of Plenum Press, New York.]

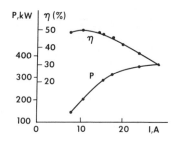

FIG. 45. Output power and efficiency of the gyrotron as functions of beam current (TE$_{521}$ mode). [From Bykov *et al.* (1975), with permission of Plenum Press, New York.]

TABLE VI
RECENT SUMMARY OF SOVIET GYROTRONS[a]

	Pulsed			cw	
Wavelength (mm)	6.7	3.0	2.8	2.0	0.9
Voltage (kV)	65	68	27	40	27
Efficiency (%)	35	34	31	22	6.2
Output power (kW)	1250	1100	12	22	1.5
Pulse length (μsec)	1000	100	—	—	—

[a] Andronov *et al.* (1978).

V. Recent Device Results in the U.S.A.

Since the 1974 announcements of Soviet breakthroughs in gyrotron development, two significant efforts in the U.S.A. have gathered momentum. Preliminary experimental results are available at this writing (Summer 1978) which we will briefly summarize here.

At the U.S. Naval Research Laboratory both an amplifier and an oscillator have been placed into operation. In preliminary tests, the amplifier, shown schematically in Fig. 7, has been shown to provide a gain of 17 dB at 35 GHz with a beam current of 7.5 A in a 13-kG superconducting magnetic field. The initial tests showed output power of about 10 kW. Steps are now being taken to measure both the saturated power level and the bandwidth of the amplifier (see Table III for the design goals). The Naval Research Laboratory oscillator is similar to the amplifier shown in Fig. 7 except that TE$_{011}$ cavity is substituted for the waveguide section. At 35 GHz this device produced an output power of 100 kW at 15% efficiency in a 1.5-μsec pulse. The electron beam parameters were according to the design: 70 kV and 9.0 A. The resonator Q was 200 and the power was coupled out through a transition from TE$_{01}$ circular guide to TE$_{10}$ rectangular guide. The spectral width of the

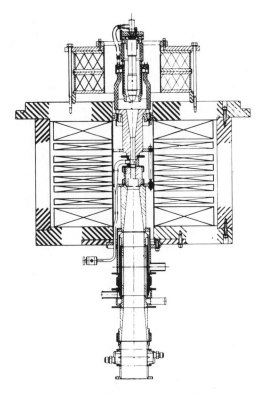

FIG. 46. Outline drawing of an amplifier for electron cyclotron resonance plasma heating, to operate at 28 GHz. [From Symons and Jory (1977).]

oscillator's output was measured to be less than 1 MHz. It is expected that operation with longer pulses (10 msec) will be possible with the same oscillator, so as to allow plasma heating experiments on long pulse fusion devices.

A 200-kW cw 28-GHz gyrotron is currently being developed by Varian Associates, Palo Alto Microwave Tube Division, Palo Alto, California, U.S.A. (Eason, 1977; Jory *et al.*, 1977). Both oscillators and amplifiers are under investigation. An outline drawing of the amplifier is shown in Fig. 46, and a photograph is shown in Fig. 47 with the solenoid removed. A photograph of the oscillator is shown in Fig. 48.

The 28-GHz gyrotron is designed to operate from a negative-polarity 80-kV beam supply at a beam current of 8 A. It employs a crossed-field electron gun which in addition requires an adjustable well regulated low-current 40-kV positive supply referenced to the cathode. The 11-kG water-cooled copper magnet required for beam focusing and for the cyclotron resonance interaction requires a well regulated low-voltage, high-current

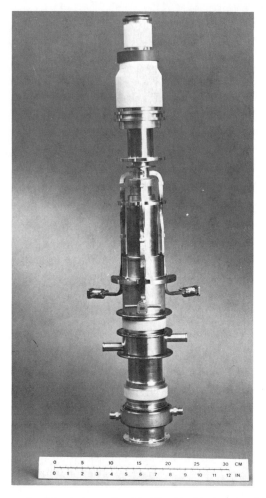

FIG. 47.   Photograph of the device outlined in Fig. 46 with the solenoid removed. [From Symons and Jory (1977).]

supply of $\approx$ 70-kW capacity. Small auxiliary coils located in the region of the electron gun provide field shaping and control over the transverse energy of the hollow electron beam. The output waveguide has a 2.5 in. diam circular cross section, and the output power propagates primarily in low-loss circular electric modes. A 200-kW, cw, rf dummy load has been developed for testing of the gyrotron. Demountable waveguide vacuum windows required for vacuum interface of the waveguide to the plasma device have been adapted from the design used on the gyrotron.

Fig. 48. Photograph of oscillator that produced peak pulsed output of 248 kW at 9 A. [From Symons and Jory (1977).]

A total of three pulsed amplifiers employed different interaction structures and one pulsed oscillator have been built and tested by Varian. Pulse testing has been used through the early stages of development to simplify construction and to permit separate consideration of the thermal design. The gain of the first two amplifiers (and hence the power output) was limited by interfering oscillations. The third amplifier, although exhibiting adequate stable gain ($\approx$ 40 dB) had a maximum peak power output of $\approx$ 75-kW corresponding to an efficiency of 9%. It is believed that the reasons for this relatively low efficiency are now understood, and a fourth modified pulsed amplifier is now under construction.

Best results have been obtained from the single cavity gyrotron oscillator which demonstrated peak 28-GHz power output up to 248-kW at an efficiency of 35%, average power output of 10.7-kW at >200-kW peak with 31% efficiency in high duty cycle operation, and long pulse operation up to 1 msec pulse duration. In addition, it was demonstrated that power output could be controlled over a range of 18 dB by adjustment of a single parameter, viz, the current in one of the auxiliary coils surrounding the electron gun which primarily determines the transverse energy of the electron beam.

VI. Prospectus

It is apparent that important strides have been made in the past decade in the theoretical understanding of the physics underlying the gyrotron. It is likewise apparent that important achievements have been made in device performance, especially regarding high power, high efficiency, and high frequency. What is not yet clear is the degree of precision with which the theory reliably guides the device design. The recent preliminary power and efficiency of the Varian gyroklystron fall below the theoretically anticipated values.

The best Soviet device performance does not appear to result from strict adherence to theoretical design guides, but depends partly upon some "fiddling around." Technical questions, such as whether or not to operate the cathode temperature—or space charge–limited, or how best to collect the spent beam are not well understood. Scientific questions, relating for example to efficient high harmonic ($n > 2$) gyrotron operation without deleterious mode competition, remain open. The design and operation of wide-band turnable low-power gyrotron local oscillators for, say, the 100–500-GHz range would have important impact on radio astronomy and electromagnetic surveillance; no such design program is underway.

What is now needed is a wider experience in device development, an intensification of the underlying theoretical work, and results from experiments designed to test the underlying physics. Given these, it may not be long before this review can be supplanted by a greatly improved second edition.

Acknowledgments

Appreciation is extended to Dr. V. L. Granatstein and his colleagues in the Electron Beam Applications Branch, Plasma Physics Division of the U.S. Naval Research Laboratory for many stimulating discussions. The warm hospitality of Professor Y. Goldstein is acknowledged for making facilities available at the Racah Institute of Physics, Hebrew University of Jerusalem, Israel, where much of the work of assembling this review was carried out. This work was supported by the U.S. Naval Electronics Systems Command, project number XF 54 581 091.

REFERENCES

Andronov, A. A., et al. (1978). Infrared Phys. **18**, 385-393.
Bekefi, G., Hirshfield, J. L., and Brown, S. C. (1961). Phys. Fluids **4**, 173-176.
Bennett, W. R. Jr. (1973). "Some Aspects of the Physics of Gas Lasers." Gordon and Breach, New York.
Borenstein, M., and Lamb, W. E. (1972). Phys. Rev. A **5**, 1298-1311.
Bratman, V. L., and Petelin, M. I. (1975). Radio Phys. Quantum Electron. **18**, 1136-1140.
Bratman, V. L., Moiseev, M. A., Petelin, M. I., and Erm, R. E. (1973). Radio Phys. Quantum Electron. **16**, 622-630.
Bykov, Yu. V., and Gol'denberg, A. L. (1975). Radio Phys. Quantum Electron. **18**, 791-792.
Bykov, Yu. V., Gol'denberg, A. L., Nikolaev, L. V., Ofitserov, M. M., and Petelin, M. I. (1975). Radio Phys. Quantum Electron. **18**, 1141-1143.
Chu, K. R., and Drobot, A. T. (1978). Memorandum Report No. 3788, Naval Research Laboratory, Washington, D.C.; also IEEE Trans. Microwave Theory Tech. (to be published).
Chu, K. R., and Hirshfield, J. L. (1978). Phys. Fluids **21**, 461-466.
Chu, K. R., Drobot, A. T., Granatstein, V. L., and Seftor, J. L. (1977). Memorandum Report No. 3553, Naval Research Laboratory, Washington, D.C.; also IEEE Trans. Microwave Theory Tech. (to be published).
Eason, H. O. (1977). Proc. Plasma Heat. Requirements Workshop pp. 421-427. U.S. Department of Energy, Gaithersburg, Maryland.
Flyagin, V. A., Gaponov, A. V., Petelin, M. I., Yulpatov, V. K. (1977). IEEE Trans. Microwave Theory Tech. **25**, 514-421.
Gaponov, A. V. (1959). Addendum, Izv. Vyssh. Uchebn. Zaved. Radiofiz. **2**, 450-462, 836.
Gaponov, A. V., and Yulpatov, Y. K. (1967). Radio Eng. Electron. Phys. **12**, 582-586.
Gaponov, A. V., Petelin, M. I., and Yulpatov, V. K. (1967). Radio Phys. Quantum Electron. **10**, 794-813.
Gaponov, A. V., Gol'denberg, A. L., Grigor'ev, D. P., Pankratova, T. B., Petelin, M. I., and Flyagin, V. A. (1975). Radio Phys. Quantum Electron. **18**, 204-211.
Gol'denberg, A. L., Pankratova, T. B., and Petelin, M. I. (1972). Soviet Patent 226044, priority July 16, 1967, granted 1972.
Granatstein, V. L. (1977). Private communication.
Granatstein, V. L., and Godlove, T. F. (1978). Memorandum Report No. 3679, Naval Research Laboratory, Washington, D.C.
Hirshfield, J. L., and Granatstein, V. L. (1977). IEEE Trans. Microwave Theory Tech. **MTT-25**, 522-527.
Hirshfield, J. L., Bernstein, I. B., and Wachtel, J. M. (1965). IEEE J. Quantum Electron. **QE-1**, 237-245.
Jory, H. R., Friedlander, F., Hegji, S. J., Shively, J. F., and Symons, R. S. (1977). Proc. Symp. Eng. Probl. Fusion Res., 7th, Knoxville, Tennessee, October 25-28.
Kisel', D. V., Korablev, G. S., Navel'yev, V. G., Petelin, M. I., and Tsimring, Sh. Ye. (1974). Radiotekh. Elektron. **19**(4), 782-797 [Engl. Transl.: Radio Eng. Electron. Phys. **19**, 781-788].
Kolosov, S. V. and Kurayev, A. A. (1974). Radiotekh. Elektron. **19**(10), 2105-2116 [Engl. Transl.: Radio Eng. Electron. Phys. **19**, 65-73].
Kurayev, A. A., and Slepyan, G. Ya. (1975). Radio Eng. Electron. Phys. **20**, 141-144.
Kurayev, A. A., Shevchenko, F. G., and Shestakovich, V. P. (1974). Radiotekh. Elektron. **19**(5), 1046-1056 [Engl. Transl.: Radio Eng. Electron. Phys. **19**, 96-103].
Manheimer, W. M., and Granatstein, V. L. (1977). Memorandum Report No. 3493, Naval Research Laboratory, Washington, D.C.
Moiseev, M. A., and Nusinovich, G. S. (1974). Radio Phys. Quantum Electron. **17**, 1305-1311.

Moiseyev, M. A., Rogacheva, G. G., and Yulpatov, V. K. (1968). All-Union Scientific Session in honor of Radio Day. Annotations and summaries of papers, Izd. NTORES im A. S. Popova, Moscow, p. 6.

Nikolayev, L. V., and Ofitserov, M. M. (1974). *Radio Eng. Electron. Phys.* **19**, 139–140.

Nusinovich, G. S. (1974). *Radio Eng. Electron. Phys.* **19**, 152–155.

Rapoport, G. N., Nemak, A. K., and Zhurakhovskiy, V. A. (1967). *Radio Eng. Electron. Phys.* **12**, 587–595.

Schneider, J. (1959). *Phys. Rev. Lett.* **2**, 504–505.

Sprangle, P., and Drobot, A. T. (1977). *IEEE Trans. Microwave Theory Tech.* **MTT-25**, 528–544.

Symons, R. S., and Jory, H. R. (1977). *Proc. Symp. Eng. Prob. Fusion Res. 7th, Knoxville, Tennessee* pp. 1–5.

Temkin, R. J., Wolfe, S. M., and Lax, B. (1977). Report Number GI, F. Bitter, National Magnet Laboratory, Massachusetts Institute of Technology, Cambridge, Massachusetts.

Twiss, R. Q. (1958). *Aust. J. Phys.* **11**, 424–446, 564–579.

Wachtel, J. M., and Hirshfield, J. L. (1966). *Phys. Rev. Lett.* **17**, 348–351.

Weibel, E. S. (1959). *Phys. Rev. Lett.* **2**, 83–84.

Wolfe, S. M., Cohn, D. R., Temkin, R. J., and Kreischer, K. (1978). Research Rep. RR-78-3, Plasma Fusion Center, Massachusetts Institute of Technology, Cambridge, Massachusetts.

Zarnitsyna, I. G., and Nusinovich, G. S. (1975a). *Radio Phys. Quantum Electron.* **18**, 223–225.

Zarnitsyna, I. G., and Nusinovich, G. S. (1975b). *Radio Phys. Quantum Electron.* **18**, 339–342.

Zaytsev, N. I., Pankratova, T. B., Petelin, M. I., and Flyagin, V. A. (1974). *Radiotekh. Elektron.* **19**(5), 1056–1061 [Engl. Transl.: *Radio Eng. Electron. Phys.* **19**, 103–107].

CHAPTER 2

IMPATT Devices for Generation of Millimeter Waves

H. J. Kuno

I. Introduction

In recent years the development of millimeter-wave systems for various applications has been increasing at a rapid rate. Applications include wide areas covering communications, radar, radiometry, radio astronomy, spectroscopy, plasma diagnostics, medical research, and many other fields.[1] The key to the recent advance of the millimeter-wave systems development is improved semiconductor technology for signal generation, modulation, and detection at millimeter-wave frequencies. In particular, IMPATT devices for millimeter-wave power generation have played an important role for the recent advancement of millimeter-wave systems development.

---

[1] See, for example, Kuno (1976).

A mechanism for achieving an oscillation at a microwave frequency resulting from avalanche breakdown and transit time delay from a somewhat complex diode structure (n^+–p–I–p^+) was first proposed by Read (1958). Experimental observation of a microwave oscillation from a simple p^+–n junction diode biased into avalanche breakdowns was reported first by Johnston et al. (1965). Since then a large amount of effort has been directed toward the theoretical understanding of the physical mechanism and the development of practical microwave and millimeter-wave power sources. In particular, the rapid progress of recent years has been pushing the frequency of operation of IMPATT devices into the upper end of the millimeter-wave region. In this chapter a review of IMPATT devices for millimeter-wave power generation is presented. Selected topics, rather than a comprehensive review, are covered. A brief summary of IMPATT device physics is first presented, followed by design considerations and general features of cw and pulsed operations of IMPATT oscillators and amplifiers at millimeter frequencies. Considerations for system applications of IMPATT devices at millimeter-wave frequencies are also discussed.

II.   Device Physics

Microwave oscillation and amplification are due to the frequency dependent negative resistance arising from the phase delay between the current and voltage wave forms due to the avalanche breakdown and transit time effect. For this reason an acronym "IMPATT" was invented which refers to *IMP*act *A*valanche *T*ransit *T*ime. For operations at millimeter-wave frequencies, simple p^+–n–n^+ type or p^+–p–n–n^+ type silicon diodes have been shown to be effective. Although the feasibility of efficient IMPATT operations from complex device structures such as the original and modified Read structures and GaAs diodes has been demonstrated at lower microwave frequencies, they are not effective at millimeter-wave frequencies. Therefore the discussion in this chapter will be limited to silicon IMPATT diodes. In a p^+–n junction, only the n region contributes to the IMPATT action. In a p–n junction, on the other hand, both p and n regions contribute to the IMPATT operation. For this reason p^+–n–n^+ type IMPATT diodes are called single-drift diodes, and p^+–p–n–n^+ type diodes are called double-drift diodes (Scharfetter et al., 1970).

A.   GENERALIZED MODEL

In order to understand the operation of IMPATT devices, let us consider a p–n junction under reverse avalanche breakdown condition. (See Fig. 1.) A set of equations that govern dynamics of electrons and holes are (Misawa, 1966):

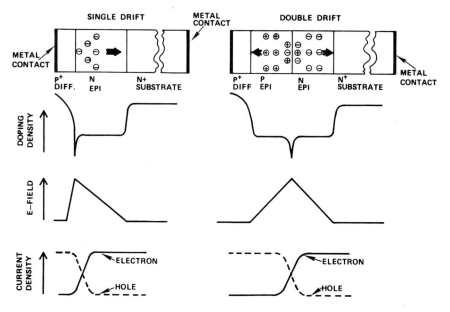

FIG. 1.   IMPATT diode model.

continuity equations for electrons and holes:

$$\partial n/\partial t = (1/q)[(\partial J_n/\partial x) + \alpha_n J_n + \alpha_p J_p], \tag{1}$$

$$\partial p/\partial t = (1/q)[(\partial J_p/\partial x) + \alpha_n J_n + \alpha_p J_p]; \tag{2}$$

and Poisson's equation

$$\partial E/\partial x = (q/\varepsilon)[N_D - N_A + p - n], \tag{3}$$

where p, n are hole and electron densities, respectively, J_p, J_n the hole and electron current densities, α_p, α_n the hole and electron ionization rates, t is the time, x the distance from the junction, and q the electron charge. The total current density is

$$J = qnv_n + qpv_p + \varepsilon \, \partial E/\partial t \tag{4}$$

where v_p, v_n are drift velocities of holes and electrons and ε is the dielectric constant. The first two terms on the right hand side of Eq. (4) represent the conduction current, while the third term represents the displacement current.

B. SMALL SIGNAL ANALYSIS

Now let us separate the electric field E and the current J into time-independent (dc) and time-varying (ac) components, i.e.,

$$E = E_0 + E_1 e^{j\omega t}, \tag{5}$$

$$J = J_0 + J_1 e^{j\omega t}, \tag{6}$$

and let

$$\alpha = \alpha_0 + \alpha' E_1 e^{j\omega t} + \frac{\alpha'' E_1^2}{2!} e^{j2\omega t} + \frac{\alpha'''}{3!} E_1^3 e^{j3\omega t} + \cdots$$

$$= \alpha_0 + [\alpha' + \tfrac{1}{8}\alpha''' E_1^2] E_1 e^{j\omega t} + \cdots, \tag{7}$$

$$n = n_0 + n_1 e^{j\omega t}, \tag{8}$$

$$p = p_0 + p_1 e^{j\omega t}, \tag{9}$$

etc., where ω = angular frequency of operation, $\alpha' = \partial\alpha/\partial E$, $\alpha'' = \partial^2\alpha/\partial E^2$, $\alpha''' = \partial^3\alpha/\partial E^3$, etc.

The subscripts "0" and "1," respectively, refer to the dc and ac components where the amplitudes of ac components are assumed much smaller than dc components. Substituting Eq. (5)–(9) into Eqs. (1)–(3) and separating the dc–ac components, we obtain

$$\partial E_0/\partial x = (1/\varepsilon)[(1/v_n) + (1/v_p)]J_{n0} + (q/\varepsilon)(N_D - N_A) - J_0/v_p, \tag{10}$$

$$\partial J_{n0}/\partial x = (\alpha - \beta)J_{n0} + \beta J_0 \tag{11}$$

where $J_0 = J_{n0} + J_{p0}$ is the total dc bias current density. This set of dc equations can be solved numerically, subject to a set of boundary conditions. The ac equations are obtained by retaining only the linear terms in the ac parts. Then we get

$$\partial E_1/\partial x = j(\omega/v_p)E_1 + (1/\varepsilon)[(1/v_n) + (1/v_p)]J_{n1} - J_1/\varepsilon v_p, \tag{12}$$

$$\partial J_{n1}/\partial x = (\alpha' J_{n0} + \beta' J_{p0} - j\omega\varepsilon\beta)E_1$$
$$+ [\alpha - \beta - j(\omega/v_n)]J_{n1} + \beta J_1, \tag{13}$$

where the derivatives with respect to t are replaced by $j\omega$ and $J_1 = J_{n1} + J_{p1} + j\omega E_1$ is the total ac current density. Having obtained the dc solutions, the ac equations can also be solved numerically for any desired frequency ω. The ac voltage V_1 across the depletion region is given by

$$V_1 = \int E_1 \, dx \tag{14}$$

and the ac impedance Z_1 or admittance Y_1 can be obtained from

$$Z_1 = 1/Y_1 = V_1/J_1A \tag{15}$$

where A is the cross-sectional area of the device. The above set of equations outline the general procedure for a small signal analysis of IMPATT devices.

Note that N_A and N_D can vary with x in a real diode and that material parameters such as α, β, v_p and v_n vary with temperature. Small signal admittance (or impedance) of various device structures can be calculated numerically as a function of frequency under various bias current and temperature conditions. The numerical analysis is normally carried out with the aid of computers.

As an example, small signal admittance of a double-drift IMPATT diode designed for operations around 100 GHz calculated for a typical bias current density is shown in Fig. 2. The small signal analysis reveals many of the important properties of IMPATT diodes. The real part of the admittance (conductance) of the diode is negative over a broad bandwidth. The reactive part goes through a resonance below which it is inductive and above which it is capacitive.

Thus small signal characteristics of an IMPATT diode may be represented by an equivalent circuit consisting of a negative conductance (or resistance), a capacitance, and an inductance, as shown in Fig. 3. An ideal equivalent circuit should have frequency independent parameters. An examination of the small signal characteristics of an IMPATT diode reveals that a parallel combination of a negative conductance $-G_n$, a junction capacitance C_j, and an inductance L_a provides fairly frequency independent values over a broad bandwidth around the normal operational frequency of interest, where the small signal negative conductance has maximum values. Although the rf output power limit and many other phenomena of an IMPATT device are determined by large signal characteristics of the diode, large signal values are closely related to small ones. Thus small signal values provide a convenient basis for comparing performance of IMPATT devices under various conditions. Physically the capacitance represents the diode junction capacitance, while the inductance and the negative conductance arise from the avalanche–transit-time mechanisms.

C. SIMPLIFIED MODEL

In order to illustrate the physical significance of the avalanche multiplication and the transit time effects in IMPATT devices, let us first consider a simplified p^+–n junction model with a uniform doping density profile as shown in Fig. 4. We follow an analysis first considered by Read (1958), then further developed by Gilden and Hines (1966). In this model the active portion (space charge region) of a diode is divided into two portions: (1) the avalanche

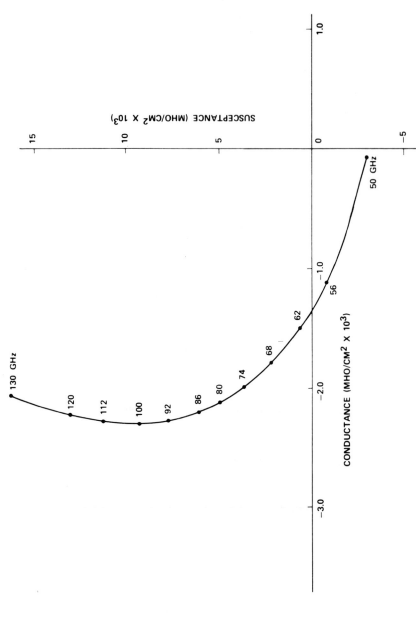

FIG. 2.   Small signal admittance of a double-drift silicon IMPATT diode for operation in the 100-GHz range. $J = 24,000$ A/cm^2, $N_A = N_0 = 2 \times 10^{17}$/cm^3, and $T_j = 250°$C.

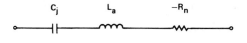

SERIES EQUIVALENT CIRCUIT

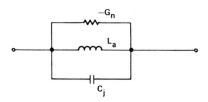

PARALLEL EQUIVALENT CIRCUIT

FIG. 3.   IMPATT diode equivalent circuits.

AVALANCHE   DRIFT

REGION   REGION

$E = E_m$ $\alpha = 0$

$\alpha = \alpha_m$ $v = v_n$

W_a W_d

FIG. 4.   Simplified IMPATT diode model: In avalanche region, $E = E_m$ and $\alpha = \alpha_m$; in drift region, $\alpha = 0$ and $v = v_n$.

region, in which the thickness is assumed to be thin so that transit time delay is negligible and the ionization rate is uniform ($\alpha_n = \alpha_p = \alpha$, $\partial \alpha / \partial x = 0$), and (2) the drift region, in which no carriers are generated and all carriers entering from the avalanche region move at their saturated (scatter limited) velocities.

In the avalanche region we let

$$\alpha'_n = \alpha'_p = \alpha'$$

$$\partial E / \partial x = 0$$

$$E = E_m$$

$$v_n = v_p = v.$$

Under these conditions, we get

$$jwq(n_1 + p_1) = 2\alpha' J_0 E_m \tag{16}$$

$$vq(n_1 + p_1) = J_1 - jw\varepsilon E_m \tag{17}$$

or

$$j(\omega/v)(J_1 - jw\varepsilon E_m) = 2\alpha' J_0 E_m. \tag{18}$$

Solving for E_m, we get

$$E_m = j(\omega/\varepsilon)J_1/(\omega_a^2 - \omega^2) \tag{19}$$

where

$$\omega_a^2 = 2\alpha' J_0 v_n/\varepsilon. \tag{20}$$

The voltage across the avalanche region is

$$V_a = E_m W_a \tag{21}$$

where W_a is the thickness of the avalanche region.

In the drift region, we have

$$\alpha = 0, \qquad J_n \gg J_p,$$

$$E = 0 \qquad \text{at} \qquad x = W_a.$$

Then we get

$$J_1 = qn_1 v_n + jw\varepsilon E_1 \tag{22}$$

$$dE_1/dx = (q/\varepsilon)n_1 = (1/\varepsilon v)(J_1 - jw\varepsilon E_1) \tag{23}$$

or

$$(dE_1/dx)J(\omega/v_n)E_1 = J_1/(\varepsilon v). \tag{24}$$

Solving for E_1 and noting that

$$V_d = \int_0^{W_d} E_1 \, dx \tag{25}$$

and

$$dE/dx = -(q/\varepsilon)(N_D - J_0/qv), \tag{26}$$

we get

$$V_d = E_m j(v_n/\omega)(-1 - e^{-j\theta}) + (J_1/\omega E_m)[(v/\omega)(e^{-j\theta} - 1) + jW_n] \tag{27}$$

where $\theta = \omega\tau = \omega W_0/v_n$. Noting that the total current I_1 is

$$I_1 = J_1 A \tag{28}$$

and the diode capacitance is

$$C_S = \varepsilon A/(W_a + W_d) \tag{29}$$

where A is the cross sectional area of the diode, we can obtain the following expression for the frequency dependent small signal impedance of the IMPATT diode:

$$Z_{SD} = \left(\frac{1}{\omega C_S}\right)\left[\frac{W_d/W}{1 - (\omega/\omega_a)^2}\right]\left\{\left[\frac{1 - \cos(\omega\tau)}{(\omega\tau)}\right]\right.$$
$$\left. -j\left[1 - \left(\frac{W}{W_d}\right)\left(\frac{\omega}{\omega_a}\right)^2 - \frac{\sin(\omega\tau)}{(\omega\tau)}\right]\right\}. \tag{30}$$

This illustrates the physical significance of avalanche ionization and transit effects in an IMPATT diode. An examination of Eq. 30 reveals that the real part (the first term) has negative values for frequencies above the critical value, viz., avalanche frequency

$$f_a = \omega_a/2\pi = (2\alpha' J_0 v_n/\varepsilon')^{1/2}/2\pi.$$

The imaginary part (the second part) is inductive at frequencies below f_a and becomes capacitive at frequencies above f_a. A similar analysis can be carried out for a simplified double-drift IMPATT diode model. We get

$$Z_{dd} = \left(\frac{1}{\omega C_D}\right)\left[\frac{1}{1 - (\omega/\omega_a)^2}\right]\left\{\left[\frac{1 - \cos(\omega\tau)}{(\omega\tau)}\right] - j\left[1 - \left(\frac{\omega}{\omega_a}\right)^2 - \frac{\sin(\omega\tau)}{(\omega\tau)}\right]\right\} \tag{31}$$

where, for a symmetrical p–n junction double-drift IMPATT diode, we let

$$\tau = \tau_n = \tau_p = W_n/v_n = W_p/v_p,$$
$$C_D = \varepsilon A/(W_n + W_p).$$

Note that $W_d/W \approx 5/6$ in Eq. (30) for a single-drift IMPATT diode and $C_D \approx (3/5)C_S$ for diodes of equal junction area designed for the same frequency range. Then, comparing Eqs. (30) and (31), we find that $Z_{DD} \approx 2Z_{SD}$.

In other words the double-drift IMPATT diode is equivalent to two single-drift IMPATT diodes connected in series. This is one of the major reasons for the fact that double-drift IMPATT diodes yield much higher output power than single-drift IMPATT diodes, particularly at millimeter-wave frequencies.

D. LARGE SIGNAL CHARACTERISTICS

In order to understand various nonlinear effects and transient responses of an IMPATT diode operated as an oscillator or an amplifier, large signal characteristics of the IMPATT diode must be taken into consideration. In the preceding section it was shown that an IMPATT diode may be characterized in terms of a parallel equivalent circuit consisting of a negative conductance $-G_n$, a junction capacitance, C_j, and an avalanche inductance L_a, as shown in Fig. 3.

Shown in Fig. 5 is a simplified equivalent circuit for a circulator-coupled IMPATT amplifier/oscillator. It consists of a circulator, an impedance transformer, a tuning element, and an IMPATT diode. Although an equivalent circuit of a practical IMPATT amplifier/oscillator in general may be more complex, it can be reduced to a simple form, as shown in Fig. 5, over a limited frequency range of specific interest. The IMPATT diode may be represented by a parallel combination of a negative conductance $-G_n$, a junction capacitance C_j, and an avalanche inductance L_a. We assume that the device is excited by a purely sinusoidal voltage waveform and that the amplitude and phase of the voltage are slowly varying functions of time in comparison with the fundamental frequency. At small signal levels, these parameters are relatively constant. However, at large signal levels, the diode parameters are dependent on the signal level. The nonlinear characteristics of the IMPATT diode parameters may be expressed by polynominal functions of the amplitude of the sinusoidal voltage across the diode terminals, i.e.,

$$-G_n = -g_n + \gamma V_d^2 + \sum_{n=2} a_n V_d^{2n} \tag{32}$$

$$(1/L_a) = (1/l_a) + \lambda V_d^2 + \sum_{n=2} b_n V_d^{2n} \tag{33}$$

$$C_j = \text{const} \tag{34}$$

where $-g_n$ and l_a are small signal parameters; γ, λ, a_n, and b_n are constants; and V_d is the amplitude of the sinusoidal voltage across the diode terminals, as shown in Fig. 5. In the region of practical IMPATT amplifier operation where the amplitude of the sinusoidal voltage is much smaller than the dc bias voltage, the higher order terms in Eqs. (32) and (33) may be neglected. Then Eqs. (32) and (33) reduce to Van der Pol-type nonlinearities (1927). The equivalent load admittance seen from the diode terminals is represented by

$$Y_L = G_L - j(1/\omega L_{ext}). \tag{35}$$

Since the imaginary component of the diode is capacitive at a normal operation frequency, an inductive load is required for tuning. The equivalent voltage and current components of the incident and reflected waves are represented by V_i, i_i, V_r, and i_r, where the subscripts i and r, respectively, refer to the incident and reflected waves. The incident and reflected waves are separated into the input and output signals by the circulator. The net current

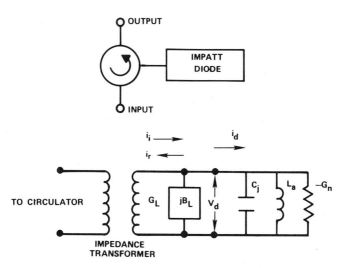

FIG. 5. A large signal equivalent circuit for a circulator-coupled IMPATT amplifier/oscillator. $-G_n = -g_n + \gamma V_d^2$, $1/L_a = 1/l_a + \lambda V_d^2$, and $B_L = -1/\omega L_{ext}$. [From Kuno (1973).]

into the diode i_d and voltage across the diode terminals V_d are given by

$$V_d = V_i + V_r, \tag{36}$$

$$i_d = i_i - i_r = G_L(2V_i - V_d). \tag{37}$$

In the equivalent circuit of Fig. 5, i_d and V_d are governed by a nonlinear integral–differential equation given by (Kuno, 1973)

$$i_d = \int_{-\infty}^{t} [(1/L_{ext}) + (1/L_a)]V_d \, dt - G_n V_d + C_j(dV_d/dt) \tag{38}$$

Substituting Eqs. (32)–(37) into Eq. (38) we get

$$\begin{aligned}(2G_L/C_j)dV_i/dt &= d^2V_d/dt^2 - (1/C_j)(g_n - G_L - 3\gamma V_d^2)(dV_d/dt) \\ &\quad + (\omega_0^2 - (\lambda/C_j)V_d^2)V_d\end{aligned} \tag{39}$$

where

$$\omega_0^2 = [(1/L_{ext}) + (1/L_a)](1/C_j) \tag{40}$$

Let us assume that the circuit Q is relatively high and that V_i, V_r, and V_d are given by

$$V_i = V_0 A_i(t) \sin[\omega t + \phi_i(t)], \tag{41}$$

$$V_r = V_0 A_r(t) \sin[\omega t + \phi_r(t)], \tag{42}$$

$$V_d = V_0 A_d(t) \sin[\omega t + \phi_d(t)], \tag{43}$$

where V_0 is an arbitrary nonzero constant and ω is the angular frequency of the input signal. $A(t)$ and $\phi(t)$ are, respectively, amplitude- and phase-modulation functions. We assume that both $A(t)$ and $\phi(t)$ are slowly varying functions of time in comparison with the carrier frequency, i.e.,

$$dA/dt \ll \omega A \qquad \text{and} \qquad d\phi/dt \ll \omega \tag{44}$$

and that the carrier frequency is close to the resonant frequency, i.e.,

$$\omega \simeq \omega_0. \tag{45}$$

Note that since V_i, V_r, and V_d are related by Eq. (36), we also have

$$A_r = [A_i^2 + A_d^2 - 2A_i A_d \cos(\phi_i - \phi_d)]^{1/2} \tag{46}$$

$$\phi_r = \tan^{-1}[(A_d \sin \phi_d - A_i \sin \phi_i)/(A_d \cos \phi_d - A_i \cos \phi_i)] \tag{47}$$

Substituting Eqs. (41)–(43) into Eq. (39), making use of conditions given by Eqs. (44) and (45), and retaining only terms involving the fundamental frequency, e.g., $\sin^3 \psi \simeq \frac{3}{4} \sin \psi$, we can derive a set of coupled differential equations that govern the amplitude and phase by the following method: Multiply the resulting equation by $\cos(\omega t + \phi_d)$ and integrate over a period. Similarly, multiply the equation by $\sin(\omega t + \phi_d)$ and integrate over a period. After simplifying the resulting equations using Eqs. (44) and (45), we obtain

$$(2/\omega_0)dA_d/dt + (A_d^2 - \varepsilon)A_d = \eta A_i \cos(\phi_i - \phi_d), \tag{48}$$

$$(2/\omega_0)A_d \, d\phi_d/dt + (\beta A_d^2 + \delta)A_d = \eta A_i \sin(\phi_i - \phi_d), \tag{49}$$

where

$$
\begin{aligned}
\varepsilon &= (g_n - G_L)/\omega_0 C_i, \\
V_0^2 &= (\tfrac{4}{9})(\omega_0 C_i/\gamma), \\
\beta &= 3\lambda V_0^2/4\omega\omega_0 C_i \simeq \lambda/3\omega_0 \gamma, \\
\delta &= \omega/\omega_0 - \omega_0/\omega \simeq (2/\omega_0)(\omega - \omega_0), \\
\eta &= 2G_L/\omega_0 C_j, \\
\omega_0^2 &= (1/L_{ext} + 1/L_a)1/C_j.
\end{aligned}
\tag{50}
$$

From Eqs. (48) and (49), A_d and ϕ_d are to be solved for a given set of input signal amplitude function $A_i(t)$ and phase function $\phi_i(t)$. Note that A_d and ϕ_d do not refer to the output signal. The output amplitude and phase functions $A_r(t)$ and $\phi_r(t)$, respectively, must be obtained from Eqs. (46) and (47). In the following section, steady-state behavior and transient response of IMPATT oscillators and amplifiers under various conditions are analyzed (Kuno, 1973).

1. Free-Running Oscillation

Let us first consider a free-running condition, i.e., $A_i = 0$ in Eqs. (48) and (49). In this case, we have

$$(2/\omega_0)dA_d/dt = -(A_d^2 - \varepsilon)A_d, \tag{51}$$

$$(2/\omega_0)d\phi_d/dt = -(\beta A_d^2 + \delta). \tag{52}$$

Note that ε may be either positive or negative depending on loading conditions. If $\varepsilon < 0$, i.e., $G_L > g_n$, dA_d/dt becomes negative for any positive values of A_d. This means that the oscillation is decaying, and only when $A_d = 0$, dA_d/dt becomes zero. Thus, when $G_L > g_n$, the IMPATT diode is stable, no free-running oscillation exists, and no output power is generated.

If $\varepsilon > 0$, i.e., $G_L < g_n$, and $A_d^2 < \varepsilon$, then dA_d/dt is positive, indicating that the oscillation is growing. If $A_d^2 > \varepsilon$, then dA_d/dt becomes negative, indicating that the amplitude of oscillation is decreasing. Steady oscillation can be found when $A_d^2 = \varepsilon$. When $A_d^2 = 0$, although dA_d/dt also becomes zero, any small disturbance such as noise can trigger growing oscillation. Thus, when $\varepsilon > 0$, the IMPATT diode is unstable, and the amplitude of the steady-state oscillation is determined by

$$A_r^2 = A_d^2 = \varepsilon. \tag{53}$$

The output power delivered to the load G_L is given by

$$P_{out} = \tfrac{1}{2}V_0^2 V_r^2 G_L = V_0^2 \varepsilon G_L = \tfrac{2}{9}[(g_n - G_L)/\gamma)]G_L \tag{54}$$

and the free-running frequency by

$$\omega_{osc} = \omega_0(1 - \beta\varepsilon/2) = \omega_0 - (\lambda/6\gamma)[(g_n - G_L/\omega_0 C_j)]. \tag{55}$$

From Eq. (54) we find that when $G_L = g_n/2$, the output power becomes maximum, i.e.,

$$P_{op} = \tfrac{1}{18}(g_n^2/\gamma) = (P_{out})_{max}. \tag{56}$$

In Fig. 6, variation of output power as a function of load, as given by Eq. (54), is plotted.

2. Stable Amplifier

For steady-state condition, i.e.,

$$dA_d/dt = d\phi_d/dt = 0,$$

Eqs. (48) and (49) become

$$A_d^2 - \varepsilon A_d = \eta A_i \cos(\phi_i - \phi_d), \tag{57}$$

$$\beta A_d^2 + \delta A_d = \eta A_i \sin(\phi_i - \phi_d). \tag{58}$$

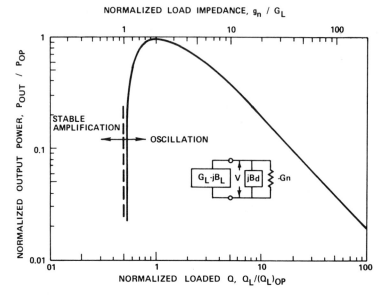

FIG. 6. Power output dependence on load impedance of an IMPATT diode oscillator. $P_{op} = g_n^2/18\gamma$, $(G_L)_{op} = g_n/2$, $Q_{op} = 2B_d/g_n$, and $-G_n = g_n + \gamma v^2$. [From Kuno (1973).]

It was shown that when $\varepsilon < 0$, i.e., $G_L > g_n$, the IMPATT diode is stable. For a small signal condition, i.e., $A_d^2 \ll \varepsilon$, we get

$$-\varepsilon(A_d/A_i) = \eta \cos(\phi_i - \phi_d), \qquad (59)$$

$$\delta(A_d/A_i) = \eta \sin(\phi_i - \phi_d). \qquad (60)$$

At the center frequency ω_0 we get $\phi_i - \phi_d = 0$, and the power gain is given by

$$(A_r/A_i)^2 = (A_d/A_i - 1)^2 = (-(\eta/\varepsilon) - 1)^2 = [(G_L + g_n)/(G_L - g_n)]^2 \quad (61)$$

as one expects from a transmission-line reflection equation.

From Eqs. (46), (47), and (57), we find that the gain at a given frequency may be given by

$$G(\omega) = (A_r/A_i)^2 = 1 + (A_d/A_i)^2 - 2(A_d/A_i)\cos(\phi_i - \phi_d)$$
$$= 1 + [(\eta/\varepsilon) + 2](\eta/\varepsilon)\cos^2(\phi_i - \phi_d).$$

Noting that from Eq. (61),

$$-\eta/\varepsilon = \sqrt{G_0} + 1,$$

we get

$$(G(\omega) - 1)/(G_0 - 1) = \cos^2(\phi_i - \phi_d).$$

If we define the amplifier bandwidth $2\,\Delta\omega$ in such a way that, at $\omega_\pm = \omega_0 \pm \Delta\omega$, gain becomes

$$(G_\pm - 1)/(G_0 - 1) = \tfrac{1}{2} \tag{62}$$

where $G_\pm = G(\omega_0 \pm \Delta\omega)$, $G_0 = G(\omega_0)$, ω_0 equals midband frequency, then, from Eqs. (59) and (60), we get

$$(2\,\Delta\omega/\omega_0)(\sqrt{G_0} + 1) = \eta = \text{const.} \tag{63}$$

For $G_0 \gg 1$, as in most amplifiers, this reduces to the conventional definition of normalized gain-bandwidth product, i.e.,

$$(2\,\Delta\omega/\omega_0)\sqrt{G_0} \simeq \eta \tag{64}$$

where $2\,\Delta\omega$ is defined by $G(\omega_0 \pm \Delta\omega) = G(\omega_0)/2$. For a range of practical interest where $G_0 \gg 1$, G_L does not vary appreciably for a large variation of gain. Thus the normalized gain-bandwidth product given by Eq. (63) remains relatively constant for a wide range of gain variation. For this reason the gain-bandwidth product is used as a measure of the amplifier quality.

3.   *Injection-Locked Oscillator*

In preceding sections it was shown that when $\varepsilon > 0$, i.e., $G_L < g_n$, the IMPATT diode becomes unstable and results in oscillation at a frequency given by

$$\omega_{\text{osc}} = \omega_0(1 - \beta\varepsilon/2) = \omega_0 - (\lambda/6\gamma)(g_n - G_L)/\omega_0 C_j \tag{65}$$

and amplitude of oscillation given by

$$A_d = \sqrt{\varepsilon} = (g_n - G_L)/\omega_0 C_j. \tag{66}$$

The oscillating IMPRATT diode can also be used for amplification of microwaves by means of injection locking. For a given A_i in Eqs. (59) and (58), a real steady-state solution for A_d can be found only in the region where

$$-\pi/2 \le (\phi_i - \phi_d) \le \pi/2.$$

Physically, this means that the oscillation cannot be locked to the input signal outside of this region. For a small input signal level, i.e., $A_i^2 \ll A_d^2$, we get $A_d^2 \simeq \varepsilon$. Then, since $\sin(\phi_i - \phi_d) \le 1$, at the edge of the locking band, vis., $\omega_\pm = \omega_{\text{osc}} \pm \Delta\omega$, we get

$$(\beta\varepsilon + \delta)(A_d/A_i) = (2\,\Delta\omega/\omega_0)(\sqrt{G} + 1) = \eta \tag{67}$$

and

$$\phi_i - \phi_d = \pm\pi/2.$$

Note that η is constant for a given loading condition. Thus the normalized locking gain-bandwidth product given by Eq. (67), which is a measurement of the injection-locked oscillator, is constant for a given loading condition.

Thus, for both a stable amplifier and an injection-locked oscillator, the normalized gain-bandwidth product is equal to $\eta = (2G_L/\omega_0 C_j)$. Comparing Eqs. (63) and (67), and noting that for a given IMPATT diode, a stable amplifier requires a larger value of G_L than an oscillator, we can see that it is easier to obtain a larger gain-bandwidth product with a stable IMPATT amplifier than with an injection-locked oscillator. It should be noted that any parasitic capacitance such as a package capacitance placed in parallel with C_j will in effect reduce the gain-bandwidth product of the amplifier.

4. Large Signal Effects

In order to analyze the large-signal effects, Eqs. (57) and (58) must be solved. The coupled nonlinear differential equations can be solved nu-

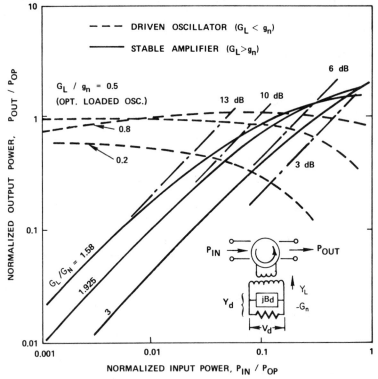

FIG. 7. Power saturation characteristics of IMPATT amplifiers. $P_{op} = g_n^2/18\gamma$, $Y_L = G_L$ $-jB_L$, $Y_d = -G_n + jB_d$, and $-G_n = g_n + V_d^2$. [From Kuno (1973).]

merically to evaluate output vs. input power characteristics of an **IMPATT** diode. Shown in Fig. 7 are the output vs. input power characteristics of IMPATT amplifiers and those of injection-locked oscillators, numerically calculated from Eqs. (57) and (58) together with Eqs. (46) and (47) at center frequencies, i.e., $\omega = \omega_0$ for stable amplifiers and $\omega = \omega_{\mathrm{osc}}$ for injection-locked oscillators. The loading conditions determine the small signal gain and bandwidth. The output and input power are normalized to the optimally loaded oscillation power given by Eq. (56). An injection-locked oscillator yields high output power with a higher gain (with a narrow bandwidth), while a stable amplifier yields high output with a low gain (with a broader bandwidth).

From Eqs. (57) and (58) together with (56) and (59), nonlinear characteristics of IMPATT amplifiers under various conditions can be calculated. Figure 8 shows calculated effects of the input signal level on the bandpass and phase characteristics of a stable IMPATT amplifier tuned to a small signal gain of 10 dB. The input and output power levels are normalized to the optimally loaded oscillator power $P_{\mathrm{op}} = g_{\mathrm{n}}^2/18\gamma$ and the frequency is normalized to $f_0 = \omega_0/2\pi$. It can be seen that as the input signal level increases the gain decreases, the bandwidth increases, and the center frequency shifts to the lower side.

The lowering of the center frequency may result in gain expansion with increasing input signal level at the lower half of the amplifier passband under certain conditions such as a high-gain narrow-band amplifier. Experimental observation of this phenomenon has been reported in many papers. It is also interesting to note the pronounced decrease in gain near the center of the amplifier band at high input signal levels, i.e., $P_{\mathrm{in}} > 0.2P_{\mathrm{op}}$. This nonlinear effect has also been observed experimentally.

Shown in Fig. 9 are locking characteristics of an IMPATT diode similarly calculated. The locking characteristics of the injection-locked IMPATT oscillator were found not to be symmetrical about the frequency of free-running oscillation.

In Figs. 10 and 11 the large signal effects are shown differently. Variations of phase shift and power gain calculated for a stable IMPATT amplifier tuned to 10-dB small signal gain are plotted as a function of input signal power level. At small signal levels (linear amplification) both power gain and phase shift remain constant. At large signal levels phase and gain will change due to the nonlinear characteristics of the diode parameter. This results in various nonlinear effects such as AM–PM conversion and intermodulation. In an injection-locked oscillator, both phase shift and output power vary with signal level even at very low input power levels except at $\omega = \omega_{\mathrm{osc}}$, where the output power and phase shift remain fairly constant at low signal levels.

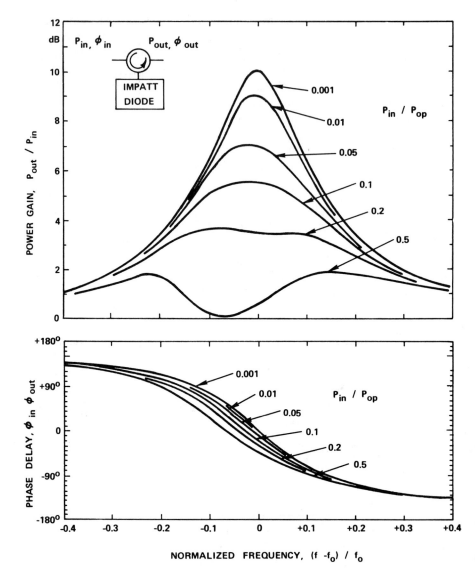

FIG. 8.　Large signal effects on gain-bandwidth and phase-shift characteristics of stabilized IMPATT amplifier. $G = P_{out}/P_{in}$, and the optimally loaded oscillator output is $P_{op} = g_n^2/18\gamma$. [From Kuno (1973).]

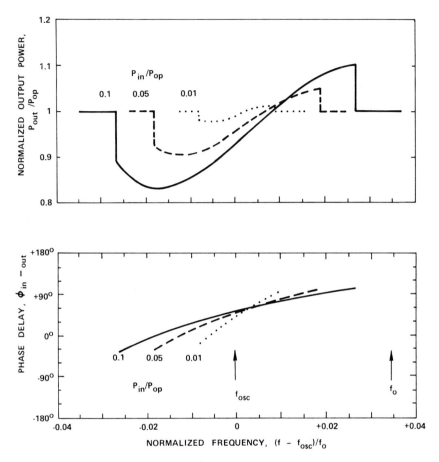

FIG. 9. Injection locking characteristics of IMPATT oscillator. [From Kuno (1973).]

5. *Transient Response*

In addition to the steady-state characteristics, the transient response of an IMPATT amplifier that determines data-rate capability is also of great importance. The transient response of an IMPATT amplifier can be analyzed by solving for output and phase responses $A_r(t)$ and $\phi_r(t)$ to given input amplitude and phase waveforms $A_i(t)$ and $\phi_i(t)$ from Eqs. (46)–(49). The analyses of transient responses of IMPATT amplifiers are, however, not covered in this chapter. Those who are interested in pursuing the transient analysis are referred to the reference cited at the end of this chapter (Kuno, 1973).

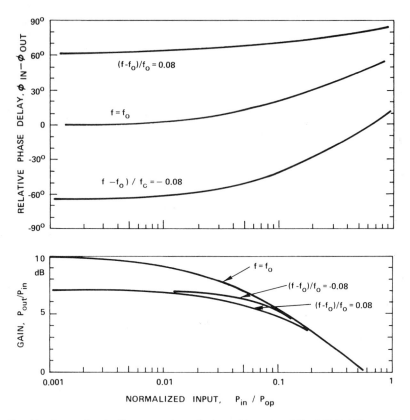

FIG. 10. Large signal effects on gain and phase delay of stabilized IMPATT amplifier. $P_{op} = g_n^2/18\gamma$ is the optimally loaded oscillator output. [From Kuno (1973).]

III. CW Oscillators

Most commonly available power sources are cw oscillators. In this section device and circuit design considerations for cw operation of millimeter-wave IMPATT devices and their performance and operational characteristics are covered.

A. Diode Design Considerations

1. p^+-n-n^+ IMPATT Diode

The simple p^+-n-n^+ type single-drift IMPATT diode structure consists of a thin layer of heavily doped p^+ region, an n-type active region, and a low resistivity n^+ type substrate as shown in Fig. 12. Critical material parameters that determine rf performance of the diode are the doping density and the

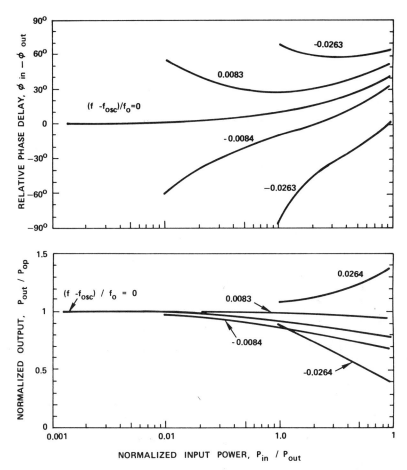

FIG. 11.   Large signal effects on power output and phase delay of injection-locked IMPATT oscillator. $P_{op} = g_n^2/18\gamma$ is the optimally loaded oscillator output. [From Kuno (1973).]

thickness of the n type epitaxial layer, the p^+–n junction, and the n–n^+ interface. The n^+–n interface should be made as abrupt as possible consistent with epitaxial growth technique. If this junction is excessively graded, a high parasitic series resistance results that reduces the output power and efficiency.

The p^+–n junction should also be made as abrupt as possible. However, there are somewhat conflicting requirements on the depth of the diffused (or ion implanted) p^+–n junction. If the junction is too shallow, it is not well isolated from the surface damage. If the junction is diffused too deeply, the abrupt p^+–n junction will not result. This also causes out-diffusion of impurities from the n^+ substrate into the epitaxial layer, resulting in a graded

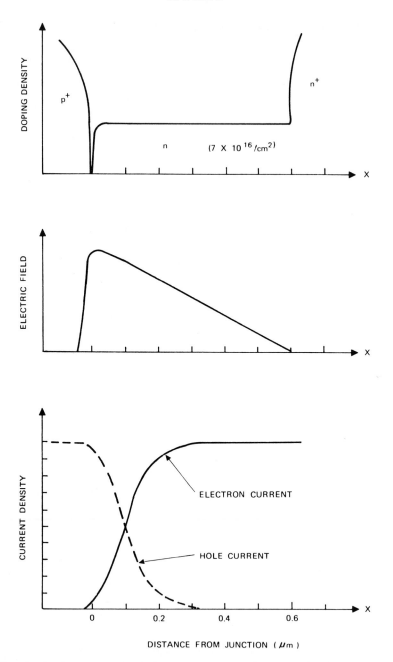

FIG. 12.   Doping density, electric field, and bias current density profiles of a typical milli-meter-wave single-drift silicon IMPATT diode.

n–n$^+$ interface. Furthermore, the increased p$^+$-region thickness also increases the thermal impedance. The depth and abruptness of the p$^+$–n and n$^+$–n junctions become more critical at millimeter-wave frequencies as the diode junction geometry becomes very small. Typically the p$^+$–n junction depth for millimeter-wave IMPATT diodes is less than 0.3 μm.

The reverse voltage applied across the diode creates a space charge region. The width of the space charge region increases as the applied voltage increases until avalanche breakdown takes place.

Electron-hole pairs are generated by the avalanche ionization in the high field region near the p$^+$–n junction.[2] Electrons then drift toward the interface, and holes to the p$^+$-region.

The ratio of the avalanche region width to the total space charge region width is an important parameter that determines the magnitude of the negative resistance (conductance) and maximum output power generated (Schroeder and Haddad, 1971). Small signal analysis has shown that the narrower the avalanche region with respect to the total space charge region, the larger the negative conductance and the greater the output power will be (Kuno, 1972).

For a uniform doping density as shown in Fig. 13a, the avalanche region width is about 33% of the total space charge region width. If the doping density is graded increasing toward the n$^+$ substrate as shown in Fig. 13b, the portion of the avalanche region will increase. If the punch-through condition occurs as in Fig. 13c, the avalanche region width also increases. If, on the other hand, the doping density decreases toward the n$^+$ substrate as shown in Fig. 13d, the avalanche region decreases. However, parasitic series resistance due to the high resistivity material at the n–n$^+$ interface will reduce the output power and efficiency. An excessive epitaxial layer thickness as shown in Fig. 13e also increases series resistance that will reduce the output power. Thus, a millimeter-wave IMPATT diode should have an epitaxial layer with a uniform doping density profile and a thickness such that the space charge region just reaches the n–n$^+$ interface under the normal operational condition.

The space charge region width at avalanche breakdown that primarily dictates the optimum frequency of operation is determined by the doping density. The space charge region width at a breakdown voltage varies as a function of doping density of the n-region as shown in Fig. 14. It should be borne in mind that the values shown in Fig. 14. are calculated for room temperature. Under normal operational condition of an IMPATT diode, the

---

[2] There are no distinct boundaries between the avalanche and drift regions; hence there is no unique definition of an avalanche region width. The boundary may be defined as the point where the hole or electron current density decreases to 10% of the total current.

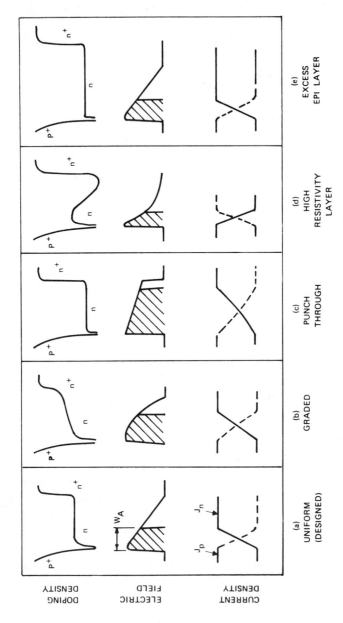

Fig. 13.   Effects of doping density profile of IMPATT diodes on effective avalanche region width.

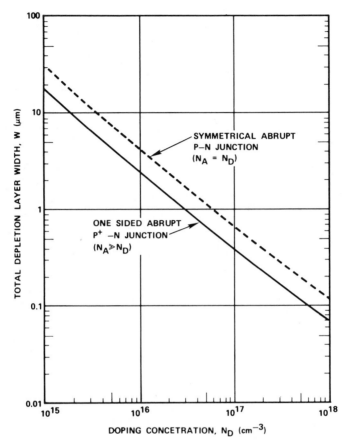

FIG. 14a. Variations of depletion layer width at breakdown (Sze, 1969; Sze and Gibbons, 1966).

junction temperature is much higher; about 500°K. As the junction temperature changes from 300 to 500°K, the space charge region width increases by about 10% for the doping density range of 5×10^{16}–10^{17}/cm², as shown in Fig. 15 (Crowell and Sze, 1966), i.e.,

$$(\Delta W/W)/\Delta T = (\Delta V_B/V_B)/\Delta T \simeq 5 \times 10^{-4}°C \tag{68}$$

The space charge region width also increases as the bias current increases. The rate of the change is given by

$$(\Delta W/W)\, \Delta J = J/qN_D v \tag{69}$$

where J is the current density and v the saturated drift velocity of electrons. Note that v also decreases as temperature increases (Gibbons, 1967). (See

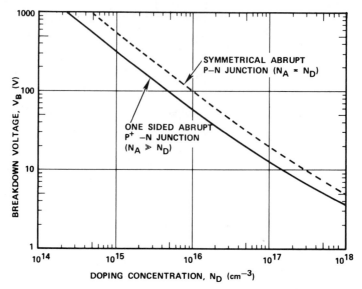

FIG. 14b.   Breakdown voltage as a function of doping density for p^+–n and p–n junction diodes ($T = 300°$K). $W = W_p + W_n$.

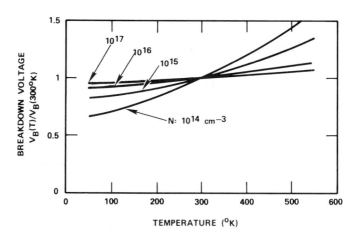

FIG. 15.   Temperature dependence of breakdown voltages in p–n junction diodes. [From Crowell and Sze (1966).]

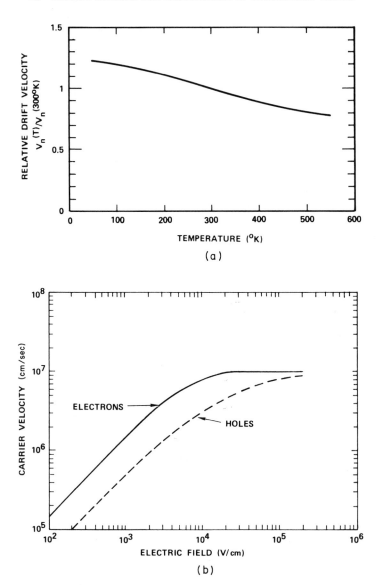

FIG. 16.   (a) Temperature dependence of saturated electron drift velocity in silicon, and (b) field dependence of carrier velocities in silicon ($T = 300°$K.) [From Gibbons (1967); see also Duh and Moll (1967), Norris and Gibbons (1967), Rodrigues *et al.* (1967).]

Fig. 16.) For a typical millimeter-wave IMPATT diode with $N_D \sim 7 \times 10^{16}/\text{cm}^3$, we have $J \sim 10^4$ A/cm^2, $v \sim 0.8 \times 10^7$ cm/sec at $T_j \sim 500°$K. The space-charge region increases by 10% due to temperature and 10% due to the current. The space charge region thickness determines the transit time of the drifting carriers. The transit time primarily determines the optimum frequency range of the IMPATT diode operation. According to large signal analysis and small signal calculation of IMPATT diodes, the optimum transit angle is given by (Mouthaan, 1969; Delagebeaudenf, 1970)

$$\theta_{op} = 2\pi f_{op}\tau = 3\pi/4, \qquad \tau = W_d/v, \tag{70}$$

where f_{op} = optimum operation frequency, τ = transit time, W_d = drift (transit) length, and v = saturated velocity. Since the avalanche ionization takes place throughout the avalanche region, the transit time is also distributed. However, it is most appropriate to define the effective (mean) transit length as the distance between the edge of the space charge region and the center of the avalanche region where the electron and hole current densities become equal, as shown in Fig. 12.

Noting that $W_a \simeq \frac{1}{3}W_n$, we find that the optimum n-region thickness for p^+–n–n^+ structure is given by

$$(W_n)_{op} = \tfrac{6}{5}(W_d)_{op} = (\tfrac{6}{5})0.75(v_n/f) = 0.45(V_n/f)$$

For example at $f = 60$ GHz, $v_n = 0.8 \times 10^7$ cm/sec, we get

$$(W_n)_{op} = 0.6 \quad \mu\text{m}.$$

In order to determine the doping density by using the data shown in Fig. 14, we must note that

$$W_{sp}\left(\frac{@\,300°\text{K}}{J_0 \simeq 0}\right) \simeq 0.8 W_{sp}\left(\frac{@\,500°\text{K}}{J_0 \sim 10^4 \text{ A/cm}^2}\right) \tag{71}$$

Then we find that

$$
\begin{aligned}
N_D &= 7 \times 10^{16}/\text{cm}^3, \\
W_n &= 0.6 \quad \mu\text{m}, \\
W_p^+ &= 0.3 \quad \mu\text{m}, \\
W_{epi} &= W_n + W_{p^+} = 0.9 \quad \mu\text{m}, \\
BV_R &= 15 \text{ v}.
\end{aligned}
$$

The design of the 60-GHz p^+–n–n^+ IMPATT diode is shown in Fig. 17. Small signal admittances of the diode, calculated by using the computer program discussed in the preceding section, as a function of frequency for various bias current densities are shown in Fig. 18. The magnitude of the negative conductance reaches peak value in the frequency range about

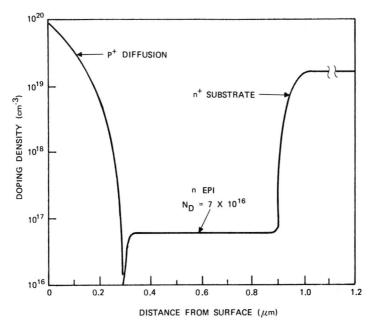

FIG. 17. Doping density profile design for a single-drift IMPATT diode for operation in the 60-GHz range.

60 GHz, indicating that the diode design is optimized for this frequency range. However, it should be noted that the negative conductance remains fairly constant over a wide frequency range, indicating a wide bandwidth operation of the device.

In Fig. 19 optimum doping density and epitaxial layer thickness similarly calculated for millimeter-wave IMPATT diodes are plotted as a function of operational frequency.

2. $p^+-p-n-n^+$ Double Drift IMPATT Diode

In the p^+-n-n^+ structure, only electrons have a proper transit angle that provides a negative conductance (or negative resistance). By adding another layer (p type) as shown in Fig. 19, it is possible to construct a diode with a single avalanche region and two drift regions, one for electrons and one for holes. Since the avalanche region is shared by both p and n type space charge regions, and electron and hole velocities are nearly equal, the transit length of the carriers are equal to each half of the total space charge region, i.e.,

$$(W_n)_{op} = (W_p)_{op} = (W_d)_{op} = \tfrac{3}{8}(v/f) \tag{72}$$

$$(W_n)_{op} = (W_p)_{op} = 0.375(v/f) \tag{73}$$

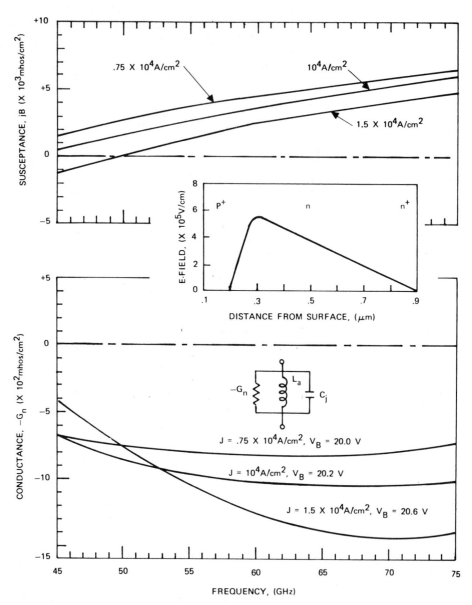

FIG. 18.   Frequency dependent small signal admittance for a 60-GHz single-drift IMPATT diode.

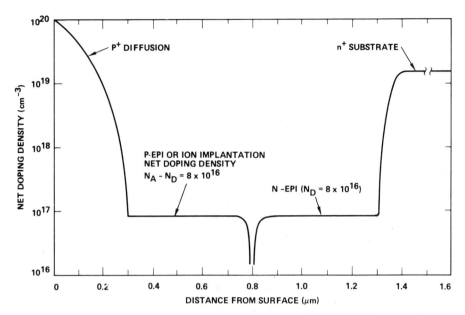

FIG. 19. Doping density profile design for a double-drift IMPATT diode for operation in the 60-GHz range.

For example, at $f = 60$ GHz, we get

$$W_n = W_p = 0.5 \quad \mu\text{m},$$

and

$$W_{sp}\left(\begin{array}{c}@300°\text{K} \\ J = 0\end{array}\right) \simeq 0.8(W_n + W_p)\left(\begin{array}{c}@500°\text{K} \\ J \sim 10^4 \text{ A/cm}^2\end{array}\right)$$

yields

$$N_D = N_A = 8 \times 10^{16}/\text{cm}^3,$$
$$W_n = W_p = 0.5 \quad \mu\text{m},$$
$$W_{p^+} = 0.3 \quad \mu\text{m},$$
$$W_{epi} = W_n + W_p + W_{p^+} = 1.3 \quad \mu\text{m},$$
$$BV_R = 23\text{v}.$$

The rf characteristics of the double-drift IMPATT diode structure are equivalent to two single-drift diodes in series. However, since the avalanche region is shared by the two drift regions, the total voltage of the double-drift diode is only about 50% higher than that of the single-drift diode (Scharfetter et al., 1970; Seidel and Scharfetter, 1970; Seidel et al., 1971). Thus, the double-drift diode yields higher efficiency and greater output power per unit area. The output power for IMPATT diodes is ultimately limited by

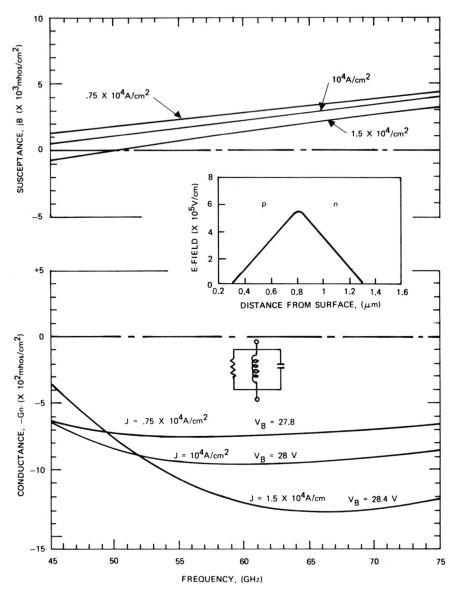

FIG. 20. Frequency dependent small signal admittance of a 60-GHz double-drift IMPATT diode.

the realizable circuit impedance, in particular at high frequencies. Since the double-drift IMPATT diode yields higher impedance for a given junction area, the junction area can be increased for a given circuit impedance level. In this way about four times the output power of a single-drift diode can be expected from a double-drift diode. Figure 20 shows calculated frequency dependent small signal admittance of the double-drift IMPATT diode designed for operation around 60 GHz under various bias current densities.

Figure 21 shows variations of optimum doping density, epitaxial layer thickness, and breakdown voltage similarly calculated for the single-drift

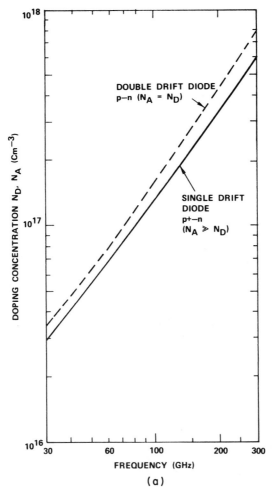

(a)

FIG. 21.   (a) Optimum doping density. (Figures 21b and 21c on following page.)

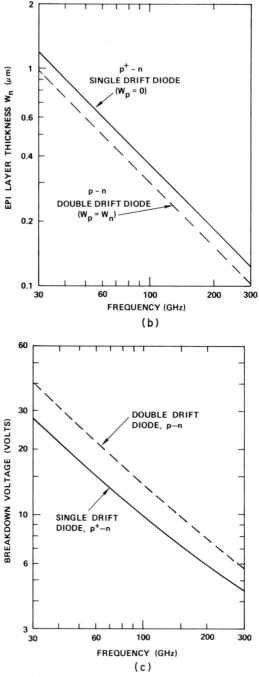

FIG. 21. (b) Epitaxial layer thickness, and (c) breakdown voltage of single- and double-drift IMPATT diodes.

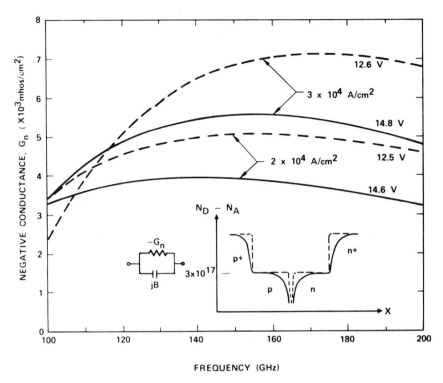

FIG. 22.   Comparison of negative conductance of 140-GHz double-drift IMPATT diodes, one with an ideal step junction (dashed line), and one with a realistic junction (solid line).

and double-drift IMPATT diodes as a function of frequency. In designing an IMPATT diode it should be noted that the negative conductance of an IMPATT diode covers a broad bandwidth and that, in a real diode, doping density transition is not an ideal step function. A real p–n junction has a doping density transition thickness of 0.1–0.3 μm. This thickness becomes a significant portion of the total active region of an IMPATT diode at millimeter-wave frequencies, particularly above 100 GHz. This doping density transition would result in reduction of the magnitude of the negative conductance. Shown in Fig. 22 is a comparison of small signal admittance of an ideal abrupt p–n junction IMPATT diode and that of a real IMPATT diode with a transition region designed for operation in the 140-GHz range. Significant reduction in the magnitude of the negative conductance can be seen. Thus, in designing millimeter-wave IMPATT diodes, effects of the nonabrupt doping density transition should be taken into consideration.

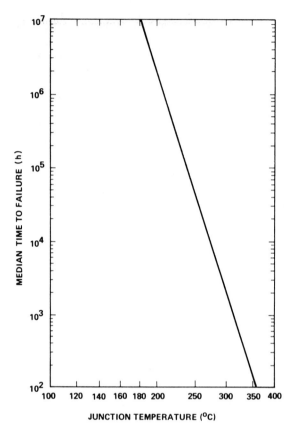

FIG. 23. Median time to failure vs. junction temperature characteristics for millimeter-wave silicon IMPATT diodes derived from accelerated life tests. [From Bernick (1973).]

B. THERMAL DESIGN

For reliable operation, IMPATT diode junction temperatures should be kept as low as possible. On the basis of life tests[3] it has been estimated that 250°C is safe for multiple-year operations as shown in Fig. 23. In order to keep the junction below 250°C and at the same time achieve the maximum output power, the heat which is dissipated in the diode must be removed efficiently. This heat removal can best be accomplished by placing a heat sink as near as possible to the junction where the heat is generated. To accomplish this goal in practice the diode is thermal-compression bonded to a copper heat sink with the p⁺ side down as shown in Fig. 24. Similarly, copper or silver

---

[3] Hughes Aircraft Company, (1974).

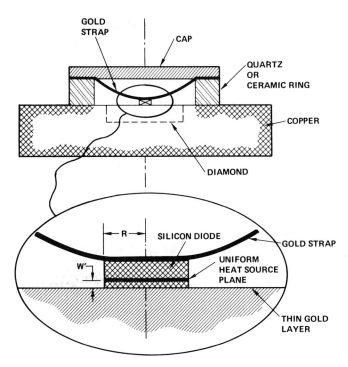

FIG. 24.   Typical package configuration and thermal model for a millimeter-wave IMPATT diode.

heat sink may also be plated on to the p^+ side of the diode. Since Type IIa diamond provides high thermal conductivity, metallized Type IIa diamond is also used as a heat sink for millimeter-wave diodes. In addition, the p^+ layer should be made as thin as possible. Since the thermal resistance due to the p^+ layer is inversely proportional to the area, this requirement is accentuated particularly at millimeter-wave frequencies, where the diode junctions necessarily become small.

Consider the thermal model of IMPATT diode as shown in Fig. 24. The model contains all the important features of actual devices, and excellent agreements have been found between calculated values of thermal resistance based on the models and experimentally determined values. The p^+ region of the cylindrical p^+–n–n^+ diode is bonded to the copper heat sink through an interface of gold. (Thin buffer metallization layers such as Cr and Pt are neglected in the analysis as their contribution to the thermal path is small.) The mechanism of heat flow is conduction from the active region of the diode through the heat sink to ambient. Detailed calculations (Bernick, 1973) indicate that the effect of corrections introduced to take into account heat flow up through the substrate and gold contact ribbon is small. The temperature for the configuration is highest along the diode center line. Since the diode radius is much larger than the length of the heat path within the diode (typically by a factor of thirty), a one-dimensional mathematical treatment for the layered diode structure may be employed to calculate the steady-state distribution of this maximum temperature. The power dissipation rate and heat flux are considered to be uniform in the radial direction. The heat sink is taken to be a semi-infinite half-space.

The heat dissipation profile follows the electric field profile in the active region since the current is constant. Thus, the power dissipation is maximum at the p^+–n interface (in a double-drift diode) and decreases linearly to zero toward the n–n^+ interface.

For the millimeter-wave IMPATT diode, it can be shown that although the power dissipation is distributed throughout the device active region, the effective heat source plane may be placed near the p–n junction without introducing a significant amount of error in the analysis. This means that the problem can be treated exactly as if all heat is generated at this plane. The thermal conductivity of the silicon k_{Si} depends on the temperature. However, detailed calculations (Bernick, 1973) show that the temperature variation across the active region is small, so that only a small error is introduced by using a constant value for k_{Si}. With this assumption, a solution of the one-dimensional linear heat flow equation for silicon leads simply to

$$R_{Si} = W'/k_{Si} A \tag{74}$$

where W' is the distance from the effective heat source plane to the heat sink and A the device area. For a p^+–n–n^+ diode structure, we have $W' = W_{p^+} + \frac{1}{3}W_n$ and for a p^+–p–n–n^+ structure, we have $W' = W_{p^+} + W_p$.

The solution for the thermal spreading resistance for a two–layer heat sink has been obtained for the case where the thermal conductivities of the material in each layer are essentially independent of temperature. Such is the case for a gold bonding layer on a copper heat sink. The result is

$$R_{HS} = \frac{1}{\pi R k_1} \int_0^\infty \frac{1 + pe^{-2UH}}{1 - pe^{-2UH}} J_1(U) \frac{dU}{U} \tag{75}$$

Where $H = t/R$, $p = (k_1 - k_2)/(k_1 + k_2)$, and where k_1 and k_2 are, respectively, the thermal conductivities of the gold layer and copper heat sink. The thermal conductivity of diamond has an inverse temperature dependence. However, an estimate of thermal resistance can be achieved by assuming a constant value of 9 W/cm °K for the thermal conductivity of diamond.

The thermal analysis for the above model involves a straightforward solution of the one-dimensional heat conduction equation. The maximum diode temperature T_{max} occurs at the n–n^+ interface. The diode thermal resistance R_{TH} is defined as

$$R_{TH} = (T_{max} - T_A)/P_{diss} \qquad (76)$$

where T_A is the ambient temperature and P_{diss} the total power dissipated. The thermal resistance may be approximately represented as a function of diode junction diameter d as

$$R_{TH} = (4/\pi)(1/k_{Cu}\, d\{1 + (t/d)[(k_{Cu}/k_{Au}) - 1]\} + (4/\pi)(w'/K_{Si}\, d^2) \qquad (77)$$

where k's are thermal conductivities of copper (or diamond), gold, and silicon; d is the diode junction diameter; w' the distance from the effective heat source plane to the diode heat sink interface; and t the thickness of the gold layer between silicon and copper. The first term of Eq. (77) is the contribution to R_{TH} from the copper heat sink with a small correction factor

$$(t/d)[(k_{Cu}/k_{Au}) - 1]$$

arising from the presence of the gold layer.

TABLE I

PARAMETERS USED IN THERMAL RESISTANCE CALCULATION

Quantity	Symbol	Value
Diode diameter	d	Variable
Thickness of p$^+$ region	w_p^+	0.3 μm
Thickness of n-region	w_n	0.6 μm
Thickness of Au layer	t	1.0 μm
Power dissipated	P	5 W
Ambient temperature	T_A	300°K
Thermal conductivity of copper	k_{Cu}	3.94 W/cm°K
Thermal conductivity of gold	k_{Au}	2.97 W/cm°K
Thermal conductivity of silicon	k_{Si}	0.8 W/cm°K
Thermal conductivity of diamond	k_{dia}	9.0 W/cm°K

The thermal resistance of millimeter-wave IMPATT diodes, calculated as a function of the junction diameter by using typical parameters as shown in Table I, is plotted in Fig. 25a. Note that the Type IIa diamond heat sink provides a thermal path twice as effective as the copper heat sink. In practice, however, it is easier to measure the junction capacitance, which is a function of the diode geometry. In Fig. 25b the calculated thermal resistance of the milli-meter-wave IMPATT diodes is plotted as a function of zero bias junction capacitance. This curve is useful for evaluating power dissipation capability for a given diode design.

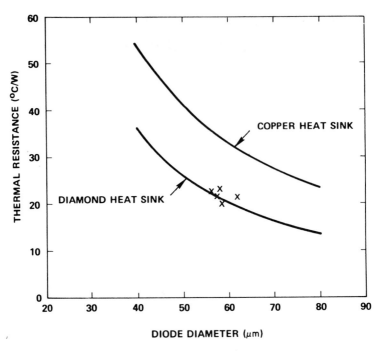

FIG. 25a.   Thermal resistance vs. diode diameter (× —experimental values for diamond heat sink).

C.   CW OSCILLATOR PERFORMANCE

Various types of oscillator/amplifier circuits have been developed for millimeter-wave IMPATT devices. Examples of basic types are depicted in Fig. 26. They may be grouped into three basic types, viz., reduced height, hat resonator, and coaxial waveguide cavity circuits. In each circuit, tuning elements are used to achieve optimum performance at a specified frequency. The frequency tuning is accomplished by varying the position of the moveable short.

Shown in Fig. 27 are typical tuning characteristics of a millimeter-wave oscillator. It can be seen that an IMPATT diode can be operated over a broad bandwidth since the negative conductance covers a wide frequency band.

Shown in Fig. 28 is state-of-the-art of cw output power achieved with IMPATT oscillators as a function of frequency. It varies from 2 W at 40 GHz to several milliwatts at 230 GHz. It appears that at frequencies below 100 GHz we have a relationship pf = const, indicating that the output

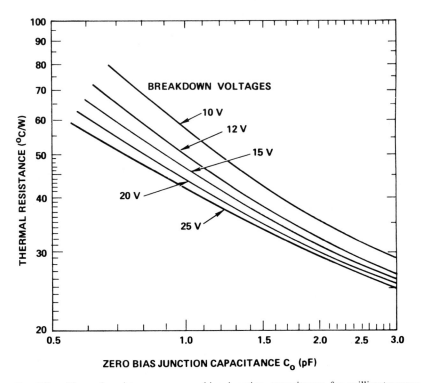

FIG. 25b. Thermal resistance vs. zero bias junction capacitance for millimeter-wave IMPATT diodes (silicone diode on copper heat sink).

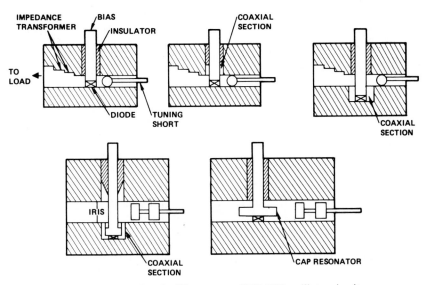

FIG. 26. Examples of millimeter-wave IMPATT oscillator circuits.

95

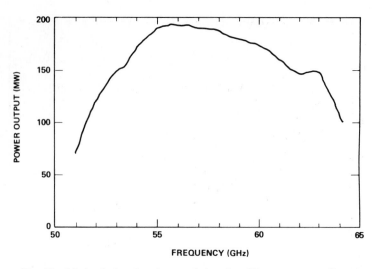

FIG. 27. Mechanical tuning characteristics of a millimeter-wave oscillator.

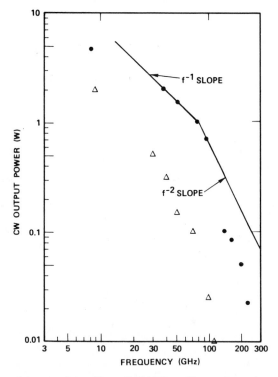

FIG. 28. State-of-the-art of cw silicon IMPATT oscillators. Legend: —●—●—, IMPATT; –△–△–, GUNN (Chao *et al.*, 1977; Hirachi *et al.*, 1976; Ishibashi *et al.*, 1976; Kuno *et al.*, 1974; Kuno, 1976; Misawa and Marinaccio, 1970; Miyakawa *et al.*, 1975; Well *et al.*, 1975; Weller *et al.* 1976).

power is determined by thermal limitation. Above 100 GHz we have $pf^2 =$ const, indicating that the power output is determined by the circuit impedance limit. The steep power fall off at high frequencies is mainly due to the increased adverse effects of diode package and mounting parasitics. With further development effort, higher output power would result, specifically at the higher part of the millimeter-wave range.

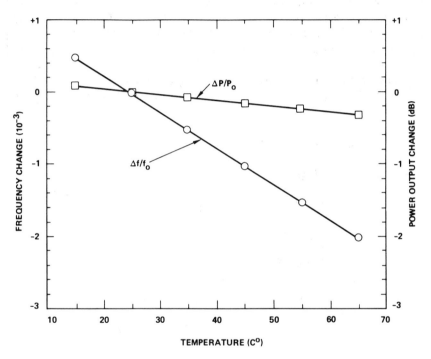

FIG. 29. Temperature effects on frequency and power output of millimeter-wave IMPATT oscillator. $f_0 = 60$ GHz, $P_0 = 425$ MW.

Small signal parameters of an IMPATT diode are strongly dependent on junction temperature. Thus the frequency of oscillation and power output of an IMPATT oscillator are temperature dependent. Shown in Fig. 29 are temperature effects on the frequency and power output of an IMPATT oscillator. Typical temperature coefficients for frequency drift and power output variation of millimeter-wave IMPATT oscillators are $-5 \times 10^{-5}/°C$ and -0.005 dB/°C, respectively.

D. FREQUENCY MODULATION

Frequency modulation is one of the important aspects of millimeter-wave sources. Frequency modulation of millimeter-wave IMPATT sources can be accomplished in a number of ways; YIG, varactor, and bias current tuning. YIG tuning is limited to the lower end of the millimeter-wave spectrum. Varactor tuning is limited to a relatively small frequency of deviation (1–2 GHz). For the millimeter-wave IMPATT oscillator, the most effective way of frequency modulation is the bias current tuning technique. Both broadband swept frequency generations and high modulation rates have been achieved with millimeter-wave IMPATT oscillators. Frequency modulation by bias current tuning of the millimeter-wave IMPATT oscillator is covered in this section.

An IMPATT oscillator may be represented by the equivalent circuit shown in Fig. 30. The small signal impedance of an IMPATT diode which is de-

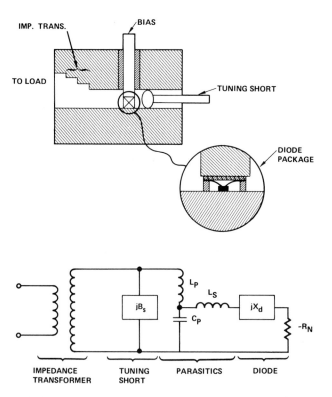

FIG. 30. Equivalent circuit of wide band millimeter-wave IMPATT swept frequency generator.

pendent on the bias current may be represented by (Gilden and Hines, 1966)

$$z_d = -R_n + jX_d = (1/\omega C)(W_d/W)[1 - (\omega/\omega_a)^2]^{-1}\{1 - \cos(\omega\tau)/\omega$$
$$-j[1 - (W/W_d)(\omega/\omega_a)^2 - \sin(\omega\tau)/(\omega\tau)]\}$$

where $\omega = 2\pi f$ is the angular frequency of oscillation, τ the electron transit time across the drift region of the diode, C the junction capacitance of the diode at the breakdown voltage, W_d the drift region thickness, and W the total depletion layer thickness. $\omega_a = (2\alpha'vJ_0/\varepsilon)^{1/2}$ is the angular avalanche resonant frequency, where $\alpha' = d\alpha/dE$ is the rate of chance of electron ionization coefficient with respect to the electric field E in the avalanche zone, ε the dielectric constant, and J_0 the bias current density.

The parasitic reactances arising from the diode package and mounting are represented by a strap inductance L_s, post inductance L_p, and package capacitance C_p. B_S is the magnitude of the susceptance due to the movable short and Y_0 the characteristic admittance of the reduced height waveguide section. In order to couple out the maximum power to the load and to eliminate spurious resonances so that a smooth, continuously swept frequency oscillation can be achieved, L_p and C_p must be made very small in comparison with L_S and $B_S\omega$, respectively. The effective load resistance R_L must be made smaller than R_N to achieve oscillation. In order to achieve low noise oscillation, we must have $B_S \gg Y_0$.

The reduced height waveguide section provides the two conditions required for a broadband FM operation, viz., small L_p and small R_L. The small diode package developed specifically for millimeter-wave operation provides the small C_p necessary for the broadband FM oscillator. By placing the short approximately one-half wavelength away from the diode, the condition $B_S \gg Y_0$ can be achieved. The impedance transformer is used to match impedance levels between the reduced height waveguide section and the standard waveguide.

The oscillation frequency is determined by the matching of the load impedance and the diode impedance. Under the conditions that $B_S \gg Y_0$, $B_S \gg \omega C_p$, and $L_p \gg L_s$, the oscillation frequency ω is determined by

$$-(1/\omega C)(W_d/W)[1 - (\omega/\omega_a)^2]^{-1}$$
$$\times[1 - (W/W_d)(\omega/\omega_a)^2 - \sin(\omega\tau)/\omega\tau] + \omega L_s = 0.$$

Using an approximation that

$$\sin(\omega\tau)/\omega\tau \cong 1 - (\omega\tau/\pi)^2$$

for normal operation, where $0 < \omega\tau < \pi$, we get

$$\omega^2 = (W/W_d)\omega_0^2 + (1 - W_d/W)(\omega_0\tau/\pi)^2\omega_a^2$$

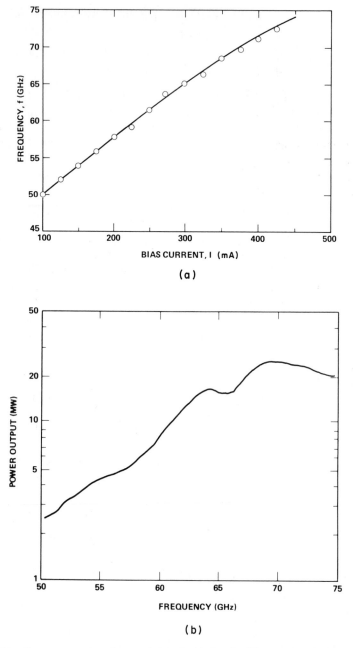

FIG. 31. Frequency tuning characteristics of wide band millimeter-wave IMPATT swept frequency generator: (a) calculated $[f = (40^2 + 8.8I)^{1/2}]$; and (b) measured.

where

$$\omega_0^2 = (L_s C_p)^{-1} \quad \text{and} \quad \omega_a^2 = (2\alpha' v J_0/\varepsilon).$$

Thus, the dependence of the oscillation frequency f on the bias current I_B may be given by

$$f = (A + BI_B)^{1/2}$$

where A and B are constants. The relationship given by the equation derived above dictates design conditions. For example, in order to achieve high frequency oscillations, ω_0^2 should be maximized by minimizing C_p and L_s. To achieve a broadband sweep, ω_0 should be made small by using a low breakdown voltage diode.

Shown in Fig. 31 are the calculated and measured variations of oscillation frequency as a function of the bias current for a typical V-band IMPATT sweeper. Good agreement between the calculation and the measured data can be seen.

In addition to the broadband swept frequency generation discussed above, frequency modulation of a relatively narrow frequency deviation but at a high modulation rate is desirable. Direct bias current controlled IMPATT oscillators can be operated at modulation rates in excess of 100 MHz. Shown in Fig. 32 is a millimeter-wave IMPATT oscillator developed for an FM communication link with a data rate capability of 100 Mbits/sec.

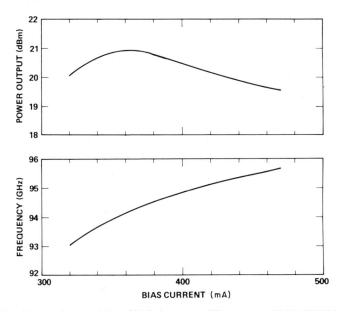

FIG. 32.   Tuning characteristics of high data rate millimeter-wave IMPATT FM source.

IV. Pulsed Oscillators

In many systems applications, high peak power pulsed oscillators are required. IMPATT devices can be operated as pulsed power sources to achieve high peak power output over a relatively short pulse width (~ 100 ns) with a low duty cycle. The device for pulsed operations requires quite different design and operational conditions from those of cw operations. The major difference results from the fact that pulsed IMPATT devices operate at much higher bias current densities than cw devices. Thus, in this section unique design considerations and operational characteristics of pulsed millimeter-wave IMPATT oscillators are covered.

Since the double-drift silicon IMPATT diode structures are much more effective in producing high peak power output than the single-drift diode structure, the discussion in this section will be concentrated on double-drift silicon IMPATT diodes.

A. Diode Design Considerations

The primary IMPATT diode design consideration for pulsed operation is the impedance-frequency characteristic of the diode as a function of current density. Because IMPATT operation is strongly current dependent, the frequency for peak negative conductance is a function of the operating current. As current density increases, the optimum frequency increases and so does the diode output power. For cw operation the maximum current density is limited thermally, but for pulsed operation this limit is extended to much higher levels depending on the pulse width and duty factor. For extremely narrow pulse widths and low duty operation, the diode is no longer thermally limited. The current density can be further extended until space charge effects cause power saturation and efficiency reduction, i.e., the ultimate diode output power is limited electronically.

To properly design an IMPATT diode for pulsed operation, it is necessary to predetermine the operating density. However, the operating current density is also dependent on other factors, such as pulse width, duty factor, power output, and device impedance. The choice of an optimum current density must therefore be derived from certain trade offs and the optimization of a set of parameters simultaneously. With a given operating current density and maximum junction temperature, the diode doping profile and epitaxial layer thickness can be optimized to obtain maximum negative conductance at the desired frequency of operation.

The following diode design is based on a symmetrical double-drift structure. The structure has equal doping densities in the n and p regions. It has been established both theoretically and empirically that the symmetrical double-drift IMPATT diodes are capable of nearly optimum performance in terms

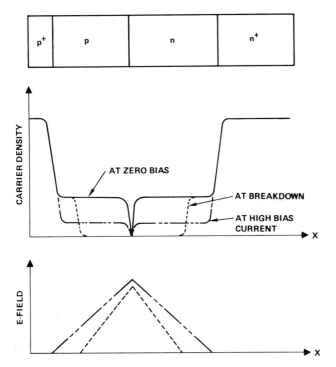

FIG. 33.   Effects of high peak bias current in double-drift pulsed IMPATT diode.

of output power and efficiency. The asymmetrical design which compensates for the unequal ionization rates of electrons and holes results in very minor improvement in the device performance. The minor improvement that can be gained from an asymmetrical design, however, can be easily lost through complication in epi growth or uncertainties in the design parameters, such as field and temperature dependence of the electron and hole ionization rates. Thus our discussion in this section follows double-drift IMPATT diodes with a symmetrical doping profile, i.e., $N_A = N_D$ as depicted in Fig. 33.

The frequency of optimum operation is dependent not only on doping density but also on bias current density. In cw IMPATT operation, the optimum transit angle is about 0.75π as discussed earlier. However, in pulsed IMPATT operation, where the bias current density is much higher, optimum transit angle increases to about π, i.e.,

$$\omega_{op}\tau = 2\pi f_{op} W/v = \pi.$$

Then we have

$$W \simeq v/2f_{op}$$

where W is the effective drift width, v the carrier drift velocity, and f_{op} the frequency of optimum operation. For example, for operation in the 100 GHz range, we find that

$$W \simeq 0.4 \quad \mu m.$$

Taking the effective drift length as the distance between the center of the avalanche region and the edge of the space charge region, we can take one side of the double-drift IMPATT diode as the effective drift length. Then we get

$$W_{sp} = W_n + W_p = 0.8 \quad \mu m.$$

Noting that the space charge region increases with increasing temperature and bias current as in cw IMPATT diodes, we must take this effect into consideration in determining the doping density.

Noting that typical bias current densities for pulsed operation of millimeter-wave IMPATT diodes in the 100-GHz range are on the order of $5 \times 10^4/cm^2$ and the doping density is in the $1-1.5 \times 10^{17}/cm^3$ range, we have

$$(\Delta W/W)_J \sim J/qN_D v \sim 30\% \quad \text{and} \quad (\Delta W/W)_T \sim 5\%.$$

Then we get

$$W_n + W_p \sim 0.6 \quad \mu m \quad \text{at breakdown.}$$

From Fig. 14 and by an iteration process, we find that

$$N_D = N_A = 1.2 \times 10^{17}/cm^3$$

The diode with this doping density has a reverse breakdown voltage of 17 V at room temperature. Under a typical operational condition of 5×10^4 A/cm² pulsed bias current density at 250°C peak junction temperature, the operating bias voltage increases to 24 V due to the junction temperature rise and space charge resistance. It should be noted that the above diode design is based on ideal abrupt junctions. In real diodes, particularly in millimeter-wave diodes, doping density transition (0.1–0.2 μm) is a significant portion of the total space charge region. Thus the epi layer should be made slightly thicker to compensate for the doping density transition.

Once diode parameters are determined, the small signal admittance can be calculated as a function of frequency for given values of bias current density. Figure 34 shows the calculated small signal admittance for the pulsed double-drift IMPATT diode with $N_A = N_D = 1.2 \times 10^{17}/cm^3$ doping density at 5×10^4 A/cm² bias current density. It can be seen that the peak negative conductance occurs in the 100-GHz range. It should be kept in mind that for large signal operation, the negative conductance decreases in magnitude and

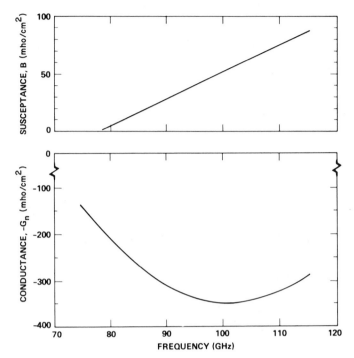

FIG. 34. Small signal admittance of pulsed double-drift IMPATT diode designed for operation in the 100-GHz range.

the device susceptance increases slightly, as discussed earlier in the large signal analysis of IMPATT devices. This causes lowering of frequency operation.

B. TRANSIENT THERMAL DESIGN

The key to proper thermal design of pulsed IMPATT diode to achieve both high peak power and low junction temperature lies in the accurate prediction of transient thermal properties of the diode, as it goes through periodic heating and cooling cycles caused by the bias pulses. Once the transient thermal resistance is known, the maximum allowable bias input power to maintain a safe operating junction temperature can be calculated. The required rf power in conjunction with maximum input power then determines the device junction area and operating current density. Thus, as a first step, the transient thermal resistance of a diode must be determined.

Consider the case of a uniform heat source with radius R on a semi-infinite heat sink of thermal conductivity k and thermal diffusivity α as shown in Fig. 35). It can be shown that the transient thermal resistance related to

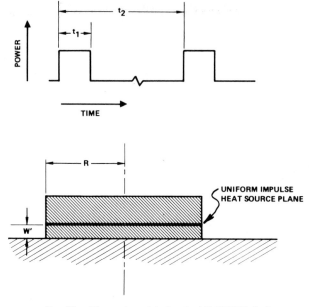

FIG. 35. Thermal model of pulsed IMPATT diode.

the maximum temperature at the center of the diode and the end of a heating cycle is given by (Jeager, 1953)

$$
\theta = \frac{2}{\pi RK} + \frac{d^{1/2}(\alpha t_1)^{1/2}}{\pi R^2 K} \left\{ \frac{2}{(\pi d)^{1/2}} (1 - e^{-R^2/4\alpha t_1}) \right.
$$

$$
\left. - \frac{R}{(\alpha t_2)^{1/2}} + I + \frac{R}{d^{1/2}(\alpha t_1)^{1/2}} \operatorname{erf} C\left[\frac{R}{2(\alpha t_1)^{1/2}}\right] \right\},
\tag{78}
$$

$$
I = \frac{2}{\pi d} \int_0^\infty \frac{e^{-dx\delta}[e^{-(1-d)x\delta} - e^{-x\delta}]\left[1 - \cos\dfrac{Rx}{(\alpha t_2)^{1/2}}\right]}{x^2(1 - e^{-x\delta})} \, dx;
$$

$d = t_1/t_2$ is the duty factor (Fig. 35). The first term is a dc term proportional to the cw thermal resistance; the remainder of the terms consist of a dc contribution as well as an ac contribution following the heat pulse.

For Type IIA diamond $k = 9.0$ W/cm°C and $\alpha = 1.52$ cm^2/S, and for copper $k = 3.96$ W/cm°C and $\alpha = 1.14$ cm^{2}S. Using these values, the transient spreading thermal resistance for both types of heat sinks can be calculated with the aid of a computer. Figures 36 and 37 present the calculated thermal resistances of both copper and Type IIA diamond for a 3% pulse duty

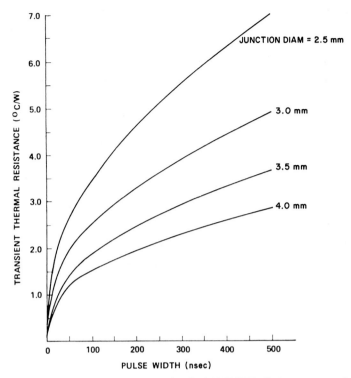

FIG. 36. Transient thermal resistance of pulsed IMPATT diode on copper heat sink. Duty = 3.0%. [From Fong *et al.* (1977).]

factor as a function of pulse width for four diode radii. For the same diode radius, the diamond is a factor of two lower in spreading thermal resistance. It should be noted that diamond thermal conductivity is a function of temperature; $k = 9.0$ W/cm°C is consistent with a junction temperature of 250°C.

In an actual diode, the heat generation is mostly confined to the active layer close to the diode junction. Since the junction for the double-drift diode is formed by a multiple epitaxy process with the diode junction located

TABLE II

CALCULATED THERMAL RESISTANCE FOR 50 μm DIAMETER DDR
DIODE ON CU AND DIAMOND HEAT SINKS

Heat sink	θ_{Si}(°C/W)	θ_{HS}(°C/W)	θ_T(°C/W)
Copper	2.86	34.29	37.15
Type IIa diamond	2.86	16.38	19.24

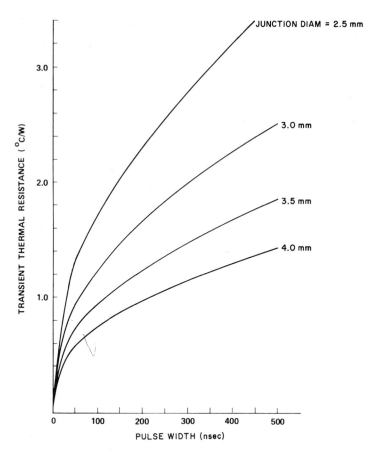

FIG. 37. Transient thermal resistance of pulsed IMPATT diode on diamond heat sink. Duty = 3.0%. [From Fong *et al.* (1977).]

at a distance from the heat sink, the silicon layer also contributes to the resistence to heat flow. For example, in a case of a typical diode for operation near 100 GHz, this layer is approximately 0.6 μm thick with a diameter of 2–4 mm. Thus its thermal mass is small as is the thermal time constant. For all practical purposes, at the pulse widths of general interest, the thermal resistance contribution from this layer approaches its cw value. We can therefore approximate the thermal resistance caused by the silicon layer by

$$\theta_{\mathrm{Si}} = W'/\pi R^2 k_{\mathrm{Si}} \qquad (79)$$

where W' is the effective thickness of the silicon layer and k_{Si} the thermal conductivity of silicon.

The total thermal resistance of an IMPATT diode, including the transient thermal spreading resistance and the silicon layer contribution, can be calculated using Eqs. 78 and 79. Given the input bias power, this thermal resistance defines the maximum junction temperature rise at the end of a heating cycle. Figures 38 and 39 show the total thermal resistance of a W-band double-drift IMPATT diode plotted as a function of junction diameter for copper and diamond, respectively. For comparison purposes, the thermal resistances for 500- and 100-nsec pulse widths are shown.

For a given input bias power, it is evident from Figs. 38 and 39 that the junction temperature decreases with increasing junction diameter. Therefore, from thermal consideration alone one would choose a large junction diameter for the diode design. However, the device impedance is inversely proportional to the diode area. In order to match the circuit impedance for efficient IMPATT operation, the diode diameter cannot be increased arbitrarily. The optimum diode diameter must therefore be chosen in conjunction with the circuit properties.

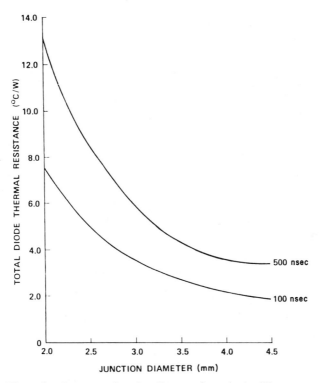

FIG. 38.   Thermal resistance vs. junction diameter for pulsed millimeter-wave IMPATT diodes on copper heat sink. Duty $= 3.0\%$. [From Fong *et al.* (1977).]

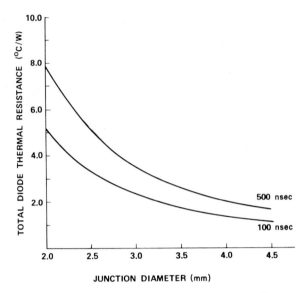

FIG. 39. Thermal resistance vs. junction diameter for pulsed millimeter-wave IMPATT diodes on diamond heat sink. Duty = 3%.

For a typical millimeter-wave oscillator, minimum resistance of a few ohms is required. This corresponds to a junction diameter of 3–5 mm for millimeter-wave IMPATT diodes, depending on frequency of operation.

C. Pulsed Oscillator Performance

As in cw oscillators, pulsed IMPATT oscillators at millimeter-wave frequencies use waveguide circuits. Shown in Fig. 40 is a photograph of a pulsed IMPATT oscillator for operation at 100 GHz. The diode is mounted in the reduced height waveguide resonator. The bias is provided through the bias pin, which is part of the low pass filter. A tuning short is provided behind the diode. An impedance transformer is also provided between the reduced height section and the full height waveguide output.

In operating pulsed oscillators, the maximum pulse width as well as the pulse duty factor is one of the most important parameters that determine the achievable peak output power. Since millimeter-wave IMPATT devices have small thermal time constants, as discussed in the preceding section, the junction temperature rises rapidly within a pulse. In order to achieve high peak power output from millimeter-wave IMPATT oscillators, the pulse width should be kept below 100 ns and the pulse duty factor below 1%. For longer pulse widths or a higher pulse duty factor, peak power output will be reduced.

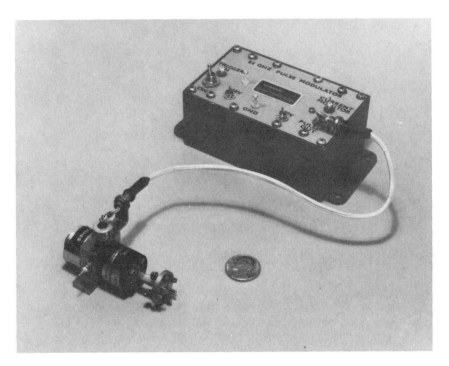

FIG. 40.   Pulsed millimeter-wave IMPATT oscillator.

Shown in Fig. 41 is the state-of-the-art peak power output achieved with pulsed IMPATT oscillators at various frequencies.[4] These values were obtained with double-drift silicon IMPATT diodes and pulse width of 100 ns or less at pulse duty factors below 1 %.

Another important property associated with the pulsed operation of an IMPATT oscillator is the frequency chirping effect. As the junction temperature increases according to the transient thermal impedance change within a pulse cycle, the diode impedance (or admittance) changes. This results in a decrease of oscillation frequency as shown in Fig. 42. Typically, frequency chirping greater than 1 GHz can be obtained with pulsed millimeter-wave IMPATT oscillators. Noting that the frequency of oscillation is also dependent on the bias current, we can control the amount of frequency chirping to meet specific systems requirements by shaping the bias pulse current waveform, as shown in Fig. 43 (Kuno and Fong, 1977).

[4] Note the high value of pf product in the 90–100-GHz range that reflects the higher level of development activities in this frequency range.

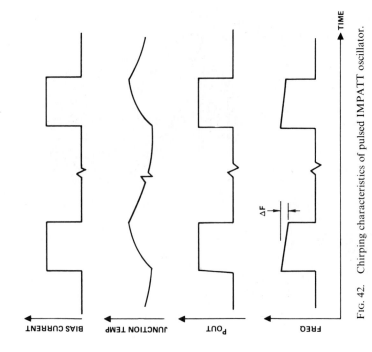

FIG. 42.   Chirping characteristics of pulsed IMPATT oscillator.

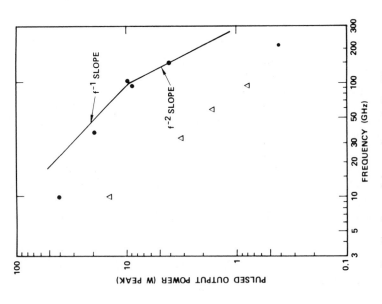

FIG. 41.   State-of-the-art for pulsed millimeter-wave IMPATT oscillators. Legend: ●●●, pulsed oscillators, ─●─, IMPATT; ─△─△─, GUNN (Chao *et al.*, 1977; English *et al.*, 1978; Fong *et al.*, 1977).

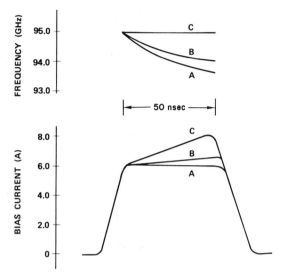

FIG. 43.   Frequency chirping control by means of bias current waveform shaping for pulsed millimeter-wave IMPATT oscillator.

V.   Power Amplifiers

In Section II, amplifier operations of IMPATT devices were analyzed in detail. IMPATT devices have been used effectively as millimeter-wave power amplifiers. For power amplification both stabilized amplifier and injection-locked oscillators have been developed (Bayuk and Rau, 1977; Fukatsu et al., 1971; Hayashi et al., 1975; Kuno, 1976; Kuno et al., 1971; Kuno and English, 1976; Kuno et al., 1972; Kuno et al., 1973; Peterson and Steinbrechher, 1973; Takayama, 1972). While injection-locked oscillators are suited for high gain (>20 dB) narrow bandwidth (<1 GHz) operation, stabilized amplifiers are for low gain (~ 10 dB/stage), broader bandwidth (>1 GHz) applications. The maximum output power achievable from an amplifier is approximately the same as that obtainable from the same device operated as an oscillator.

In order to achieve optimum performance at a desired frequency, both real and imaginary parts of the circuit impedance must be matched properly to those of the device impedance, as described in Section III. In order to achieve this impedance matching, an amplifier circuit must have two degrees of tuning freedom. Shown in Fig. 44 are a cross sectional view and equivalent circuit representation of waveguide circuit developed for millimeter-wave IMPATT amplifiers (Kuno and English, 1973). In the circuit, parallel and series tuning elements are provided by means of a tuning short and an adjustable coaxial section. Figure 45 shows a photograph of a two stage 60-GHz IMPATT amplifier. Each stage consists of a circulator coupled

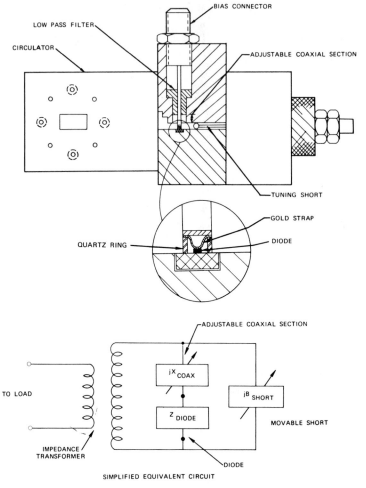

FIG. 44. Cross sectional view and equivalent circuit representation of millimeter-wave IMPATT amplifier. [From Kuno and English (1973).]

reflection type IMPATT amplifier. Additional isolators are used to provide interstage isolations. Shown in Figs. 46–49 are various small and large signal characteristics of the IMPATT amplifier (Kuno and English, 1973, 1977). It is of interest to note the quite good agreement between these data and the analysis presented in Section II.

Pulsed IMPATT amplifiers are also feasible. However, due to the rapid impedance change within a pulse caused by a temperature change, it is difficult to achieve stabilization with millimeter-wave IMPATT devices. Injection locking of pulsed millimeter-wave IMPATT oscillators however,

FIG. 45.   Three stage IMPATT amplifier. [From Kuno and English (1973).]

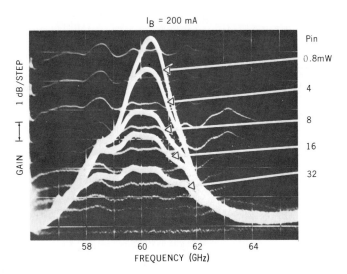

FIG. 46.   Measured large signal effects on gain–bandwidth characteristics of a stabilized IMPATT amplifier. [From Kuno and English (1973).]

have been achieved successfully (Bernues and Ying 1978). One of the important factors in a pulsed injection-locked oscillator is that the locking bandwidth must be larger than that of the chirping bandwidth. Shown in Fig. 50 are the test results of pulsed injection-locked oscillators.

In order to achieve high power output, a number of devices can be combined in an amplifier cavity or a number of amplifiers can be combined by means of hybrid couplers. The feasibility of millimeter-wave power combining techniques has been demonstrated.

For example, Fig. 51 shows a combiner consisting of eight diodes in a TM_{010} mode cavity (Harp and Russel, 1974). With this technique 10-W cw output power has been achieved at 35 GHz. Another example is shown in Fig. 52 where four amplifiers are combined to produce 1-W cw output power at 60 GHz over a bandwidth of 8 GHz (Kuno and English, 1976). Millimeter-wave power combiner/amplifier technique development is currently pursued very actively. The power combining techniques should prove useful in many millimeter-wave systems.

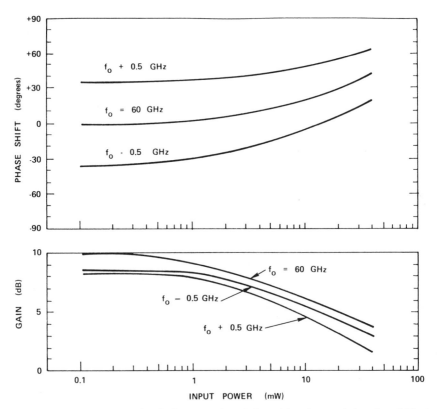

FIG. 47.   Measured large signal effects on gain and phase delay characteristics of a stabilized IMPATT amplifier. [From Kuno and English (1973).]

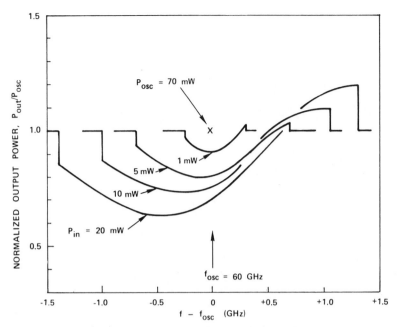

FIG. 48.   Measured locking characteristics of an injection-locked IMPATT oscilator. [From Kuno and English (1973).]

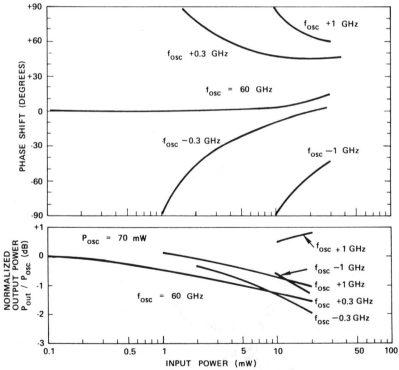

FIG. 49.   Measured large signal effects on power output and phase delay of an injection-locked IMPATT oscillator. [From Kuno and English (1973).]

117

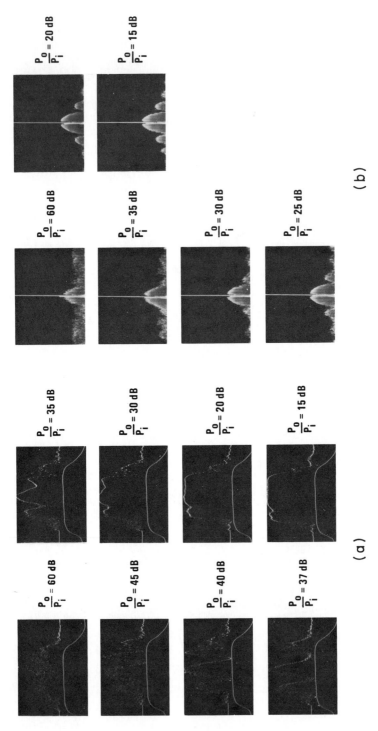

FIG. 50. Injection locking characteristics of pulsed IMPATT oscillators. Values given are for output power/projected power (P_o/P_i), in decibels. (a) Video output from phase detector. Phase width is 50 nsec, and frequency is 10 kHz. (b) Spectrum analyzer of signal from phase detector. Phase width is 50 nsec, frequency is 10 kHz.

118

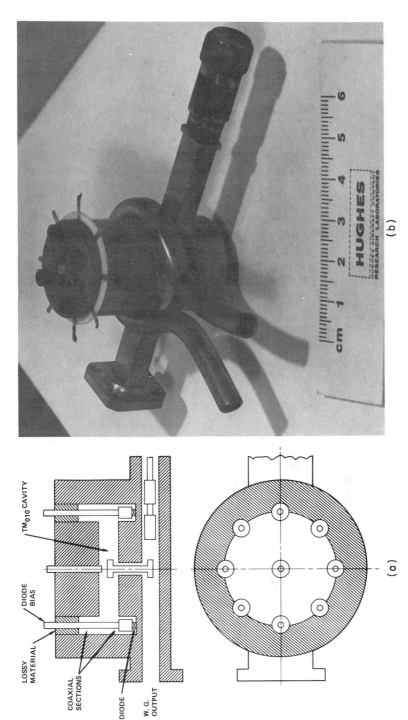

(b)

(a)

FIG. 51. Single cavity IMPATT power combiner/amplifier. [From Harp and Russell (1974).]

119

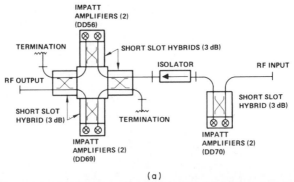

POWER COMBINER OUTPUT STAGE   DRIVER STAGE

(a)

(b)

FIG. 52.   Hybrid coupled IMPATT power amplifier/combiner. [From Kuno and English (1976).]

VI.   AM and FM Noise Characteristics

In addition to power output, noise characteristics of oscillators are important properties for system applications. Shown in Fig. 53 are measured

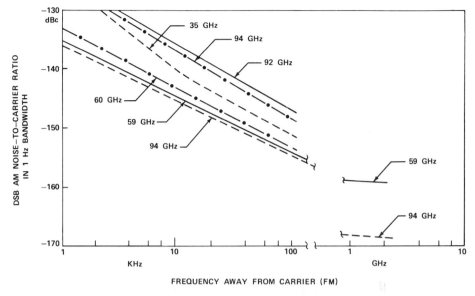

FIG. 53. AM noise characteristics of millimeter-wave oscillators. Legend: ——, IMPATT, ---, GUNN; -•-•-, Klystron.

AM noise characteristics of typical millimeter-wave oscillators. For comparison, those of Gunn oscillators and Klystrons are also shown. It is interesting to note that AM noise characteristics of IMPATT oscillators near the carrier are similar to those of Gunn oscillators and Klystrons. At higher modulation frequencies (higher than several hundred MHz) however, the IMPATT oscillator noise is higher than that of a Gunn oscillator or a Klystron. For this reason IMPATT oscillators are in general difficult to use as local oscillators for mixers in receiver applications.

Shown in Fig. 54 is the calculated effect of local oscillator (LO) on receiver noise figures for various values of LO noise suppression factor. For a typical IMPATT local oscillator with carrier-to-noise ratio of 150 dB/Hz, LO noise suppression of 30 dB is required to keep the adverse effect of the LO on the receiver noise figure negligible. The required LO noise suppression can be accomplished either by a bandpass filter for the LO frequency or by a balanced mixer configuration. With proper LO noise suppression techniques, IMPATT oscillators can be used effectively as local oscillators for low noise millimeter-wave mixers.

Shown in Fig. 55 are measured FM noise characteristics of a millimeter-wave IMPATT oscillator. FM noise characteristics on the oscillator strongly depend on the circuit Q. Values of circuit Q for typical millimeter-wave IMPATT oscillators range between 20 and 100. These values are based on the injection locking gain–bandwidth characteristics measurements.

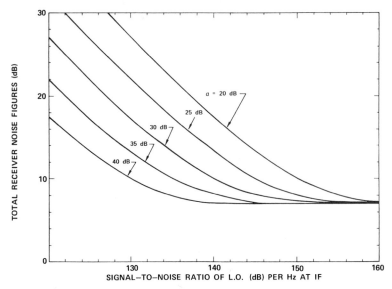

FIG. 54.   Effects of local oscillator noise on mixer noise figure. $NF = L_C(N_{IF} + N_R - 1) + P_{LO}(S/N)_{LO}/KT_0\alpha)$, $L_c(N_{IF} + N_R - 1) = 7$ dB, and $P_{LO} = 5$ dBm.

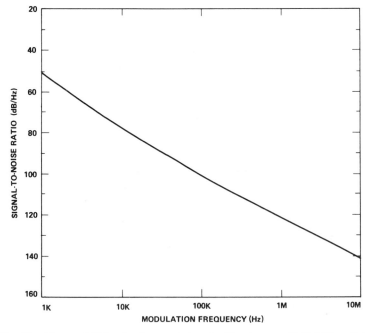

FIG. 55.   Measured FM noise characteristics of millimeter-wave IMPATT oscillator.

FM noise characteristics can be improved by means of injection or phase locking (see the following section) to a signal with low FM noise.

VII.   Frequency and Phase Stabilization

In Section III.C it was pointed out that free running IMPATT oscillators typically have a frequency stability of $-0.5 \times 10^{-4}/°C$. (See Fig. 29.) This means that the frequency drift rate is approximately -2 MHz/°C at 40 GHz and -5 MHz/°C at 100 GHz, for example. For applications where temperature variations cause excessive frequency drifts, a number of techniques have recently been developed for controlling the frequency. The simplest method is to control the oscillator cavity temperature by means of a small heater and a control circuit. (See Fig. 56.) Since solid state millimeter-wave oscillator cavities have small masses, it is relatively easy to control their temperature in this way. The temperature variation can be kept to less than 1°C.

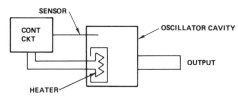

FIG. 56.   Cavity temperature controlled frequency stabilization.

The frequency drift of an IMPATT oscillator can be reduced significantly by means of a high Q cavity. Shown in Fig. 57 is a schematic diagram for the high Q cavity frequency stabilization technique (Gray *et al.*, 1969).

Another approach to frequency stability is to use a frequency discriminator such as an invar cavity filter with an AFC loop (see Fig. 58). This method does not require heater power but does require a more complex control circuitry.

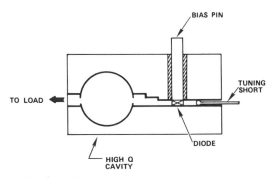

FIG. 57.   High Q cavity for frequency stabilization.

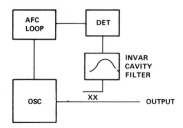

FIG. 58. AFC loop controlled frequency stabilization.

In addition to long term frequency stability, many systems require phase stabilities of crystal controlled oscillator quality. For such applications two basic approaches have recently been developed for millimeter-wave sources. One is a phase-lock loop approach, and the other is an injection locking approach using a multiplier chain as shown in Figs. 59 and 60. In the phase-locked oscillator, sampled power of the millimeter-wave frequency is converted down to an IF by a harmonic mixer, the phase is compared with the phase of the reference crystal oscillator, and the phase error is then corrected by means of a feedback loop to the millimeter-wave oscillator. This technique has successfully been applied to millimeter-wave IMPATT oscillators (Baprawski *et al.*, 1976). Shown in Fig. 59 is a comparison of phase noise of a free running and phase-locked millimeter-wave IMPATT oscillators. Significant reduction in phase noise can be seen within the locking band. The locking bandwidth is limited by the locking loop bandwidth, which is typically 1–10 MHz.

Similar improvement in phase stability can be achieved by means of an injection-locked oscillator with a frequency multiplier chain as shown in Fig. 61. Since trade off between locking gain and bandwidth can be made, a broader locking bandwidth can be achieved in an injection-locked than in a

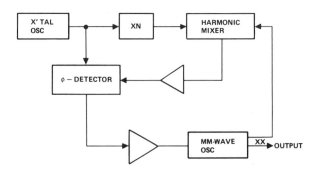

FIG. 59.   Phase-locked millimeter-wave IMPATT oscillator.

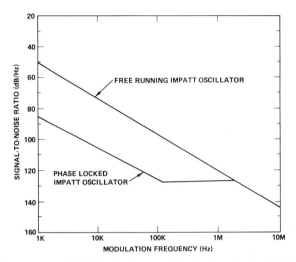

FIG. 60.   Improvement of FM noise characteristics in phase-locked IMPATT oscillator. $f_0 = 60$ GHz.

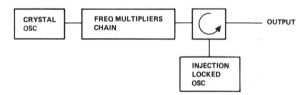

FIG. 61.   Injection locking with frequency multiplier chain for phase stabilization.

phase-locked oscillator. However, an injection-locked oscillator using a multiplier chain is considerably more complex than a phase-locked oscillator at millimeter-wave frequencies.

VIII.   Conclusion

After a long, slow start the rate of millimeter-wave technology development and progress has rapidly increased in the last several years. This is due partly to the changing requirements for millimeter-wave systems and partly to the recent progress achieved in the solid state devices technology for various millimeter-wave system applications, such as communications, radar, missile guides and seekers, radiometers, and instrumentation.

Millimeter-wave systems offer many advantages over both microwave and electro-optical systems. In comparison with electro-optical systems, milli-meter waves offer penetration capabilities through clouds, fog, smoke, and dust. In addition, as compared with lower frequency microwave systems, millimeter-wave systems offer advantages of smaller antenna size, lighter weight, more compact size, increased system accuracy, and unique frequency dependent atmospheric attenuation characteristics. With further technological development, millimeter-wave systems should lead to lower systems production cost.

In each millimeter-wave system, millimeter-wave power generation devices are required. At this moment silicon IMPATT devices are the most powerful solid state power sources at millimeter-wave frequencies. In particular, silicon IMPATT devices are the only viable solid state devices for power generation that have been developed to date. The frequency coverage of the IMPATT devices is currently reaching the 300 GHz range.

At millimeter-wave frequencies one can combine conventional microwave techniques with optical techniques. The development of millimeter-wave IMPATT devices and circuits can take advantage of both.

Perhaps the quasi-optical approach may be the only way to extend the frequency coverage of IMPATT devices into submillimeter waves. There is no doubt that IMPATT devices will invade the submillimeter-wave region.

In parallel with the currently expanding millimeter-wave systems development activities, further experimental and theoretical work on IMPATT source development will certainly enhance their potential values and capabilities. The IMPATT devices will then find a preeminent place in various millimeter-wave systems applications and will play a key role in the development of submillimeter-wave technology.

REFERENCES

Baprawski, J., Smith, C., and Bernues, F. (1976). *Microwave J.* 41–44.

Bayuk, R. J., and Raue, J. E. (1977). Ka-band solid state power amplifier, *IEEE Int. Microwave Symp., San Diego, California*.

Bernick, R. L. (1973). Thermal Analysis for IMPATT Diodes. Hughes Aircraft Company Research Rep. No. 1003.

Bernues, F., and Ying, R. S. (1978). Recent Advances in Millimeter-Wave Radar Technology. Electro 78, Boston, Massachusetts.

Chao, C., Bernick, R. L., Ying, R. S., Weller, K. P., Lee, D. H., and Nakaji, E. M., (1977a). Pulsed IMPATT diode oscillators above 200 GHz, *IEEE Int. Solid-State Circuits Conf., Philadelphia, Pennsylvania*.

Chao, C., Bernick, R. L., Nakaji, E. M., Ying, R. S., Weller, K. P., and Lee, D. H., (1977b). *IEEE Trans. Microwave Theory Tech.* **MTT-25**, 985–991.

Crowell, C. R., and Sze, S. M. (1966). *Appl. Phys. Lett.* **9**, 242.

Delagebeaudeuf, D. (1970). *Proc. IEEE* **58**, 1140–1141.

Duh, C. Y., and Moll, J. L. (1967). *IEEE Trans. Electron Devices* **ED-14**, 46–49.

English, D. L., Nakaji, E. M., Fong, T. T., and Kuno, H. J. (1978). Millimeter-wave pulsed oscillators, Presented at the *IEEE Int. Microwave Symp. (Late-News Items), Ottawa, Canada*.

Fong, T. T., Lane, W. R., Kuno, H. J., and Kramer, N. B., (1977). High power W-band pulsed IMPATT oscillators, *IEEE Int. Microwave Symp. San Diego, California.*

Fukatsu, Y., Akaike, M., and Kato, H. (1974). Amplification of high speed PCM phase-shift keyed millimeter-wave signals through an injection locked IMPATT oscillator. IEEE ISSCC, Philadelphia, Pennsylvania.

Gibbons, J. F. (1967). *IEEE Trans. Electron Devices*, **ED-14**, 27.

Gilden, M., and Hines, H. F. (1966). *IEEE Trans. Electron Devices* **ED-13**, 169–175.

Gray, W. W., Kikushima, L., Morenc, N. P., and Wagner, R. J. (1969). Applying IMPATT Sources to Modern Microwave Systems, IEEE ISSCC, Philadelphia, Pennsylvania.

Harp, R. S., and Russel, K. J. (1974). Improvements in Bandwidth and Frequency Capability of Microwave Power Combinatorial Techniques, IEEE ISSCC, Philadelphia, Pennsylvania.

Hayashi, H., Iwai, F., and Fujita, T. (1974). 80 GHz IMPATT Amplifier, IEEE ISSCC, Philadelphia, Pennsylvania.

Hirachi, Y., Nakagami, T., Toyama, Y., and Fukukawa, Y. (1976). *IEEE Trans. Microwave Theory Tech.* **MTT-24**, 731-737.

Hughes Aircraft Company (1974). Engineering Evaluation Life Tests on IMPATT Diodes, Rep. No. 3.

Ishibashi, T., and Ohmori, M. (1976). *IEEE Trans. Microwave Theory Tech.* **MTT-24**, 858–859.

Jeager, J. C. (1953). *Aust. J. Q. Math.* **11**, 123–137.

Johnston, R. L., Loach, B. C., Jr., and Cohen, G. B. (1965). *Bell Syst. Tech. J.* **44**, 369–372.

Kuno, H. J. (1973). *IEEE Trans. Microwave Theory Tech.* **MTT-21**, 694–702.

Kuno, H. J. (ed.) (1976). *IEEE Trans. Microwave Theory Tech.* **MTT-24**, 683–895.

Kuno, H. J., and English, D. L. (1973). *IEEE Trans. Microwave Theory Tech.* **MTT-21**, 703–706.

Kuno, H. J., and English, D. L., (1976). *IEEE Trans. Microwave Theory Tech.* **MTT-24**, 744–751.

Kuno, H. J., and English, D. L., (1976). *IEEE Trans. Microwave Theory Tech.* **MTT-24**, 758–767.

Kuno, H. J., and Fong, T. T., (1977). Solid State Millimeter-Wave Source, EASCON-77, Washington, D.C.

Kuno, H. J., Bowman, L. S., and English, D. L. (1971). Millimeter-Wave Silicon IMPATT Power Amplifiers for Phase-modulated Signals, IEEE ISSCC, Philadelphia, Pennsylvania.

Kuno, H. J., English, D. L., and Pusateri, P. H. (1972). *IEEE Int. Microwave Symp., Chicago, Illinois.*

Kuno, H. J., English, D. L., and Ying, R. S., (1973). High Power Millimeter-Wave IMPATT Amplifiers, IEEE ISSCC, Philadelphia, Pennsylvania.

Kuno, H. J., Weller, K. P., Ying, R. S., and Lee, D. H., (1974). Pumps and Local Oscillators, Technical Rep. AFAL-TR-76-421.

Misawa, T. (1966). *IEEE Trans. Electron Devices*, **ED-13**, 137–151.

Misawa, T., and Marinaccio, L. P. (1970). *Proc. Symp. Submillimeter Waves* pp. 53–67.

Miyakawa, T. *et al.* (1975). *IEEE Int. Microwave Symp., Palo Alto, California.*

Mouthaan, K. (1969). *IEEE Trans. Electron Devices* **ED-16**, 935–945.

Norris, C. B., and Gibbons, J. F. (1967). *IEEE Trans. Electron Devices* **ED-14**, 38–43.

Peterson, D. F., and Steinbrecher, D. H. (1973). *IEEE Trans. Microwave Theory Tech.* **MTT-21**, 19–27.

Read, W. T. (1958). *Bell Syst. Tech. J.*, **37**, 401–446.

Rodrigues, V., Ruegg, H., and Nicolet, M. A. (1967). *IEEE Trans. Electron Devices* **ED-14**, 44–46, January 1967.

Scharfetter, D. L., Evans, W. J., and Johnston, R. L. (1970). *Proc. IEEE* **58**, 1131–1133.

Schroeder, W. E., and Haddad, G. I. (1971). *Proc. IEEE* **59**, 1245–1248.

Seidel, T. E., and Scharfetter, D. L. (1970). *Proc. IEEE* **58**, 1135–1136.

Seidel, T. E., Davis, R. E., and Inglesias, D. E. (1971). *Proc. IEEE* **59**, 1222–1228.

Sze, S. M. (1969). "Physics of Semi-Conductor Devices." Wiley (Interscience), New York.

Sze, S. M., and Gibbons, G. (1966). *Appl. Phys. Lett.* **8**, 111–112.

Takayama, Y. (1972). *IEEE Trans. Microwave Theory Tech.* **MTT-20**, 266–272.

Van del Pol, B. (1927). *Phil. Mag. Ser. 7* **3**, 65–80.

Weller, K. P., Ying, R. S., and Nakaji, E. M. (1975). Millimeter-Wave 94 GHz Silicon IMPATT Amplifiers, IEEE ISSCC, Philadelphia, Pennsylvania.

Weller, K. P., Ying, R. S., and Lee, D. H. (1976). *IEEE Trans. Microwave Theory Tech.* **MTT-24**, 738–743.

CHAPTER 3

Pulsed Optically Pumped Far Infrared Lasers*

Thomas A. DeTemple

I. Introduction

The production of intense far infrared (FIR) radiation via stimulated emission presents a difficult and challenging task using conventional laser excitation techniques. The difficulty resides with the basics of the excitation processes and the small energy difference of less than kT between initial

* Supported by National Science Foundation, Army Research Office—Durham, and United States Air Force.

and final laser states. Since inversion requires a selective excitation into the upper laser level, the equivalent selectivity by electron impact excitation ($\sim 10^2 kT$), chemical excitation ($\sim 2kT$) and V–V energy transfer ($\sim 2kT$) precludes the use of these except in special circumstances (for example, HF and HCN lasers). In contrast, the selectivity of a laser pumping source can be quite high ($< 10^{-4} kT$), offering a state selective excitation without parallel.

Another difficulty in producing intense FIR is in the lack of energy storage. Intense conventional lasers such as CO_2, Nd, and ruby are characterized by having relatively long-lived (1–1000 μsec) upper state lifetimes permitting slow pumping followed by a rapid depletion by a Q-switch action. For FIR lasers, specifically those involving pure rotational states, the lifetimes are typically 10 nsec at 1 Torr, a value too short to permit a Q-switch enhancement. Because of this, the most intense FIR signals have been generated using lasers as the excitation source with FIR output powers simply proportional to input powers. With the state selectivity in excitation, the overall conversion has approached the Manley–Rowe limit, i.e., one FIR photon per pumping photon, suggesting that the approach may be more efficient than, or at least as efficient as a direct-electrically excited FIR laser.

The success of the optically pumped FIR lasers can best be appreciated from the performance characteristics summarized in Table I. Intense FIR lines spanning over a decade in wavelength with powers in the 0.1–10-MW range have been achieved as indicated in this table. These data thus indicate that the technique is not an inefficient curiosity but rather a successful and viable approach for the generation of intense, step tunable lines in the FIR. It is the purpose of this chapter to discuss the details and characteristics of optically pumped FIR lasers. As a prelude to the sections which follow, a

TABLE I

OPTICALLY PUMPED FAR INFRARED LASER PERFORMANCE SURVEY

Molecule	$\lambda(\mu m)$	Power	Reference
$C^{13}H_3F$	1222		Hacker et al. (1976)
	1207		Hacker et al (1976)
	1006	1 MW	Hacker et al. (1976)
	862		Hacker et al. (1976)
$C^{12}H_3F$	496	1 MW	Evans et al. (1975)
CH_3I	447	100 kW	Hutchinson (1977)
D_2O	385	10 MW	Evans et al. (1976)
	113	10 MW	Evans et al (1977b)
	66	10 MW	Evans et al. (1977b)

simplified model of the FIR laser will be presented which illustrates the major features of these systems.

A. System Dynamics

The salient dynamic features of an optically pumped FIR source can best be illustrated with the case of CO_2 laser pumped $C^{12}H_3F$ (Chang and Bridges, 1970). Shown in Fig. 1 is a partial energy level diagram near the low K, Q branch transitions in the v_3 mode. The $P_9(20)$ transition of CO_2 is nearly resonant with the $^{Q}Q(12, 2)$ transition (Freund et al., 1974). Absorption of the pump will create population in the $J = 12$, $K = 2$ vibrationally excited state with a particular molecular velocity v and magnetic quantum state M. If the rate of production of molecules into these states exceeds the rate of decay, an inversion will be established on the $^{Q}R(11, 2)$ transition at 496 μm. Cascade inversions in the ground and excited state are also possible.

From a simple rate equation treatment, the equivalent production rate is $\sigma_{13} I_p$ where σ_{13} is the absorption cross section for the $^{Q}Q(12, 2)$ transition and I_p is the pump flux (cm^{-2} sec^{-1}). The decay processes are rich in number: ΔJ, ΔK, ΔM, Δv, and Δv_3 changing collisions may exist separately or in any combination (Brewer and Shoemaker, 1971; Frenkel et al., 1971; Oka, 1973; Schmidt et al., 1973; Weitz and Flynn, 1973; Shoemaker et al., 1974; Johns et al., 1975; Earl et al., 1976). For $C^{12}H_3F$ in a high J low K state, solely ΔM and Δv changing collisions are not probable relative to ΔJ changing collisions (Shoemaker et al., 1974; Johns et al., 1975). In addition the rate of ΔJ changing collisions is approximately one order of magnitude faster than the rate of

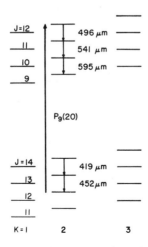

FIG. 1.   Partial energy level diagram of the v_3 band absorption and FIR emission in $C^{12}H_3F$. Only the $K = 2$ lines are shown.

ΔK changing collisions, (Frenkel *et al.*, 1971) with Δv_3 collisions being the slowest (Weitz and Flynn, 1973). Thus, the dominant collision process is the hard or ΔJ collision, the rate of which will be denoted by T_1^{-1}. Population inverse thus requires $\sigma_{13} I_p T_1 > 1$ or $I_p > kW/cm^2$ Torr^2 for the $^QQ(12, 2)$ transition. Intensities of this magnitude are easily available from high pressure (TEA) lasers.

The equivalent gain which can be established on these pure rotational transitions is quite large. Setting ρ_{11}^e to be the equilibrium fractional population in the ground absorbing state (typically 1 %), the gain can be estimated as

$$G_f \simeq \tfrac{1}{2}\rho_{11}^e N\sigma_{32} = \tfrac{1}{2}\rho_{11}^e N(\lambda_{32}^2 A_{32}/4\pi^2 \Delta v_H)$$

where N is the molecular number density, σ_{32} the FIR emission cross section on line center in the homogeneously broadened limit with line width (FWHM) Δv_H typically 40 MHz/Torr, and A_{32} the reciprocal radiative lifetime. Representative values of σ_{32} are 10^{-14}–10^{-15} cm^2 resulting in gain coefficients of $G_f \sim 0.1$–1 cm^{-1}, values which are quite large.

With these large gains and the large pumping intensities, the absorption and emission transitions are easily saturated, leading to a simplified picture of the conversion dynamics. With a strong delta function pulse, the amount of energy which can be extracted in the FIR is proportional to $\rho_{11}^e N$. The rate of extraction is determined by the refilling rate of the ground absorbing state and the emptying rate of the upper laser states, both being T_1^{-1}. Hence, the extractable or saturated power density is proportional to $\hbar\omega_{32}\rho_{11}^e N/T_1$. Taking λ_{32} to be 500 μm and $T_1 = 10$ nsec Torr, this value is about 10 kW/1 torr^2, indicating that large FIR powers require large pumping volumes. Operating pressures are limited to a few torr due to the competing effect of ΔJ collisions.

B. Topic Restrictions

From the previous section, the pulsed FIR system is characterized by having a large gain in a low pressure, large volume configuration. The remainder of this chapter will be devoted to the details of these systems. In particular: in Section II, the density matrix treatment appropriate to the optically pumped FIR is discussed, formalizing the dominant contributions to the gain: laser and stimulated Raman emission; in Section III, the gross spectral characteristics of molecules which have produced at least 1 kW of power in the FIR are reviewed; in Section IV, the various geometrical configurations, experimental details, and techniques of these systems are discussed; in Section V, secondary effects such as the transient behavior and multiple wave interactions are outlined; and Section VI summarizes the chapter. Since many of the detailed collision rates are not known for the molecules under consideration—indeed some are only now becoming available (Leite, 1977a,b)—we will assume that T_1 is dominated by ΔJ collisions and that $T_1 = (\pi \Delta v_H)^{-1} = 8$ nsec Torr, or $\Delta v_H \simeq 40$ MHz/Torr for numerical estimates (Murphy and Boggs, 1967). In addition, Doppler broadening on either the FIR or pump transition will not be treated explicitly. This is valid because at a few torr pressure, both are approximately in the homogeneously broadened regime.

II. Theoretical Considerations

The general treatment of multiple coherent waves interacting with an atomic or molecular system is of considerable interest from the standpoint of parametric interactions including optical pumping, optical ionization phenomena, laser induced chemistry, laser isotope separation, and multimode or multiline lasers. For the specific case of optically pumped lasers, one has at least two coherent waves interacting with at least a three-level system. Except under special circumstances, the standard rate equation approach in describing the various interactions with the gain material is not general enough in that transient coherent effects and multiphoton effects, such as stimulated Raman emission, are not readily treated. In contrast, a semiclassical treatment based on Maxwell's equations and the density matrix equations for the material is capable of describing all interactions excluding spontaneous emission phenomena and questions associated with electromagnetic field statistics (Pantell and Puthoff, 1969; Loudon, 1973; Sargent et al., 1974). In the following sections, selected solutions of this approach will be reviewed and interpreted for situations appropriate to this chapter.

A. Two-Wave Three-Level System

The simplest interaction occurs for two coherent waves interacting with a three-level system and is diagrammed in Fig. 2. In this figure, (a) is the normally inverted system with the FIR transition occurring in the vibrationally excited state, and (b) is the compliment to (a) with the FIR occurring on a ground state transition; both transitions have been observed.

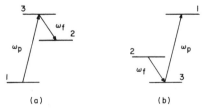

(a)   (b)

FIG. 2.   Possible absorption/emission combinations in a three-level system with

$$\omega_p \simeq |E_3 - E_2|/\hbar \quad \text{and} \quad \omega_f \simeq |E_3 - E_2|/\hbar.$$

Assuming that the transition dipole moment between states 1 and 2, μ_{12}, is zero, the density matrix equations of evolution appropriate to non-degenerate levels shown in Fig. 1a are, for the diagonal elements ρ_{ii}:

$$\dot{\rho}_{11} = -(\rho_{11} - \rho_{11}^e)/T_1 + \mathbf{\mu}_{13} \cdot \mathbf{E}(\rho_{13} - \rho_{31})/i\hbar, \tag{1}$$

$$\dot{\rho}_{22} = -(\rho_{22} - \rho_{22}^e)/T_1' + \mathbf{\mu}_{23} \cdot \mathbf{E}(\rho_{23} - \rho_{32})/i\hbar, \tag{2}$$

$$\dot{\rho}_{33} = -(\rho_{33} - \rho_{33}^e)/T_1'' - \mathbf{\mu}_{13} \cdot \mathbf{E}(\rho_{13} - \rho_{31})/i\hbar - \mathbf{\mu}_{23} \cdot \mathbf{E}(\rho_{23} - \rho_{32})/i\hbar; \tag{3}$$

and for the off-diagonal elements ρ_{ij}:

$$\dot{\rho}_{13} = -(i\Omega_{13} + 1/T_2)\rho_{13} + \mathbf{\mu}_{13} \cdot \mathbf{E}(\rho_{11} - \rho_{33})/i\hbar + \mathbf{\mu}_{23} \cdot \mathbf{E}\rho_{12}/i\hbar, \tag{4}$$

$$\dot{\rho}_{12} = -(i\Omega_{12} + 1/T_2')\rho_{12} + (\mathbf{\mu}_{23} \cdot \mathbf{E}\rho_{13} - \mathbf{\mu}_{13} \cdot \mathbf{E}\rho_{32})/i\hbar, \tag{5}$$

$$\dot{\rho}_{32} = -(i\Omega_{32} + 1/T_2'')\rho_{32} + \mathbf{\mu}_{23} \cdot \mathbf{E}(\rho_{33} - \rho_{22})/i\hbar - \mathbf{\mu}_{31} \cdot \mathbf{E}\rho_{12}/i\hbar, \tag{6}$$

where $\Omega_{ij} = (E_i - E_j)/\hbar$, the T_1's are longitudinal or energy relaxation times, and ρ_{ii}^e is the equilibrium value (Javan, 1957; Javan and Szöke, 1965). The T_2 parameters are the transverse or polarization dephasing times which are defined in terms of the homogeneously broadened linewidth Δv_H^i (FWHM) by $\Delta v_H^i = (\pi T_2^i)^{-1}$. The diagonal elements are identified with fractional populations since $\text{Tr}(\rho) = 1$, and the off-diagonal elements represent the induced macroscopic polarization since $\mathbf{P} = N \, \text{Tr}(\rho\mathbf{\mu})$ (Pantell and Puthoff, 1966; Sargent et al., 1974); Tr is the matrix trace operation (Fano, 1957).

Although the general solutions to these equations are not known (MacGurk et al., 1974; Brewer and Hahn, 1975; Fenillade, 1976; Fenillade and Botteher, 1977; Leite et al., 1976), they can be solved with approximations and assumptions appropriate to the optically pumped FIR systems. First, a reasonable simplifying approximation is that all T_1 are equal and are approximately equal to T_2. Second, since T_2 is about 10 nsec at a typical operational pressure of 1 Torr, a quasi-static solution may be sought because the pumping time scale is usually $\gtrsim 100$ nsec. Finally, all wave detunings are assumed suboptical, i.e., terms containing $\Omega_{ij} - \omega$ dominate terms containing $\Omega_{ij} + \omega$, where ω is the optical frequency.

With these approximations and setting the total field, pump plus FIR, to be $\mathbf{E} = E_p\hat{\mathbf{\varepsilon}}_p \cos \omega_p t + E_f\hat{\mathbf{\varepsilon}}_f \cos \omega_f t$, where the subscripts p and f refer to the pump and FIR, respectively, and $\hat{\mathbf{\varepsilon}}_i$ is a polarization unit vector; the dominant Fourier coefficients of the off-diagonal elements, the "driving" terms for population changes in Eqs. (1)–(3), can be identified as

$$\rho_{13} \simeq \tilde{\rho}_{13}e^{i\omega_p t}, \qquad \rho_{32} \simeq \tilde{\rho}_{32}e^{-i\omega_f t}, \qquad \text{and} \qquad \rho_{12} \simeq \tilde{\rho}_{12}e^{i(\omega_p - \omega_f)t}$$

with $\tilde{\rho}_{ij}$ being the time independent (or slow varying) amplitudes. These amplitudes can readily be found by substitution into Eqs. (4)–(6) and using a short time average on the right hand side of these equations.

The resulting solutions can be used to recast the population Eqs. (1)–(3) into the following standard form:

$$\dot{\rho}_{11} = -(\rho_{11} - \rho^e_{11})/T_1 + I_p G_p, \tag{7}$$

$$\dot{\rho}_{22} = -(\rho_{22} - \rho^e_{22})/T_1 + I_f G_f, \tag{8}$$

where the intensity is $I_i = c\varepsilon_0 |E_i|^2/(2\hbar\omega_i)$ and the "gain" coefficients are

$$G_f = \sigma_{32}\{S_1(\rho_{33} - \rho_{22}) + S_2(\rho_{11} - \rho_{22})\}, \tag{9}$$

$$G_p = \sigma_{13}\{S_3(\rho_{33} - \rho_{11}) - S_4(\rho_{11} - \rho_{22})\}, \tag{10}$$

where

$$\sigma_{32} = \frac{\omega_f(\mathbf{\mu}_{32} \cdot \hat{\varepsilon}_f)^2 T_2}{\hbar\varepsilon_0 c}, \tag{11}$$

and

$$\sigma_{13} = \frac{\omega_p(\mathbf{\mu}_{13} \cdot \hat{\varepsilon}_p)^2 T_2}{\hbar\varepsilon_0 c} \tag{12}$$

are the homogeneously broadened line center cross sections for the $3 \to 2$ and $1 \to 3$ transitions. The form of G_f in Eq. (9) shows that in contrast to the usual laser, there are two contributions to the FIR gain: a laser-like term proportional to $(\rho_{33} - \rho_{22})$ and a Raman-like term proportional to $(\rho_{11} - \rho_{22})$. The cross section multipliers S_i are conveniently expressed in terms or normalized variables defined to be

$$x = T_2(\Omega_{31} - \omega_p), \qquad y = T_2(\Omega_{32} - \omega_f) \quad \text{(detunings)},$$

$$P = \mathbf{\mu}_{13} \cdot \hat{\varepsilon}_p E_p T_2/2\hbar, \qquad Q = \mathbf{\mu}_{23} \cdot \hat{\varepsilon}_f E_f T_2/2\hbar \quad \text{(fields)},$$

$$L(z) = z + i \quad \text{(lineshape)},$$

as

$$S_1 = \text{Im}\{[1 + (P^2 - Q^2)/L(x - y)L(x)]/L^*(y)\,\Delta\}, \tag{13}$$

$$S_2 = \text{Im}\{-P^2/L(x - y)L(x)L^*(y)\,\Delta\}, \tag{14}$$

$$S_3 = -\text{Im}\{[1 + (P^2 - Q^2)/L(x - y)L^*(y)]/L(x)\,\Delta\}, \tag{15}$$

$$S_4 = \sigma_{32} I_f S_2/\sigma_{31} I_p, \tag{16}$$

where

$$\Delta = 1 + [P^2/L^*(y) - Q^2/L(x)]/L(x - y).$$

Although algebraically complicated, these expressions are general, containing laser and Raman terms and AC Stark shifts due to both the pump and FIR (Javan and Szoke, 1965; Panock and Temkin, 1977). These results are also valid for the inverted configuration in Fig. 2b, requiring only minor sign changes.

To illustrate these results further, consider very weak fields ($P \ll 1$, $Q \ll 1$) such that $\Delta \simeq 1$. Then

$$S_1 \simeq \text{Im}[1/L^*(y)] = 1/(y^2 + 1) = 1/[T_2^2(\Omega_{32} - \omega_f)^2 + 1],$$

which is recognized as a Lorentzian with a maximum value of 1 at $\omega_f = \Omega_{32}$, the laser resonance. In addition, if $x \gg 1$,

$$S_2 \simeq P^2/x^2[(x - y)^2 + 1] = P^2/x^2[T_2^2(\Omega_{21} - \omega_p + \omega_f)^2 + 1],$$

which is identified with a Raman–Stokes process since S_2 has a maximum value of P^2/x^2 when $\omega_f = \omega_p - \Omega_{21}$, the Raman resonance condition.

Next, consider $x \gg 1$, $Q \simeq 0$, and $P^2/x^2 < 1$. In this case

$$S_1 \simeq 1/[(y + P^2/x)^2 + 1], \qquad S_2 \simeq P^2/x^2[(x - y + P^2/y)^2 + 1)],$$

which results in a laser resonance at $y = -P^2/x$ and a Raman resonance at $y \simeq x + P^2/x$. Since the normal resonances would occur at $y = 0$ and x respectively, both the laser and Raman transitions have been AC Stark shifted by the presence of the pump in opposite directions by an amount P^2/x. Finally consider $x = 0$, $P > 1$. For this case

$$S_1 = \frac{[1 - P^2/(y^2 + 1)](1 + P^2)}{y^2[1 - P^2/(y^2 + 1)]^2 + [1 + P^2/(y^2 + 1)]^2},$$

which results in $S_1 < 0$ when $y = 0$, $S_1 = 0$ when $y = \pm(\sqrt{P^2 - 1})$; $S_1 > 0$ when $y > \pm\sqrt{P^2 - 1}$. In this case, the strong pump splits the laser contribution (S_1) into a doublet, sometimes called Autler–Townes or AC Stark splitting (Autler and Townes, 1955; Heppner and Weiss, 1977; Heppner et al., 1977). The Raman contribution (S_2) is also split into a doublet under the same conditions with both being graphed in the lower part of Fig. 3 for $P = 3$.

Thus because of the strong near resonance interaction, not only a laser-like behavior but also a Raman-like behavior may occur. With these processes are associated AC Stark shifts, and for sufficiently intense fields, strong line splittings may occur, at which point it becomes nebulous to talk about a laser versus Raman process since they merge together as a single process.

The steady state population and gains can now be found by solution of Eqs. (1)–(3). For the particular case of $\rho_{22}^e = \rho_{33}^e \simeq 0$ characteristic of a

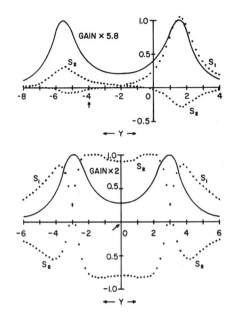

FIG. 3. Representative profiles for S_1 and S_2 and the net gain $G_f/\sigma_{32}\rho_{11}^e$ versus FIR detuning $y = T_2(\Omega_{32} - \omega_f)$. The pump detuning $x = T_2(\Omega_{31} - \omega_p)$ is indicated by the arrow and $\rho_{22}^e = \rho_{33}^e = 0$. (Top: $P = 3, x = -4$; bottom: $P = 3, x = 0$.)

$3 \rightarrow 2$ transition in the vibrationally excited state, the populations are found to be

$$\rho_{11} = \rho_{11}^e(1 + f_1 + 2F_1 + f_1F_1 + f_1F_2 + F_1F_2 + F_2)/D, \qquad (17)$$

$$\rho_{22} = \rho_{11}^e(f_1F_1 + F_2 + f_1F_2 + F_1F_2)/D, \qquad (18)$$

$$\rho_{33} = \rho_{11}^e(F_1F_2 + f_1 + f_1F_1 + f_1F_2)/D, \qquad (19)$$

$$D = 1 + 2f_1 + 2F_2 + 2F_1 + 3(f_1F_1 + F_1F_2 + f_1F_2),$$

where

$$f_1 = \sigma_{13}I_pT_1S_3, \qquad F_1 = \sigma_{32}I_fT_1S_1, \qquad F_2 = \sigma_{32}S_2I_fT_1.$$

Having these results, numerical examples appropriate to the FIR systems will now be considered.

B. SOLUTION DISCUSSION

Typical of most molecules under consideration, $\mu_{13} \simeq 0.05D$, $\mu_{32} \simeq 1D$ and $0 < |\Omega_{31} - \omega_p|/2\pi \lesssim 1$ GHz. For a pump value of $I_p \simeq 1$ MW/cm^2, the normalized pump field is $P \simeq 25$ Torr, and the normalized pump detuning x ranges from $0 < T_2|\Omega_{31} - \omega_\mathrm{p}| < 60$ Torr, resulting in an AC Stark shift due to the strong pump of $P^2/x > 10$ (>5 homogeneously broadened linewidths) and a Raman cross section multiplier S_2 of $P^2/x^2 > \frac{1}{6}$ which is not negligible.

Shown in Fig. 3 are S_1, S_2, and $G_\mathrm{f}/\sigma_{32}\rho_{11}^\mathrm{e}$ obtained from Eqs. (9) and (17)–(20) versus y for various fixed P and x. Negative S_i values should be interpreted as an interference with the dominant positive S_j term. Clearly shown are the separate processes, laser and stimulated Raman emission, contributing to the net gain profile as well as the AC Stark shifts and large field interference effects. Other representative profiles can be found elsewhere (Panock and Temkin, 1977; Takami, 1976; Temkin, 1977). From these results, it follows that the full range of behavior, from weak to very strong fields, is accessible in the optical pumping approach used to generate FIR.

Actual gain estimates can be made after generalizing the previous equations to a degenerate three-level system. With level degeneracy, the AC Stark shifts become M dependent since μ_{ij} will depend on this quantum number (Schalow and Townes, 1955; Skribanowitz et al., 1972a). In addition, because of the M variation in μ_{ij}, the populations which are excited or depleted will be tensorial, which may favor a preferential polarization of the FIR. The polarization rules are quite simple; a net $(1 \rightarrow 2)$ $\Delta J = 0, \pm 2$ yields parallel polarized FIR, while $\Delta J = \pm 1$ yields perpendicularly polarized FIR (Henningsen, 1977); however, the amount of gain anisotropy depends on the J value of the transitions and the strength of the pumping field. For moderate pumping fields such that the AC Stark shifts may be ignored, summing over the various M states results in an M averaged gain of

$$G_\mathrm{f} \simeq \beta_\varepsilon \sigma_{32}\{S_1[\rho_{33} - (g_3/g_2)\rho_{22}] + S_2[\rho_{11} - (g_1/g_2)\rho_{22}]\} \qquad (20)$$

with $P = \mu_{13} E_\mathrm{p} T_2/2\hbar$ and

$$\sigma_{32} = \omega_\mathrm{f}\mu_{32}^2 T_2/3\hbar\varepsilon_0 c \equiv \lambda_\mathrm{f}^2 A_{32}/4\pi^2 \Delta\nu_\mathrm{H} \qquad (21)$$

with μ_{32} the M averaged emission dipole moment, μ_{13} the M averaged absorption dipole moment, and g_i the level degeneracy. The polarization dependent β_ε coefficients are listed in Table II in the high J limit. For more accurate gain estimates, the explicit M dependence can be inserted in the previous equations and the results summed over all M, a task best suited for machine computations, particularly with strong AC Stark shifts and saturation.

One further parameter of interest is the saturated power density U_i which is the large field limit of the product $G_i I_i$ (Gerry, 1966). This parameter is useful for estimating the fully saturated extracted or absorbed power density and is obtained from the gain expressions and steady-state popula-

TABLE II

HIGH J LIMIT OF THE POLARIZATION DEPENDENT COEFFICIENT β_e^a

Pump transition $1 \rightarrow 3$	FIR transition $3 \rightarrow 2$	Net ΔJ	$\beta_\parallel^b$	$\beta_\perp^b$	Dominant polarization
(Linearly polarized)	(Linearly polarized)	—	—	—	—
Q	Q	0	$\frac{3}{5}$	$\frac{1}{5}$	$\parallel$
R, P	Q	± 1	$\frac{1}{5}$	$\frac{2}{5}$	$\perp$
Q	R	-1	$\frac{1}{5}$	$\frac{2}{5}$	$\perp$
R, P	R	$0, -2$	$\frac{2}{5}$	$\frac{3}{10}$	$\parallel$
(Unpolarized)		—			
P, Q, R	Q, R	—	$\frac{1}{3}$	$\frac{1}{3}$	Any

^{a} Broyer (1976).
^{b} Direction with respect to the pump polarization.

tions, Eqs. (17)–(19). In the Raman (R) dominated case, let F_2 become large, then in the large field limit the saturated power densities are found to be

$$U_{R,f} = I_f G_f \simeq I_f \sigma_{32} S_2(\rho_{11} - \rho_{22}) \rightarrow \rho_{11}^e/2T_1 \qquad \text{(FIR)}, \qquad (22)$$

$$U_{R,p} = I_p G_p \simeq I_f \sigma_{32} S_2(\rho_{22} - \rho_{11}) \rightarrow -\rho_{11}^e/2T_1 \qquad \text{(PUMP)}. \qquad (23)$$

In the laser (L) dominated case, let the product F terms be much larger than the single F terms; then for a similar limit

$$U_{L,f} \rightarrow \rho_{11}^e/3T_1 \qquad \text{(FIR)}, \qquad (24)$$

$$U_{L,p} \rightarrow -2\rho_{11}^e/3T_1 \qquad \text{(PUMP)}. \qquad (25)$$

From these, the Raman case is seen to be twice as efficient as the laser case since $U_{L,f}/U_{L,p} = 0.5 < U_{R,f}/U_{R,p} = 1$.

These results can best be appreciated with some numerical examples. For the case of the 496 μm $K = 2$ line in CH_3F, $\rho_{11}^e \simeq 0.66\%$ and $\sigma_{32} \simeq 7.4 \times 10^{-15}$ cm^2 Torr (Eq. 21). The laser gain for a saturating pump is approximately $\sigma_{32}\rho_{11}^e/2 = 0.8$ cm^{-1} with saturated power densities of $U_{L,f} = 3.2$ and $U_{L,p} = 319$ W/cm^3 Torr^2. Because of the very high gain, saturation may occur easily but, because of the low saturated power density, rather large volumes ($> 1l$) are required for high power operation. For the case of the 385-μm transition in D_2O which has been identified as a Raman line, $\rho_{11}^e \simeq 1.8\%$, $x/2\pi T_2 \simeq 300$ MHz, $\sigma_{32} \simeq 7.3 \times 10^{-15}$ cm Torr, and setting $P^2/x^2 = \frac{1}{6}$ as an example, the Raman gain is calculated to be 0.43 cm^{-1}. The saturated power densities for this example are $U_{R,f} = 17$ and $U_{R,p} = 652$ W/cm^3 Torr^2. Although the gain is slightly lower than the $C^{12}H_3F$ example, the saturated power density is almost a factor of 6 higher, indicating a larger potential power for the same volume and pressure.

These numerical examples illustrate that potentially large laser and Raman gains are available and that a required large FIR power necessitates the use of a large volume. Fortunately, since the FIR cell is passive, the latter presents no great handicap in using these systems as laboratory sources.

C. SECONDARY FEATURES

In concluding this section, brief mention will be made of Doppler broadening, saturation intensity, and a vibrational bottlenecking effect.

1. *Doppler Broadening*

The previous formalism can be generalized to include Doppler broadening by replacing the molecular frequencies by $\Omega_{31} \rightarrow \Omega_{31} - \mathbf{k}_p \cdot \mathbf{v}$ and $\Omega_{32} \rightarrow \Omega_{32} - \mathbf{k}_f \cdot \mathbf{v}$, where $\mathbf{k}_i$ is the wave vector and $\mathbf{v}$ the molecular velocity of a homogeneously broadened subgroup (Feld and Javan, 1969; Hänsch and Toschek, 1970; Feldman and Feld, 1972). In addition ρ_{ii}^e should be subscripted with $\mathbf{v}$ to reflect the equilibrium Maxwell–Boltzman distribution. When this approach is taken, one finds that a gain anisotropy due to Raman contributions exists between co-propagating and counter-propagating waves with the co-propagating case having the higher gain (Hänsch and Toschek, 1970; Feldman and Feld, 1972; Skribanowitz et al., 1972b; Heppner and Weiss 1977; Seligson et al., 1977). The degree of anisotropy is dependent on the ratio λ_f/λ_p and the degree of homogeneous broadening, generally decreasing the higher the pressure and the larger λ_f/λ_p (Seligson et al., 1977).

2. *Saturation Intensity*

The saturation intensity is a steady-state parameter, defined to be the intensity at which a population difference, and hence gain, falls to $\frac{1}{2}$ the unsaturated value in the homogeneously broadened limit (Pantell and Puthoff, 1969). It is always of the form $I_{SAT} \sim \Gamma_{eff}/\sigma_{ij}$ where Γ_{eff} is either a single dominant relaxation rate or an algebraic expression containing many different relaxation rate processes. Since Γ_{eff} typically is dominated by the slowest rate, e.g., diffusion (Tucker, 1974; Hodges and Tucker, 1975; Weiss, 1976), the true steady state value of I_{SAT} requires additional kinetic equations in order to be extracted from the simple analysis presented here (DeTemple and Danielwicz, 1976). In the quasistatic limit, a working and sometimes meaningful saturation intensity can be extracted from Eqs. (17)–(19) valid for time $t > T_1$ but less than other characteristic decay times. For the Raman case with large F_2,

$$\rho_{11} - \rho_{22} \rightarrow \rho_{11}^e/(1 + 2F_2)$$

from which $I_{SAT,R,f} = (T_1 \sigma_{32} S_2)^{-1}$ may be obtained. Noting that $\sigma_{32} S_2$ is equivalent to a stimulated Raman emission cross section σ_{12}, then

$I_{\text{SAT,R,f}}$ is of the expected form. For the pure laser case with small F_2 and f_1,

$$\rho_{33} - \rho_{22} \to \rho^{\text{e}}_{11} f_1/(1 + 2F_1)$$

from which $I_{\text{SAT,L,f}} = (T_1 2\sigma_{32})^{-1} < I_{\text{SAT,R,f}}$ on line center ($y = 0$, $S_1 = 1$). Other cases may be obtained directly from Eqs. (17)–(19).

3. Bottlenecking

If the pump duration is long, a sufficiently large vibrational population may be excited such that the overall ground state population may be depleted or simply bleached. Since vibrational relaxation is very slow compared with the typical pump duration of 100 nsec, an equivalent vibrational saturation or bottleneck may occur which has the effect of terminating the interaction (Tucker, 1976; Temkin and Cohn, 1976). This can be treated by writing separate equations of motion for the ground and first vibrational level and setting $\rho^{\text{e}}_{ii} N = \alpha_{ii} N_j$ where α_{ii} is the equilibrium fraction of the total population N_j associated with state i. Treating the saturated laser case, setting $\alpha_{11} = \alpha_{22} = \alpha_{33}$ and including the vibrational saturation, result in a modified form of the saturated power density of

$$U_{\text{L,p}} = -(2\rho^{\text{e}}_{11}(0)/3T_1)\,e^{-\Gamma t}, \qquad U_{\text{L,f}} = (\rho^{\text{e}}_{11}(0)/3T_1)(2\,e^{-\Gamma t} - 1),$$

where $\Gamma = 4\alpha_{11}/3T_1$, time is measured from the start of an assumed step function pump, and $\rho^{\text{e}}_{11}(0)$ is the thermal equilibrium value of the initial population. The exponential terms are indicative of vibrational bleaching and when $\Gamma t > 0.368$, $U_{\text{L,f}} < 0$, indicating absorption on the FIR transition. As an example, for $C^{12}H_3F$ the ΔK relaxation rate is approximately a factor of 10 slower than the rotational relaxation rate, so one may assume that the K rotational manifolds are partially decoupled. For this case $\alpha_{11} \simeq 6\%$ (i.e., state $J = 12$ contains 6% of the total $K = 2$ rotational manifold population, which is 11% of the total ground state population) leading to $\Gamma^{-1} \simeq 110$ nsec Torr and FIR laser absorption after 40 nsec Torr. Including ΔK relaxation will lengthen these somewhat, but the basic results and implications are inescapable; only a finite amount of power *and* energy density may be extracted on a per pulse basis.

D. SUMMARY

Because of a near resonant enhancement, it is possible to have a Raman-like as well as laser-like contribution to the gain in the optically pumped FIR systems, the major difference between the two being the FIR frequency. The value of the gain for the separate laser and Raman components is similar, suggesting that other features, such as ground state absorption or pump modes, may favor the growth of one over the other. The saturated power density and maximum conversion efficiency are both higher for the Raman case than for the laser case, suggesting that the former is to be preferred over the latter if efficiency is a consideration.

With this somewhat abbreviated background, specific molecular systems will now be discussed.

III. Specific Molecular Systems

At present there are close to one thousand FIR lines generated by the optical pumping technique from some thirty molecules spanning the region from 30 μm to 2 mm (Yamanaka, 1976; Rosenbluh et al., 1976; Gallagher et al., 1977; Strumia, 1975; Beck, 1976). Of these, less than a dozen truly high powered lines have been found in only a few molecules with the most notable being $C^{12}H_3F$, $C^{13}H_3F$, and D_2O. Since the latter are all light molecules, one may speculate that the strongest lines may exist only for these because light also implies large rotational and centrifugal distortion constants, which will have the effect of maximizing a given ground state population by minimizing the rotational partition function, and will minimize the possibility of any ground state FIR absorption by creating a sparse spectrum, widely shifted from the equivalent excited state spectrum (DeTemple and Lawton, 1978). Whether these are the only considerations remains to be seen.

In what follows, gross spectral features of the stronger lines will be presented in tabular form, an example of which is Table III. In this table, the CO_2 pumping line is subscripted with the band, 9 or 10 μm, the rotational states are denoted either as J, K for symmetric tops or as J_τ ($\tau = K_- - K_+$) for asymmetric tops, the frequency offset is $\nu_{laser} - \nu_{molecule}$, and order-of-magnitude emission intensities are indicated. Also, the most accurate transition frequency, λ_f^{-1}, is listed based either on heterodyne frequency measurements (M) or calculations from the best available spectroscopic data (C); the listed wavelength λ_f is rounded off to the nearest μm.

A. $C^{12}H_3F$

One of the stronger FIR emission lines occurs in $C^{12}H_3F$ at 496 μm. This line, being the first FIR line to be generated by the optical pumping technique (Chang and Bridges, 1970), has been actively studied in the continuous and pulsed modes of operation (DeTemple and Danielewicz, 1976; Chang, 1974; Evans et al., 1975). Cascade lines have also been observed in the ground and excited states, as indicated in Table III and Fig. 1. The Q branch absorptions from a $J + 1$-fold K multiplet lacking the $K = 0$ transition since $\mu_{13}^2 \sim K^2$. The multiplet structure is vividly illustrated in Fig. 4, in which is show the $^QQ(12, K)$ absorption spectra obtained with the use of a tunable diode laser (Sattler and Simonis, 1977). Also shown is a calculated spectra obtained using the K^2 dependence of μ_{13}^2, nuclear spin

TABLE III

SYMMETRIC TOP TRANSITIONS

| Species | Line | Absorption | | | | λ_f (μm) | $1/\lambda_f$ (cm^{-1})^a | Mode | Emission | | | |
		Mode	Lower	Upper	Offset				Lower	Upper	Pol^b	Int^c
C^{12}H$_3$F	P$_9$(20)	ν_3	12,K (12,2)	12,K 12,2	−44 MHz	496	20.157M	ν_3	11,2	12,2	⊥	1 MW
						541	18.479M	ν_3	10,2	11,2	⊥	—
						595	16.801C	ν_3	9,2	10,2	—	—
						452	22.128M	0	12,2	13,2	⊥	—
						419	23.827C	0	13,2	14,2	—	—
C^{13}H$_3$F	R$_{10}$(32)	ν_3	32,5	31,5	69 MHz	193	51.861C	ν_3	30,3	31,3	=	1 kW
	P$_9$(32)	ν_3	4,K (4,3)	5,K 5,3	24 MHz	1222	8.184C	ν_3	4,3	5,3	=	1 MW
						1207	8.291C	0	4,3	5,3	—	—
						1006	9.949C	0	5,3	6,3	—	—
						862	11.607C	0	6,3	7,3	—	—
CH$_3$Br^{79}	P$_{10}$(28)	ν_6	9,2	8,1	~20 MHz	1965	5.088C	ν_6	7,1	8,2	=	1 kW
CH$_3$I	P$_{10}$(18)	ν_6	44,5	45,6	>50 MHz	447	22.364M	ν_6	44,6	45,6	=	100 kW
	P$_{10}$(32)	ν_6	15,5	16,6	−30 MHz	1254	7.973C	ν_6	15,6	16,6	=	—
NH$_3$	R$_{10}$(6)	ν_2	s(5,4)	a(5,4)	−568 MHz	292	34.339C	ν_2	s(5,4)	a(5,4)	=	—
	P$_{10}$(32)	ν_2	a(5,3)	s(5,3)	−960 MHz	152	65.985C	ν_2	a(4,3)	s(5,3)	⊥	1 kW
						291	34.373C	ν_2	s(4,3)	a(4,3)	=	—
PH$_3$	P$_9$(18)	ν_4	10,2	9,3	400 MHz	121	82.339C	ν_4^{+1}	8,3	9,3	=	1 kW

^{a} Transition frequency in cm^{-1}. M = measured [C^{12}H$_3$F (Chang and Bridges, 1970); CH$_3$I (Radford et al., 1977)], C = calculated [C^{12}H$_3$F (Freund et al., 1974); C^{13}H$_3$F (Graner, 1975), CH$_3$I (Freund et al., 1974); CH$_3$Br (Deroche and Betrencourt–Stirneman, 1976)]; CH$_3$I (Arimondo et al., 1978); NH$_3$ (Udea and Shimoda, 1975); PH$_3$ (Malk et al., 1978).

^{b} Pol: Polarization of FIR with respect to the pump polarizer.

^{c} Int: Order-of-magnitude FIR intensity.

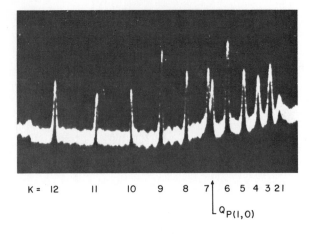

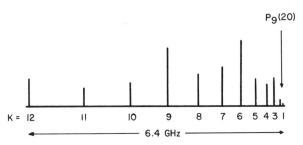

FIG. 4.   The $^Q Q(12,K)$ absorption spectra in the v_3 band of $C^{12}H_3F$ obtained with a tunable diode laser (Sattler, 1977) and calculated from spectroscopic constants. The location of the $P_9(20)$ laser line center is shown by the arrow.

statistical weights which populate levels with K a multiple of three twice as heavy as other K levels, and the K dependent Boltzmann factors (Freund et al., 1974; Schalow and Townes, 1955). Since a free running CO_2 TEA laser may have a spectral bandwidth of $\gtrsim 1$ GHz, a number of K levels may be populated, which may lead to a frequency spread in the emitted FIR; up to 1.7 GHz if all K levels were populated.

B. $C^{13}H_3F$

Closely related to $C^{12}H_3F$ is the isotopic species $C^{13}H_3F$, present as a naturally occurring impurity (1.1 % concentration) in $C^{12}H_3F$ or in iso-topically enriched mixtures. The dominant emission is at 1222 μm, with strong cascade transitions in the ground state (Hacker et al., 1976). The K multiplet absorption spectra is shown in Fig. 5 which, if all were pumped, could lead to an FIR frequency spread of only 75 MHz for the 1222-μm

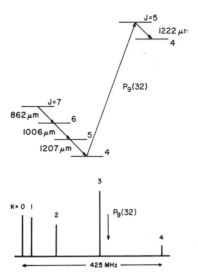

FIG. 5.   The $^Q R(J,3)$ absorption and emission lines in the v_3 band of $C^{13}H_3F$ pumped by the $P_9(32)$ line. Also shown is the $^Q R(4,K)$ absorption spectra with the location of the $P_9(32)$ laser line center.

line (Freund et al., 1974; Hacker et al., 1976). The major disadvantage of this molecule is the high cost, roughly ten times the cost of $C^{12}H_3F$, necessitating the need for a gas recovery system on large cells.

C. D_2O

One of the more interesting FIR molecules is D_2O, the spectral properties of which are listed in Table IV (Plant et al., 1974; Keilmann et al., 1975; Evans et al., 1976; Woskoboinikow et al., 1976; Lipton and Nicholson, 1977). The FIR wavelengths range from 50 to 385 μm with the highest powers observed at 66, 113, and 385 μm, all exceeding 1 MW (Evans et al., 1976, 1977b). In addition, the various pump lines are known to be detuned from the D_2O absorption by 6–40 infrared Doppler widths, implying the existence of off-resonant pumping and stimulated Raman emission (Lin and Shaw, 1977; Petuchowski et al., 1977; Temkin, 1977). Shown in Fig. 6 is a partial energy level diagram near the $P_9(32)$ pumping transition, with the insert showing the absorption profile obtained from tunable diode spectroscopy (Worchesky, 1978). For this pump line, both the 66- and 50-μm lines have been identified as being produced by stimulated Raman emission (Petuchowski et al., 1977). Shown in Fig. 7 is a partial energy level diagram near the $R_9(22)$ pump line (Keilmann et al., 1975). The 385-μm line has been found to exhibit a rather interesting behavior in that at low pressures it is laser-like, while at higher

TABLE IV

ASYMMETRIC TOP TRANSITIONS

Species	Line	Mode	Lower	Upper	Offset	λ_f (μm)	$1/\lambda_f$ (cm^{-1})[a]	Mode	Lower	Upper	Pol[b]	Int[c]
			Absorption					Emission				
D$_2$O	P$_9$(32)	ν_2	$6_6, 6_5$	$5_4, 5_5$	1.2 GHz	66	151.697C	ν_2	$4_3, 4_4$	$5_4, 5_5$	∥	10 MW
						83	120.55 C	ν_2	$3_2, 3_3$	$4_3, 4_4$	∥	1 kW
						50	198.771C	0	$6_5, 6_6$	$7_6, 7_7$	∥	100 kW
	P$_9$(32)	ν_2	7_{-1}	6_{-3}	−554 MHz	116	85.850C	ν_2	5_{-3}	6_{-3}	∥	1 MW
	R$_9$(12)	ν_2	10_{-8}	9_{-6}	565 MHz	95	105.692C	ν_2	8_{-6}	9_{-6}	∥	10 kW
						113	88.827C	ν_2	9_{-8}	9_{-6}	⊥	10 MW
						144	70.885C	0	10_{-8}	10_{-6}	⊥	1 kW
	R$_9$(22)	ν_2	5_0	4_0	−326 MHz	385	25.980M	ν_2	4_{-2}	4_0	⊥	10 MW
						358	27.906C	ν_2	4_{-4}	4_{-2}	⊥	1 MW
						240	41.736C	0	5_0	6_{-2}	⊥	10 kW
						276	36.188C	0	6_{-2}	6_0	⊥	10 kW
	R$_9$(30)	ν_2	10_{-7}	10_{-9}	−651 MHz	99	101.104C	ν_2	9_{-9}	10_{-9}	∥	100 kW
	R$_9$(32)	ν_2	9_{-9}	8_{-7}	589 MHz	121	82.305C	ν_2	7_{-7}	8_{-7}	∥	10 kW
						98	101.726C	0	9_{-9}	10_{-9}	∥	10 kW
	R$_9$(34)	ν_2	4_2	3_0	−2.3 GHz	253	39.452C	ν_2	3_{-2}	3_0	⊥	—
D$_2$S	P$_{10}$(32)	ν_2	6_3	7_5	—	136	73.7M	ν_2	6_3	7_5	∥	1 kW
HCOOH	R$_9$(18)	ν_6	35_{-18}	34_{-17}	—	394	25.404M	ν_6	33_{-16}	34_{-17}	∥	—
	R$_9$(20)	ν_6	34_{-34}	33_{-33}	—	433	23.114M	ν_6	32_{-32}	33_{-33}	∥	1 kW
	R$_9$(28)	ν_6	27_{-18}	26_{-17}	—	513	19.493M	ν_6	25_{-16}	26_{-17}	∥	1 kW

[a] Transition frequency in cm^{-1}. M = measured [D$_2$O (Fetterman, 1979); D$_2$S (Danielewicz and Keilmann, 1979); HCOOH (Dangoisse et al., 1979)]; C = calculated [D$_2$O (Lin and Shaw, 1977)].

[b] Pol: Polarization of FIR with respect to the pump intensity.

[c] Int: Order-of-magnitude FIR intensity.

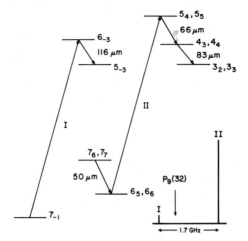

FIG. 6. Partial energy level diagram of the v_2 band of D_2O near the $P_9(32)$ CO_2 laser line showing the FIR emission lines. Also shown is an absorption spectra based on tunable diode laser spectroscopy.

pressures it is Raman-like (Temkin, 1977). An example of this is shown in the FIR Fabry–Perot interferometer scans of the resulting output in Fig. 8. The shift of the peak by 390 MHz between high and low pressure is quite dramatic, exceeds the known pump detuning of 326 MHz, and partially resolves the AC Stark shift. The separate laser and Raman contributions to the gain of this transition have been resolved in tunable oscillator-amplifier measurements; the results for two FIR polarization directions are displayed in Fig. 9 (Drozdowicz *et al.*, 1978). The remaining D_2O lines are undoubtedly

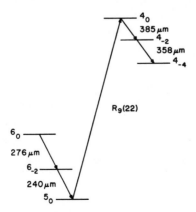

FIG. 7. Partial energy level diagram of the v_2 band in D_2O near the $R_9(22)$ CO_2 laser line showing the FIR emission lines.

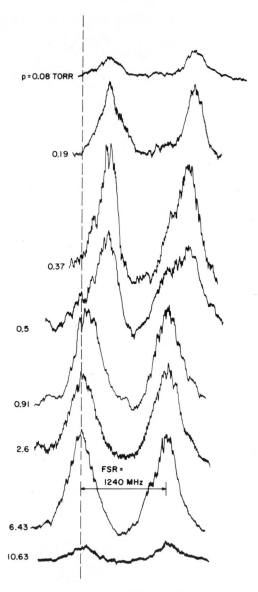

FIG. 8. Fabry–Perot interferometer scans of the superfluorescent output from a 3-m D_2O cell at 385 μm showing a frequency shift between low and high pressure operation (Temkin, 1977).

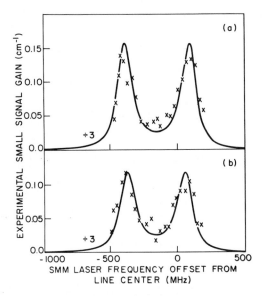

FIG. 9. Experimental (x) and theoretical (solid) small signal gain vs. FIR laser frequency detuning from the D_2O 385-μm line center. Conditions: single mode CO_2 laser of 500 kW/cm² power, 3 Torr of D_2O and (a) perpendicular and (b) parallel polarization. The Raman contribution would be identified at a detuning of -420 MHz, which is the sum of the known pump detuning plus the AC Stark shift (Drozdowicz *et al.*, 1978).

produced in part by a Raman process or a resonant cascade, a conclusion which follows from the previous behavior and the recently available high resolution conventional and laser spectroscopic infrared data (Lin and Shaw, 1977; Worchesky *et al.*, 1978; Lipton and Nicholson, 1978).

D. CH_3Br, CH_3I, HCOOH, NH_3, D_2S, AND PH_3

The remaining molecules in Tables III and IV, while being somewhat weaker, contribute some information to the emission processes and candidate molecular characteristics. For CH_3I, the absorption and subsequent 447 μm emission is due to an isolated J, K transition with the emission identified as laser rather than Raman (Graner, 1975; Hutchinson, 1977; Wiggins *et al.*, 1978). Since ρ_{11}^e, and hence $U_{L,f}$, is a factor of 10 smaller than the strong transition in $C^{12}H_3F$, somewhat larger volumes may be required to produce the same output power. A number of the FIR lines from NH_3 have been identified as being produced by stimulated Raman emission with a behavior similar to the 385-μm D_2O line (Fetterman *et al.*, 1972; Gullberg *et al.*, 1973; Wiggins *et al.*, 1978). Both isotopes of CH_3Cl and CH_3Br emit rather weakly—an observation which can be correlated with the existence of nearby ground state absorption (Plant *et al.*, 1974; Semet and Luhmann, 1976;

TABLE V

SUMMARY OF TRANSITION PARAMETERS FOR THE STRONGER LASER LINES

Molecule	μ_{IR} (D)	μ_{FIR} (D)	Pump	Lower state	ρ_{11}^e (%)	$-G_p^a$ (cm^{-1})	λ_f (μm)	G_f^b (cm^{-1})	$U_{L,f}^c$ (W/cm^3 Torr^2)
$C^{12}H_3F$	0.2	1.79	$P_9(20)$	12,2	0.66	0.053	496	0.8	3.0
$C^{13}H_3F$	0.2	1.79	$P_9(32)$	4,3	0.74	0.49	1222	0.24	0.9
CH_3Br^{79}	0.04	1.82	$R_{10}(28)$	9,2	0.3	0.20	1965	0.08	0.3
CH_3I	0.04	1.65	$P_{10}(18)$	44,5	0.07	0.047	447	0.16	0.3
PH_3	0.06	0.58	$P_9(18)$	10,2	0.38	0.24	121	0.14	7.0
D_2O	0.12	1.87	$P_9(32)$	$6_6,6_5$	0.67	0.26	66	4.3	22.0
			$R_9(12)$	10_{-8}	0.47	0.16	113	4.9	9.0
			$R_9(22)$	5_0	1.8	0.45	385	2.2	10.0
HCl^{35}	0.064	1.095	$\sim 3\ \mu m$	R(3)	18.0	>10	118	>10	331.0

[a] $\sigma_{13}\rho_{11}^e N$
[b] $\sigma_{32}\rho_{11}^e N/2$
[c] $\hbar\omega_f \rho_{11}^e N/3T_1$

Deroche and Betrencourt–Stirneman, 1976; Deroche 1978). The PH_3 lines are interesting in that they appear to be as strong as the D_2O lines operating in the same experimental system even though the permanent dipole moment, and hence σ_{32}, is smaller (Malk et al., 1978). This molecule may be suitable as a candidate zero or low gain, high saturated power density amplifier, useful for producing spectrally narrow and intense FIR signals.

E. HCL

In an entirely different approach, a tunable 3-μm pump has been used to produce tunable stimulated Raman emission in HCl in the 60–160-μm range at the 100-kW power level (Frey et al., 1977). Using a pulsed, tunable dye laser which is doubly Stokes shifted to 3 μm in an H_2 cell, tunable nanosecond pulses of 10 MW were used to pump HCl on and near the $R(J)$, $J = 2$–$J = 7$, transitions of the fundamental absorption ($v = 0$–$v = 1$). Tunable Raman emission on a number of $v = 1$ $R(J)$ transitions was observed as well as $v = 1$ $R(J - 1)$ cascade laser transitions. The observed tuning of ± 1.1 cm^{-1} about each $R(J)$ resonance represents an equivalent bandwidth of about 1 % of the FIR frequency, comparable to conventional microwave sources. The basic approach may be extended to other HX molecules, HF being a recent addition (deMartino, 1978), and to isotopic species to generate quasi-tunable, intense signals, perhaps down to 500 μm.

F. SUMMARY

In summary, the strong FIR emitters, while being few in number, are quite spectacular in that overall photon conversion efficiencies approaching the Manley–Rowe limit have been achieved in certain cases. Selected parameters for the stronger emitters are listed in Table V. Here, the infrared band and permanent dipole moments are listed as well as the saturated power densities in the laser mode and the small signal, homogeneously broadened, line center FIR gain (G_f) and pump (G_p) absorption coefficients. The strongest lines are notable in that they have large saturated power densities which correlate with the large ground state populations ρ_{11}^e.

IV. Operational Configurations

The geometrical configurations which have been employed to generate intense FIR are numerous, but conveniently group into three categories: (a) single pass cells; (b) cavities; and (c) oscillator–amplifiers.

A. SINGLE PASS CELLS

The earliest FIR configurations were simply long cells operating in the mirrorless or superfluorescent mode, sometimes referred to as amplified spontaneous emission (Brown et al., 1973; DeTemple et al., 1973).[1] The large gain obtained with the use of TEA CO_2 lasers and the long path lengths are both conducive to a saturated behavior with efficient conversion.

The starting field in the superfluorescent mode is noise, both blackbody and spontaneous emission, at an intensity $\sim 10^{-15}$ W/cm^2 at 500 μm. The FIR saturation intensity is of order 5 W/cm^2 Torr^2 for $C^{12}H_3F$ at 496 μm, indicating that a gain $[\exp(G_f z)]$ of 10^{16} would be needed before the noise initiated FIR wave is amplified to a level sufficient to saturate and efficiently extract energy from the sample. For $G_f \sim 0.37$ cm^{-1}, this gain requires a path length of only 1 m. Thus in a long cell, a simplified model is that the first meter of the cell acts as a small signal amplifier with the remainder a saturated linear amplifier, with efficient conversion occurring in the latter.

In operation, the pulsed systems differ somewhat from the simple model. For example, graphed in Fig. 10a is the output energy versus cell length for $C^{12}H_3F$ at 496 μm showing a scaling behavior as $\exp(z^{1/2})$ (DeTemple et al., 1973) which is in contrast to the e^z scaling behavior in Fig. 10b, $\exp(z^{1/2})$ being one characteristic of a small signal lethargic amplifier (to be discussed in the next section) or a small signal Raman amplifier (Carman et al., 1970). All scaling forms have been observed in a number of different experiments (DeTemple et al., 1973; Brown et al., 1973, Plant et al., 1974), from which one may surmise that the dynamical evolution of the noise driven FIR systems is a rather complex process, possibly compounded by the presence of a multimode pump, Raman and AC Stark shifts and the usual saturation effects.

Although the superfluorescent configuration has resulted in the most intense signals from $C^{12}H_3F$, D_2O, and other molecules, (Evans et al., 1975, 1976, 1977b; Semet and Luhmann, 1977) the FIR spectral characteristics were far from the ideal laser source. The source of the problem is twofold in origin: the pump spectral purity and the large FIR gain–bandwidth product. The spectral purity of a free running TEA CO_2 laser is comprised of many longitudinal modes (50–75 MHz spacing) spread over a few GHz. Shown in Fig. 11 are the equilibrium populations (ρ_{ii}^e) for $C^{12}H_3F$ and the spectral distribution of a TEA CO_2 laser. Considering a simple saturation of the available $^QQ(12, K)$ transitions by this pump, the emitted FIR at 496 μm might have a spectral distribution as shown in Fig. 11b, which should be compared with one measurement shown in Fig. 11c (Brown et al., 1974). Even with a single mode pump, stimulated Raman emission off the various K multiplets may increase the effective gain bandwidth, as illustrated in the

---

[1] We use the laser physics definition of superfluorescence, a high gain noise amplifier, as opposed to superradiance, a high gain, noise driven, coherent radiator, and remark that the definitions may be different in other areas and in earlier laser related literature.

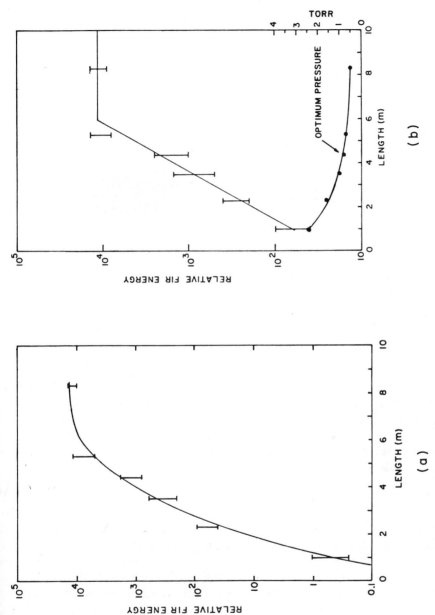

FIG. 10. Total superfluorescent energy at 496 μm versus cell length at (a) constant pressure (0.5 Torr, $C^{12}H_3F$) and (b) optimum pressure (DeTemple, 1973). The CO_2 laser was multimode with a power of ~0.5 MW with resulting FIR power of ~ 1 kW.

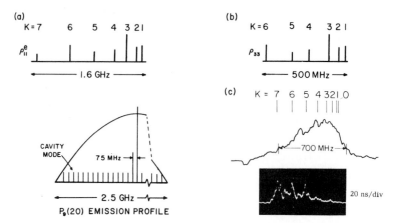

FIG. 11. (a) Ground state population for the $^{Q}Q(12,K)$ v_3 absorption in $C^{12}H_3F$ vs. relative infrared frequency superimposed on a TEA CO_2 laser emission profile. (b) Relative $^{Q}R(11,K)$ emission strength at 496 μm, assuming a simple saturation of the K levels in (a). (c) Measured superfluorescent emission spectrum. Also shown is the equivalent time behavior of the emission. (Brown et al., 1974).

calculated gain profiles in Fig. 12 for $C^{12}H_3F$ (Chang, 1977). For isolated lines, the spectral width may be determined by the spectral purity of the pump due to the Raman contributions, suggesting that the narrowest signals may occur when the pump is single frequency and the transition is isolated (no K multiplet). With a multimode pump, representative FIR spectral widths in the superfluorescent mode are $C^{13}H_3F$ (1222 μm) $\lesssim$ 100 MHz (Hacker et al., 1976), $C^{12}H_3F$ (496 μm) $\sim$ 700 MHz (Brown et al., 1974), and D_2O (385 μm) $\sim$ 500 MHz (Woskoboinikow et al., 1976). With a single mode pump, the FIR spectral purity under various conditions may be transform limited, determined by quasi-cavity modes caused by insipient feedback sources (Plant and DeTemple, 1976), or may exhibit gain narrowing (Casperson and Yariv, 1972), all of which may be as low as 1–10 MHz.

B. CAVITY OPERATION

Although indicative of the potential of the technique, the superfluorescent mode has two disadvantages; lack of good transverse mode profile and poor spectral purity. These have been overcome with the use of an FIR cavity which provides natural frequency filtering. Even though the original optical pumping experiments (Chang and Bridges, 1970) and the first TEA laser experiments (Brown et al., 1972) utilized hole-coupled FIR cavities, the resulting FIR output was quite low—an observation one may now attribute to the small active volume and poor mirror coupling. Subsequently, with the use of mesh type mirrors (Ulrich et al., 1970) and larger volumes, the FIR

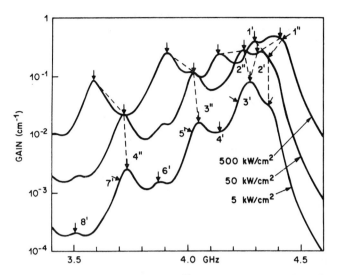

FIG. 12. Calculated FIR gain spectrum in $C^{12}H_3F$ for single-mode pump intensities indicated in the figure. Numbers and arrows indicate the shift of some of the laser/Raman contribution from the various K multiplets due to the AC Stark shift. The $^QR(11,2)$ line center is at 4.3 GHz on this scale. (Chang, 1977).

power and spectral characteristics were considerably improved over the superfluorescent and original cavity experiments.

One of the earlier cavity experiments utilized transverse optical pumping (TOP) in $C^{12}H_3F$ at 496 μm and is shown in Fig. 13a (Brown et al., 1974). Pulses of $\sim$0.5 kW with a spectral purity of $\sim$30 MHz were obtained from a short (33-cm) cavity using a variable mesh coupler. A novel feature of this short laser with a multimode pump is that the FIR could be tuned over $\sim$500 MHz as shown in Fig. 13b.

Using a one-round-trip pump configuration in $C^{12}H_3F$ at 496 μm as shown in Fig. 14a, pulses of 10 kW with a spectral purity of 150 MHz corresponding to 3–4 cavity longitudinal modes were obtained with a beam splitter coupler (Sharp et al., 1975). From a zig-zag configuration (Fig. 14b), pulses of 9-kW, 28-MHz line width were obtained from $C^{12}H_3F$ at 496 μm on a single longitudinal and transverse mode, using a multimode pump (Cohn, 1975). From another zigzag system, 17-kW, 2.5-μsec long pulses at 66 μm were obtained from D_2O with a multimode spectral width of a few hundred MHz (Dodel and Magyar, 1978). In addition, an interesting angular correlation was observed; a FIR superfluorescence pulse was detected co-propagating with the pump (i.e., zigzag) in addition to a FIR laser pulse colinear with the cavity (Allen et al., 1979). Pulses at the kW level with a narrow linewidth ($<$60 MHz) were achieved using a single pass, IR

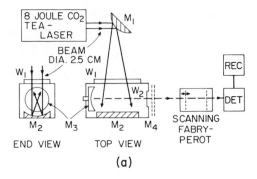

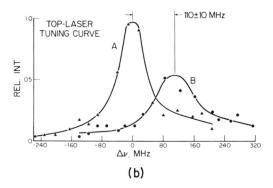

FIG. 13.   (a) Arrangement for TOP experiment using an 8J CO_2 TEA laser: M_1 and M_2 are cylindrical mirrors, M_3 and M_4 form the FIR cavity of length 33 cm, and W_1 and W_2 are windows. (Brown, 1974). (b) Output intensity versus cavity tuning for the TOP laser; labels A and B refer to two transverse modes (Brown et al., 1974). Legend: ——▶—, pump beam; – –▶– –, FIR beam.

etalon coupled cavity shown in Fig. 14c, again from $C^{12}H_3F$ at 496 μm (Brown et al., 1976). Similar FIR performances have also been obtained using dielectric waveguide structures, (Hodges and Hartwick, 1973; Plant and DeTemple, 1976; Semet and Luhmann, 1976) as opposed to conventional open resonators, and unstable resonators (Hutchinson, 1977; Weber, 1978; Ewanizky et al., 1979).

In order to suppress multilongitudinal mode behavior inherent in longer cavities, additional frequency filtering in the form of either a Fox–Smith (Smith, 1972) or Michelson (Kumar et al., 1977) mode selector has been used, an example of the former being shown in Fig. 15 (Evans et al., 1977a). This approach assures single mode behavior which is also tunable.

The benefits of a cavity in obtaining single frequency, tunable FIR are clear and encouraging except that the typical conversion efficiencies are a factor of ten smaller than the Manley–Rowe limit or $h\omega_f/h\omega_p$. Most of this

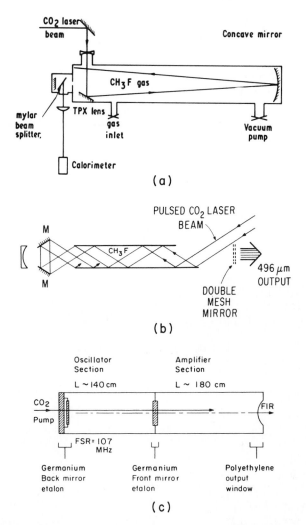

FIG. 14.   Various FIR cavity configurations: (a) conventional open resonator with a beam splitter coupler (concave mirror radius is 3.76 m) (Sharp, 1975); (b) zigzag laser with a mesh coupler (Cohn, 1975); and (c) coaxial etalon coupled laser (Brown *et al.*, 1976).

may be attributed to the use of a spectrally broad pump, only part of which is effective in generating a spectrally narrow signal.

There are a number of techniques for producing a spectrally narrow or single-mode pump, which include the use of an intracavity absorption cell (DeTemple and Nurmikko, 1971), an intracavity CO_2 gain cell (Gondhalekar *et al.*, 1975) an etalon (Weiss and Goldberg, 1972), and injection

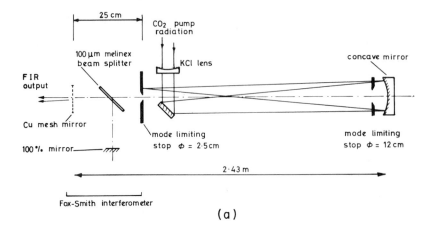

(a)

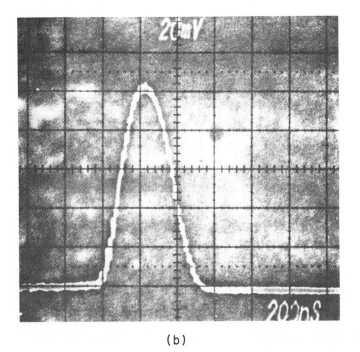

(b)

FIG. 15. (a) FIR open resonator with a Fox–Smith coupler which suppresses all but one longitudinal mode (Evans *et al.*, 1977a), and (b) time dependent output from the laser in (a) showing a complete lack of structure associated with multimode behavior. (Au-coated concave mirror has 2.46 m radius.)

locking (Lachambre *et al.*, 1976; Izatt *et al.*, 1977; Hutchinson and Vander Sluis, 1977). With the exception of the etalon, the various approaches result in a fixed frequency output.

Single-mode pumping has been used in the stimulated Raman, super-fluorescent and cavity configurations, resulting in temporally smooth FIR pulses (Plant and DeTemple, 1976; Petuchowski *et al.*, 1977). More recently, an etalon tuned pump has been used which has the advantage of optimizing the FIR output in terms of the IR frequency. Shown in Fig. 16a is the transmission in D_2O near the $P_9(22)$ line, responsible for the 66- and 116-μm FIR, obtained with an etalon tuned TEA laser (Lipton *et al.*, 1977; Nicholson and Lipton, 1977). The FIR emission in the SF mode is shown in Fig. 16b, while the cavity emission is shown in Fig. 16c; both indicating an optimum IR detuning. A similar behavior has been found for the $R_9(22)$ pump line shown in Fig. 17 (Woskoboinikow *et al.*, 1979). Here, the optimum detuning of -0.4 GHz results in 385-μm Raman emission with a somewhat higher output than the equivalent laser emission obtained by tuning to the D_2O line center. Since a similar behavior has been found for other lines (Lipton, 1978), one may conclude that the use of a tunable single-mode pumping laser offers considerable promise in obtaining an intense, high spectral purity FIR signal.

C. OSCILLATOR–AMPLIFIERS

In an effort to increase the FIR power up to that obtained in the super-fluorescent mode, oscillator–amplifier configurations have been used. This is particularly effective because of the high amplifier gain and large driver power available from FIR cavity sources. Conceptually, the amplifier is simply a long cell terminated with a suitable exit window and containing an input coupler to "mix" the FIR driver with the pump.

Shown in Fig. 18 are various coupling schemes which have been used in amplifiers: (a) free standing mesh (Evans *et al.*, 1977a), (b) Ge Brewster coupler (Semet and Luhmann, 1976), (c) Si (Plant and DeTemple, 1976) and SiO_2 (Woskoboinikow *et al.*, 1976; Drozdowicz *et al.*, 1977) Brewster-mirror couplers, (d) AR etalon coupler (Brown *et al.*, 1976), and (e) an open retroreflector (Drozdowicz *et al.*, 1976). The free standing mesh has the advantage of being highly reflective in certain FIR regions while being only semitransparent to the pump. The etalon coupler is FIR frequency sensitive and has optical damage problems associated with the coatings. The open retroreflector is limited to a zigzag configuration as in Fig. 14b. The Ge Brewster coupler is effective only for orthogonally polarized FIR, while the Brewster-mirror coupler can be made polarization independent, although having an optical damage limitation. The most versatile appear to be the Brewster-mirror couplers which can also be used for the cell window, thus providing a minimum feedback necessary to suppress parasitic oscillation or superfluorescence.

The oscillator–amplifier approach has been applied mainly to $C^{12}H_3F$ at 496 μm and D_2O at 385 μm. At 496 μm, a FIR power in the range (Brown et al., 1977; Evans et al., 1977b; Semet and Luhmann, 1976) 0.25–0.5 MW with <60 MHz spectral purity has been obtained. At 385 μm, a FIR power in the range 0.5–1 MW with a spectral purity of better than 50 MHz was observed (Woskoboinikow et al., 1977). In addition, a broad low level background, presumably due to superfluorescence was observed on both

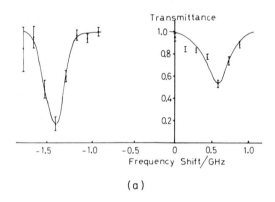

(a)

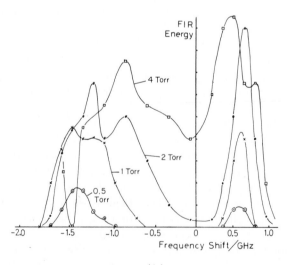

(b)

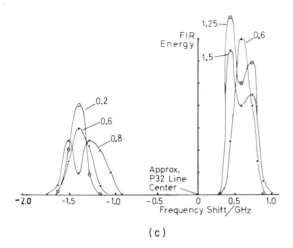

(c)

FIG. 16. (a) Measured transmission through D_2O vapor using a tunable single mode TEA laser on the $P_9(32)$ transition (Lipton *et al.*, 1977); the left and right transmission minima corresponds to transitions II and I in Fig. 6 responsible for the 66 and 116 μm lines. (b) Relative total superfluorescent D_2O emission vs. $P_9(32)$ detuning at various pressures (Lipton *et al.*, 1977). (c) Relative FIR cavity output from D_2O vs. $P_9(32)$ detuning at various pressures (Lipton *et al.*, 1977); the curves to the left and right of line center correspond to the 66- and 116-μm transitions. (Figures 16a,b on facing page.)

systems, which may be caused by the use of a multimode pump and the existence of spurious feedback regions (Brown *et al.*, 1977; Drozdowicz *et al.*, 1977; Evans *et al.*, 1977a; Woskoboinikow *et al.*, 1977). This background is evident in the time domain (Fig. 19a) (Brown *et al.*, 1977), and in the optical frequency domain (Fig. 19b) (Evans *et al.*, 1977a). In addition, the overall energy conversion efficiency was still somewhat lower than the Manley–Rowe limit, a fact attributable to the use of a multimode pump. Aside from these, the oscillator–amplifier approach has produced some of the most intense and spectrally pure signals in the FIR region.

To summarize, for a select few lines ($C^{12}H_3F$ at 496 μm, $C^{13}H_3F$ at 1.222 mm, and D_2O at 66, 113, and 385 μm) conversion efficiencies in the superfluorescent mode have approached the Manley–Rowe limit to within a factor of 2. In comparing all approaches, the efficiencies rank as: 0.5 $h\omega_f/h\omega_p \sim$ SF > oscillator–amplifier $\gtrsim$ cavity mode. Certain improvements in spectral purity and efficiency may be anticipated with the use of a tunable single-mode pumping source. Reduction or elimination of the persistent superfluorescence background may occur with the use of an intrinsically lower gain, narrow bandwidth transition, such as PH_3 at 121 μm, which need not necessarily have a low saturated power density (Table V).

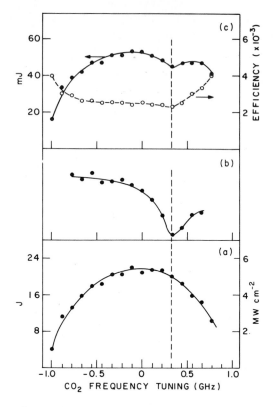

FIG. 17.   Output from D_2O at 385 μm vs. single mode $R_9(22)$ laser detuning; 3-m FIR cavity with a mesh mirror and Brewster input coupler of Fig. 18c. (a) CO_2 laser tuning curves, (b) small signal $R_9(22)$ laser transmission through cell, and (c) total FIR output energy and efficiency (Woskoboinikow et al., 1979).

D.   EXPERIMENTAL TECHNIQUES

In concluding this section, a short synopsis of experimental materials and techniques will be outlined.

The emission wavelengths are measured either with a grating spectrometer or Fabry–Perot interferometer, or by cavity scans if the source is a laser cavity. The emission frequency can be measured by Fabry–Perot interferometry if a reference line is available, heterodyne spectroscopy with an up-converted microwave source, or pressure dependence of self-absorption if the molecular spectra of the absorber is known.

Detectors group into two categories: cryogenic and room temperature, with cryogenic detectors having somewhat higher responsivity in the video

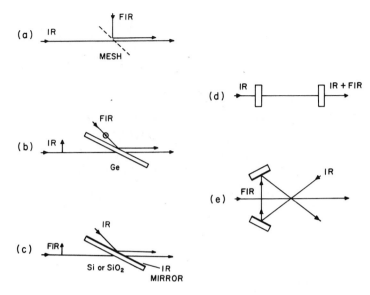

FIG. 18. Various amplifier coupling geometries: (a) free standing mesh (Evans *et al.*, 1977a), (b) Ge Brewster window (Semet and Luhmann, 1976), (c) Si or SiO$_2$ Brewster window with IR mirror (Plant and DeTemple, 1976; Woskoboinikow *et al.*, 1976), (d) etalon couplers (FIR etalons with IR AR coatings (Brown *et al.*, 1976), and (e) open retroreflector (Drozdowicz *et al.*, 1976).

mode than the room temperature detectors. The gross features of some FIR detectors are summarized in Table VI. Energy measurements are performed using either calibrated Golay cells or joulemeters (pyroelectric or thermopile).

Materials for windows, couplers, and lenses include high resistivity Si and high purity Ge, z-cut SiO$_2$, high density polyethylene, TPX, and Teflon—all readily available and fabricated. Dielectric reflectors for the IR must also be low loss in the FIR if the latter is to propagate through them. For example, ZnS is low loss for $\lambda_f > 50$ μm but ThF$_4$, a common IR coating material, has moderate FIR loss. Interferometer mirrors are typically free standing inductive meshes, while FIR polarizers are either film supported grids or a "pile-of-plates" polarizer.

Cell base pressures in the millitorr range appear adequate. Pressure should be sensed with gauges insensitive to gas species: McCloud gauges, capacitive manometers, and mechanical gauges. Long-term operation with the halide-bearing molecules may cause corrosion or deterioration of some materials as well as contamination and decomposition of pump oils. For obvious reasons, the pump exhaust should not be vented into the laboratory.

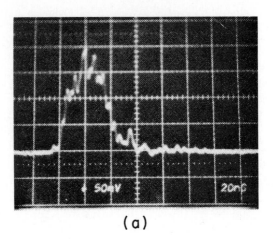

(a)

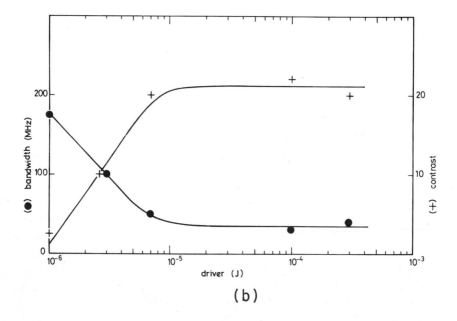

(b)

FIG. 19.   (a) Output pulse from a 496-μm oscillator–amplifier showing residual super-fluorescence beating with the main frequency component (Brown *et al.*, 1977), and (b) emitted bandwidth and contrast ratio from a CH_3F oscillator–amplifier vs. oscillator energy (Evans *et al.*, 1977a). The contrast ratio is the transmitted intensity ratio through a Fabry–Perot interferometer at tuning maxima and minima; it is a measure of the spread and persistence in the background noise, ideally tending to infinity with a perfect source.

TABLE VI
FIR DETECTORS (VIDEO)

Detector	°K	Wavelength range	Speed[a]	Responsivity[a]
InSb[b]	4.3	500 μm–4 mm	$\frac{1}{4}$ μsec	>1V/W
Si:P[c]	2.0	75–600 μm	nsec	∼V/W
GaAs[d]	4.3	100–400 μm	nsec	∼V/W
Ge:Ga	4.3	20–130 μm	nsec	∼V/W
Ge:Cu	4.3	1–30 μm	<nsec	>1V/W
Pyroelectric[e]	300	uv–mm	nsec	∼mV/W
Photon drag[f]	300	IR–mm	nsec	∼μV/W
Schottky[g]	300	FIR–mm	<nsec	∼V/W
MOM[g]	300	IR–mm	<nsec	∼mV/W

[a] Order of magnitude, responsivity into 50 Ω.
[b] Low bias voltage; tuning achieved with magnetic field.
[c] Range is temperature and doping sensitive.
[d] Weak response to beyond 1 mm.
[e] Small area, range dependent on absorbing coating.
[f] Not useful near crystal absorption.
[g] Short wavelength response determined by diode contact area; responsivity decreases with increasing FIR frequency.

V. Additional Effects

There are four additional effects associated with the experimental and theoretical aspects of optical pumped FIR systems which will now be discussed. These are (a) chirping, (b) three-photon processes, (c) phase coherent or transient effects, and (d) short pulse generation.

A. CHIRPING

The formalism developed in Section II, while being quite general, indicates that the overall conversion of IR into FIR is a dynamical process complicated by the space–time dependence of the various waves, which is one reason why only qualitative comparisons have been made with experiments. The dynamical nature, even in the quasi-static limit, can be seen in the gain profile calculation in Fig. 12 and the fact that pumping pulses are not rectangular in time. This means that as the pump enters the interaction region, the gain and the resonant frequencies are evolving in time as well as space. Similarly, the countering AC Stark shift due to the FIR is present and not negligible (Panock and Temkin, 1977). From Eqs. (13) and (14), the AC Stark shift of the pump is cancelled out when $Q^2 \sim P^2$ or $I_f \sim \mu_{13}^2 I_p/\mu_{32}^2 \sim 10^{-3}I_p$, a value easily approached in practice.

One manifestation of the dynamics of the interaction would be a chirp $(\partial\omega_f/\partial t \neq 0)$ on the FIR again suggesting that the highest spectral purity would come from the cavity configuration because of the built-in frequency filter. Evolution in the superfluorescent and Raman modes may have the most chirp, with some preliminary experimental indications in the latter case (Petuchowski *et al.*, 1977).

If the Raman and AC Stark effects are ignored, the formalism developed in Section II reduces to a simple rate equation approach. This limit has been used to describe cw and pulsed systems with some nominal success (Bluyssen *et al.*, 1975; DeTemple and Danielewicz, 1976; Henningsen and Jensen, 1975; Yamanaka *et al.*, 1974; Pichamuthu and Sinha, 1978).

B.   THREE-WAVE INTERACTIONS

Another complicating, but interesting, feature is the presence of a third FIR wave, either a cascade line in the vibrationally excited state or a separate line in the ground state. Possible three-wave processes are graphed in Fig. 20: (a) a simple cascade (e.g., the 385- and 358-μm lines in D_2O), (b) isolated lines (50- and 66-μm lines in D_2O), (c) ground state cascades (1207- and 1006-μm lines in $C^{13}H_3F$), (d) parallel transitions (95- and 113-μm lines in D_2O), and (e) two-photon pumping, both laser and hyper-Raman (Bischel *et al.*, 1975; Cotter *et al.*, 1977; Vrehen and Hikspoor, 1977b; Jacobs *et al.*, 1976; Leap *et al.*, 1978).

Conceptually, an induced emission process can be thought of as simply transferring population between states, in which case the normal cascade behavior in Fig. 20a follows naturally from the dynamics implied in Fig. 2a.

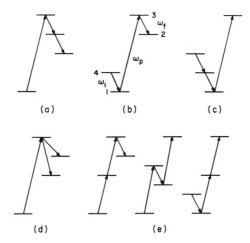

FIG. 20.   Possible absorption/emission combinations involving three waves in a four-level system; (b) is labeled according to Fig. 2a.

What is of particular interest is the possibility of a three-photon interaction, in addition to the normal one- and two-photon interactions. To extract the conditions for this to occur, the density matrix treatment has been applied to the three-wave, four-level system in the near resonant approximation, much as in Section II.A (Petuchowski *et al.*, 1979), resulting in four diagonal or population equations and six off-diagonal equations. For the case of Fig. 20b, the net gain for a wave near the $3 \rightarrow 2$ transition can be grouped as

$$G_f = \sigma_{32}[S_1(\rho_{33} - \rho_{22}) + S_2(\rho_{11} - \rho_{22}) + S_3(\rho_{44} - \rho_{22})] \quad (26)$$

where σ_{32} is defined in Eq. (11), S_1 is the equivalent laser or one-photon cross section multiplier, S_2 the Raman or two-photon multiplier, and S_3 the three-photon multiplier. All S_i contain AC Stark shift contributions due to all three waves. Since the equations for S_i with all the Stark shift terms are particularly nontransparent, only the weak field limits for the S_i are presented for nondegenerate levels and are

$$S_1 = \text{Im}[1/L^*(y)],$$

$$S_2 = \text{Im}\left\{ -\frac{P^2}{L^*(y)L(x)L(x-y)} \right.$$

$$\left. \times \left[1 + \frac{R^2}{L(x-z)}\left(1 + \frac{L(x)}{L(x-y-z)}\right)\left(\frac{1}{L(x)} - \frac{1}{L^*(z)}\right)\right]\right\},$$

$$S_3 = \text{Im}\left\{ -\frac{P^2 R^2}{L^*(y)L^*(z)L(x-z)L(x-y)}\left[\frac{1}{L(x)} + \frac{1}{L(x-y-z)}\right]\right\},$$

where $z = T_2(\omega_{41} - \omega_l)$, $R = \boldsymbol{\mu}_{41} \cdot \hat{\boldsymbol{\varepsilon}}_l E_l T_2/2\hbar$, the subscript *l* refers to the second FIR wave near the $1 \rightarrow 4$ transition, and the remaining terms are defined as in Section II. S_1 and the leading term of S_2 are the same as before in the same limit. The new term S_3 is the three-photon interaction, proportional to P^2R^2 or $I_p I_l$, as is the correction to the S_2 term. The key feature is that S_3 maximizes under conditions of two- and three-photon conservation of energy, or $y + z = x$ ($\Omega_{24} = \omega_p - \omega_f - \omega_l$) and y or $z = x$. If $x \neq 0$ (off resonant pumping) a maximum three-wave interaction occurs when one of the FIR waves satisfies the Raman resonance with the other satisfying the laser resonance, while for $x = 0$, the maximum interaction occurs for $y = z = 0$ or all waves on resonance. The latter condition may occur in $C^{12}H_3F$ and $C^{13}H_3F$ while the former condition may occur in D_2O, since the data obtained for Fig. 17 indicated that the 385-μm frequency offset tracked the pump offset, indicating a Raman-like behavior, while the cascade 358 μm remained fixed in frequency—a laser line (Woskoboinikow *et al.*, 1979). A similar behavior has also been observed in the HCl Raman experiment (Frey *et al.*, 1977). These results also imply that the 50 and 66 μm D_2O Raman lines (Fig. 6) are coupled predominately through two-wave interactions and AC Stark shifts and not through a three-photon interaction. Noting that $R^2 = 1$ corresponds only to $I_l \sim 10$ W/cm^2 Torr^2 and that FIR intensities $> $ kW/cm^2 have been typically generated, one may conclude that one-, two- and three-photon processes may be present in these systems.

For reference purposes, the AC Stark shift term appropriate to S_3, the simplest term is

$$
\Delta = \left(1 + \frac{P^2}{L^*(z)L(x)} - \frac{R^2}{L(x-z)L(x)} - \frac{Q^2}{L(x-z)L(x-y-z)}\right)
$$

$$
\times \left(1 + \frac{P^2}{L^*(y)L(x-y)} - \frac{R^2}{L(x-y)L(x-y-z)} - \frac{Q^2}{L(x)L(x-y)}\right)
$$

$$
- \frac{R^2Q^2}{L(x-z)L(x-y)}\left(\frac{1}{L(x)} + \frac{1}{L(x-y-z)}\right)^2
$$

which should be inserted by replacing R^2 by R^2/Δ (Petuchowski *et al.*, 1979). When $R \to 0$ and $z \to \infty$, this Δ reduces to the previous form as in Eq. (14).

C. COHERENT EFFECTS

When dealing with time scales which are comparable to or less than T_2, the intrinsic response time of the material becomes important. Phenomena such as optical nutation, optical free induction decay, self-induced transparency, and superradiance are indigenous to this regime (Allen and Eberly, 1975). Originally predicted in 1954, superradiance is a cooperative emission process from a system which radiates an intensity as $\bar{N}^2$ for a time $A_{32}^{-1}/\bar{N}$, where $\bar{N}$ is the number of molecules (Dicke, 1954). Because A_{32}^{-1} can be as short as 1 nsec and $\bar{N}$ large, severe pumping requirements have precluded the observation of superradiance for many candidate systems. Since for pure rotational FIR transitions A_{32}^{-1} can be quite long and $\bar{N}$ sufficiently low such that a substantial gain can still be produced, it is not surprising that the first observations of superradiance were on these transitions (Skribanowitz *et al.*, 1973).

1. Superradiance

Before discussing the experiments, the key features of superradiance can be derived from the density matrix equations for one wave present in a two-level system or equivalently, the Maxwell–Bloch equations. For these, a traveling wave, delta function pump creating an oriented, fully inverted population and central tuning ($\omega_f = \Omega_{32}$) are assumed for simplicity. The normalized equations of motion in retarded time $\tau = t - z/c$ and normalized variables are

$$\partial P/\partial \tau = -P/T_2 + EN, \tag{27}$$

$$\partial N/\partial \tau = -N/T_1 - EP, \tag{28}$$

$$\partial E/\partial z = \alpha P - \tfrac{1}{2}\kappa E, \tag{29}$$

where $P = -2 \operatorname{Im} \rho_{32}$, $N = (\rho_{33} - \rho_{22})/\rho_{33}(\tau = 0)$, $E = \mu_{32} E_f/\hbar$, κ is a power loss, α is related to the initial gain created by the pump as $\alpha = G_f/2T_2 = \sigma_{32}\rho_{33}(\tau = 0)/2T_2$, and the pump is applied at $\tau = 0$. Because the system undergoes a natural decay process, the equivalent initial spontaneous decay has to be inserted artificially by assuming some small initial value of $P = \sin \theta_0 \simeq \theta_0$ in the spirit of the Bloch model (Stroud et al., 1972; Bonifacio et al., 1975).

In the so-called "mean-field" limit where $\partial E/\partial z \simeq 0$, (Bonifacio and Lugiato, 1975; Rehler and Eberly, 1971) the solution of Eqs. (27)–(29) for $T_1 = T_2 = \infty$ is

$$I_f = (B/\tau_s^2) \operatorname{Sech}^2[(\tau/\tau_s) - \ln(2/\theta_0)] \tag{30}$$

where B is some constant and the superradiant time $\tau_s \sim \bar{N}^{-1}$, $\bar{N}$ being the total number of radiating atoms. This evolution indicates that there are a characteristic pulse delay $t_0 = \tau_s \ln(2/\theta_0) \sim \bar{N}^{-1}$, pulse width $\Delta\tau = 1.76\tau_s \sim \bar{N}^{-1}$, and peak intensity $I_f \sim \bar{N}^2$, which may be taken together as one signature of superradiance. Other characteristics are detailed elsewhere (MacGillivray and Feld, 1976; Bonifacio and Lugiato, 1975; Allen and Eberly, 1975; MacGillivray and Feld, 1977b).

With a finite value of $T_1 = T_2$, the solution in the mean field limit becomes

$$I_f = (B/\tau_s^2) e^{-2\tau/T_2} \operatorname{Sech}^2[(T_2/\tau_s)(1 - e^{-\tau/T_2}) - \ln(2/\theta_0)] \tag{31}$$

which is asymmetric relative to the previous solution (Bonifacio et al., 1975). Both forms are graphed in Fig. 21. The conditions for a well defined pulse are simply that $\tau_s \ll T_2$ and $G_f z \gg 1$, with the latter condition assuring a strong signal. Both conditions can easily be satisfied for some FIR transitions.

The first convincing observation of superradiance was in HF (Skribanowitz et al., 1973; Herman et al., 1974), with subsequent observations in $C^{12}H_3F$ (Rosenberger et al., 1977) and the metal vapors Tl, Cs, Li, and Na (Gross et al., 1976; Flushberg et al., 1976; Vrehen et al., 1977; Vrehen, 1977; Gibbs, 1977; Okada et al., 1978). Using an HF laser as the pumping source, superradiance was observed on a number of Doppler broadened $R(J)$ transitions in the vibrationally excited state. An example of the behavior is shown in Fig. 22, along with a numerical solution of the

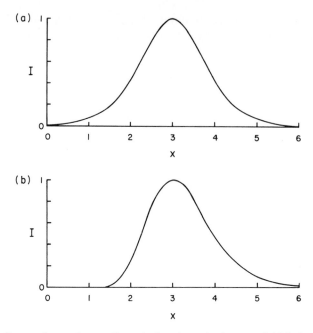

FIG. 21.   Superradiant pulses vs. dimensionless time x in the mean field limit: (a) Eq. (30): $\text{Sech}^2(x - x_0)$ with $x_0 = 3$, and (b) Eq. (31): $e^{-2x}\,\text{Sech}^2[\beta(1 - e^{-x}) - x_0]$, with $\beta = 40$ and $\beta - x_0 > 3$. Vertical scales are relative. [(a) $T_1 = T_2 = \infty$; (b) $T_1 = T_2 \neq \infty$.]

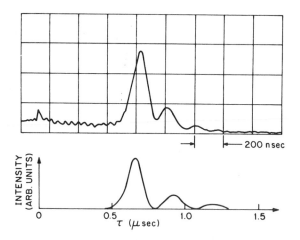

FIG. 22.   Observed and predicted superradiant pulses from HF at 84 μm. The small peak in the left side of the upper trace is the pump pulse. The conditions were $p = 1.3$ m Torr, $L = 1$ m, and R(2) pump with $\tau_R = 2T_2^*/G_f L$, where T_2^* is the equivalent Doppler dephasing time. Lower graph for $T_2^* = 330$ and $\tau_R = 6.1$ nsec. (Skribanowitz et al., 1973).

Maxwell–Bloch equations for the conditions appropriate to the experiment (Skribanowitz *et al.*, 1973). The system was not in the mean field limit, implying that propagation effects were important, one manifestation of which was the ringing in Fig. 22 caused by different portions of the cell radiating at different times (MacGillivray and Feld, 1976). The $\bar{N}^2$ dependence of the FIR intensity was also verified (Herman *et al.*, 1974).

A complimentary experiment in the homogeneously broadened limit was performed in $C^{12}H_3F$ at 496 μm (Rosenberger *et al.*, 1977). To provide good separation between the pumping pulse and superradiant pulses, the former was truncated with a plasma shutter resulting in pump and super-radiant pulses shown in Fig. 23 (Kwok and Yablonovitch, 1975; Rosenberger *et al.*, 1977). The asymmetry on the FIR pulses is evident in this figure

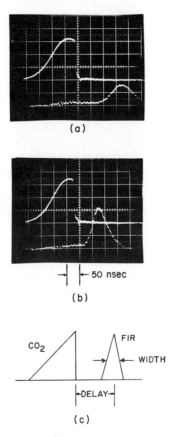

FIG. 23. Superradiant pulses from $C^{12}H_3F$ at 496 μm at two pressures: (a) 0.078 Torr; and (b) 0.122 Torr. In (c) the experimental parameters of Fig. 24 are defined (Rosenberger *et al.*, 1977).

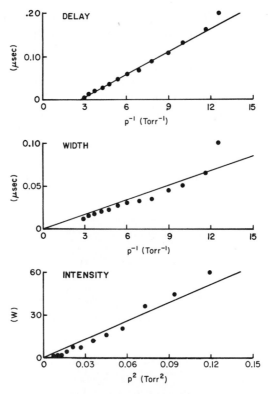

FIG. 24. Scaling behavior of the superradiant pulse parameters with the definitions in Fig. 23c. The assumption is that $\bar{N} \sim \rho_{11}^e p$, with p the convenient variable to use in this figure (Rosenberger, 1978).

and is similar to Fig. 21b) even though the system was not in the mean field limit. The observed delay, pulse width, and intensity are shown in Fig. 24, clearly illustrating the expected behavior (Rosenberger, 1978).

2. Swept-Gain Steady State Superradiance

In most of the previous experiments, low intensity superradiant pulses of approximately equal intensity were observed in the backward direction. As the cell is made longer, the superradiant pulse co-propagating with the pump will continue to grow while the counter-propagating wave will diminish; these features have been observed in $C^{12}H_3F$ (Bowden et al., 1977; Ehrlich, 1978). For a sufficiently long cell, the co-propagating wave may approach a steady state, a condition in which the power gained from the material is balanced by the power lost by the κ processes (Bonifacio et al., 1975). The solution of Eqs. (27)–(29) for this case, a solitary wave (Scott, 1973), is

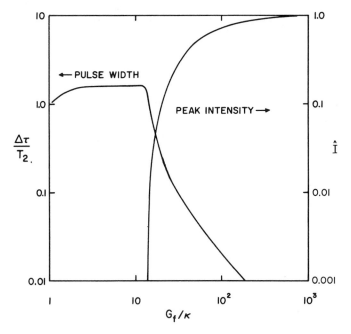

FIG. 25. Steady-state superradiant pulse intensity $\hat{I}$ and pulse width versus G_f/κ. Pulse width (FWHM) is in units of T_2 and pulse intensity in units of $_J B/\tau_s^2$, $\theta_0^2 = \tau_s A_{32}$, and $A_{32}T_2$ was taken to be 10^{-10} (Bonifacio et al., 1975).

identical to Eq. (31) with $B = c\varepsilon_0 \hbar^2/2\mu_{32}^2$ and $\tau_s = T_2\kappa/G_f$. The behavior of the steady-state pulse width and pulse intensity are graphed in Fig. 25. For $G_f/\kappa < 10$, the pulses are weak, with pulse width $\sim T_2$. Above this region, the pulse sharply contracts in width, the delay becomes very short, and the intensity becomes quadratic in G_f. In this contracted regime, the peak intensity is approximately B/τ_s^2 with a pulse width of $\Delta\tau \simeq 1.76T_2\kappa/G_f$; i.e., the pulse is approximately given by Eq. (30).

In contrast to the experiments in the bidirectional superradiance regime, the steady-state pulses can be very short and intense. An example: $T_2 = 10$ nsec and $G_f/\kappa = 100$ would yield a steady-state pulse of $I_f \sim 10$ kW/cm^2 with $\Delta\tau \sim 176$ psec. From computer solutions of Eqs. (27)–(29) the steady-state is achieved in a distance $\kappa z \gtrsim 20$, indicating that very long path lengths, tens of meters, would be required to achieve these ultrashort pulses.

3. Lethargic Gain

Conceptually, superradiance may be thought of as radiation from a high gain, lethargic, noise driven amplifier. This viewpoint, while somewhat removed from the original concepts of superradiance, (Dicke, 1954) is consistent with the Maxwell–Bloch treatment, which contains stimulated emission and propagation effects (MacGillivray and Feld, 1976). The amplifier is lethargic because the pulses are evolving on a sub-T_2 time scale (Hopf *et al.*, 1976). To see the lethargic effect more clearly, the small signal amplifier solution of Eqs. (27)–(29) under the assumption of large G_f/κ, a weak injected FIR field E_0 at $\tau = 0$, and swept delta function excitation is (Hopf *et al.*, 1976)

$$E(z, \tau) \simeq E_0 \, e^{-\kappa z/2} \, e^{-\tau/T_2} I_0[\sqrt{2G_f z(1 - e^{-\tau/T_2})}],$$

where I_0 is a modified Bessel function. Interestingly, as $z \to \infty$, $E(z, \tau) \to 0$, leaving only any steady-state superradiant pulse which may evolve simultaneously with the injected field. Taking $\kappa \to 0$, and $G_f z$ large, the asymptotic form of I_0 results in

$$E(z, \tau) \simeq E_0 e^{-\tau/T_2} \exp[\sqrt{2G_f z(1 - e^{-\tau/T_2})}]/\{2\pi[\sqrt{2G_f z(1 - e^{-\tau/T_2})}]\}^{1/2}$$

which is less than the expected small signal field gain of $\exp(G_f z/2)$. Since the denominator is slowly varying compared with the exponential term, $E(z, \tau) \sim \exp(z^{1/2})$, a most unusual behavior. The time evolution of various swept amplifiers is graphed in Fig. 26, showing the gain reduction in the lethargic case.

Clearly separate from superradiance, the lethargic gain should dominate swept, low pressure, high gain FIR amplifiers for which $T_1 \simeq T_2$. Even though the data in Fig. 10a are of the form $\exp(z^{1/2})$, one can only claim a mathematical analogy with lethargic gain because of the experimental time scale ($> 10T_2$), and possible Raman contributions to the gain and to the existence of a multimode pump. Nevertheless, coherent effects such as superradiance and lethargic gain may be investigated in the FIR, providing insight into the dynamics of high gain amplifiers, a topic of interest for both FIR and x-ray pulse generation (Scully *et al.*, 1975; MacGillivray and Feld, 1977a).

D. Short Pulse Generation

The standard technique used to produce short pulses, mode locking, relies on the existence of multilongitudinal mode operation and a means to phase lock the modes together: typically, AM modulators or saturable absorbers. Since the latter optical elements are not available in the FIR and since multimode operation is obtainable for only a few high gain transitions, coherent phenomena such as steady-state superradiance appear attractive. In addition to superradiance, there are alternate approaches which might be employed to generate short pulses which include the use of a mode locked pump and the utilization of backward stimulated Raman scattering.

Pulse narrowing in stimulated backward Raman emission has been observed in the visible and uv regions from ruby, Nd, and KrF pumped

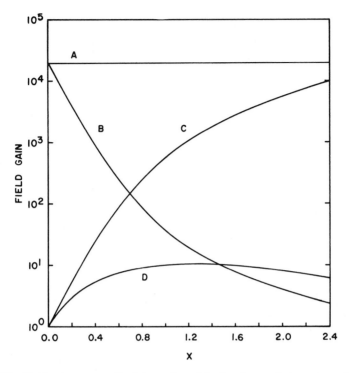

FIG. 26. Field gain vs. normalized time x for a delta function excited swept amplifier with $G_f z = 20$: (A) conventional amplifier, $\exp(0.5G_f z)$; (B) T_1 dominated amplifier, $\exp(0.5G_f z e^{-x})$; (C) T_2 dominated amplifier, $\exp[0.5G_f z(1 - e^{-x})]$; and (D) a lethargic amplifier,

$$e^{-x}I_0[\sqrt{2G_f z(1 - e^{-x})}].$$

samples (Maier *et al.*, 1969; Kachen and Lowdermilk, 1977; Murray *et al.*, 1978). Conceptually, a co-propagating Stokes wave is limited in intensity by the instantaneous intensity of the pump since both propagate with the same velocity, the pulse width being proportional to the pump duration. In contrast, a counter-propagating Stokes wave always propagates into regions where little or no pump depletion has occurred. For large intensities, the leading edge of the Stokes wave can become very intense and short, extracting energy from the total pump pulse. This behavior is graphed in Fig. 27 for various spatial distances into a backward Raman amplifier (Maier *et al.*, 1969), clearly showing the steepening and contraction of the leading edge. For sufficiently long amplifiers, the backward wave may evolve into a steady-state pulse with a functional form and pulse width identical to the steady-state superradiant pulses [Eq. (30)] in the fully contracted regime (Maier *et al.*, 1969).

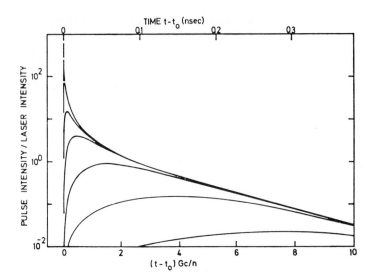

FIG. 27. Calculated backward Raman pulse waveforms at different spatial steps into the amplifier at length intervals $\Delta l = 2.77/G$, where G is the Stokes gain (Maier *et al.*, 1969). The initial conditions were a square pumping pulse and a starting spontaneous Raman noise field, which was a linear ramp in time. The reshaping and contraction proceeds from right to left, corresponding to greater propagation distances into the amplifier. The bottom scale is a dimensionless time scale while the top refers to a particular experiment.

Since a number of strong FIR lines have been produced by stimulated Raman emission, this technique may prove useful in generating short pulses. The virtues of the approach can simply be estimated by assuming a saturated condition and square pump of duration τ_p. The backward intensity and pulse width are approximately given by

$$I_f \simeq (\hbar\omega_f/\hbar\omega_p)(\tau_p/\Delta\tau)I_p, \qquad \Delta\tau \simeq I_p/cU_{R,p},$$

where $U_{R,p}$ is the saturated power density for the pump in the Raman case. For $I_p = 100$ kW/cm^2 and $\tau_p = 100$ nsec (385-μm line of D$_2$O), $I_f \simeq 100$ kW/cm^2 Torr^2 and $\Delta\tau \simeq 2$ nsec Torr^2; i.e., the backward Raman intensity can actually exceed the pumping intensity for a short duration! A number of obvious difficulties remain however; the interaction length has to be $c\tau_p/2$ (~ 15–30 m) for efficient conversion, the AC Stark shifts and any co-propagating Raman waves may effectively reduce the conversion, (Murray *et al.*, 1978) and $\Delta\tau$ may be less than T_2, placing the growth in a lethargic gain regime (Akhmanov *et al.*, 1974). Nevertheless, as an approach it is attractive because the effect leads to a pulse compression which may produce FIR pulses of the same intensity as the pump in seeming violation of the Manley–Rowe limit.

The large observed superfluorescence linewidths suggests that an effective gain bandwidth larger than Δv_H exists due to either an equivalent inhomogeneous broadening associated with the various $R(J, K)$ transitions in a K-multiplet or Raman contributions from each frequency component of a multimode pump; a combination of both can also be present. These imply that sub-T_2 pulses might be generated by using mode locked or short pulse pumping lasers. Precedence for this comes from synchronously mode locked pumped dye laser experiments (Chan and Sari, 1974; Heritage and Jain, 1978) and from early FIR observations using free running or unlocked TEA lasers (Plant, 1974; Evans, 1975). Recent experiments using a mode-locked TEA laser have indeed resulted in mode-locked FIR pulses of duration of $< T_2/8$ from $C^{12}H_3F$ and D_2O as shown in Fig. 28 (Lee, 1979). The interactions are clearly in the lethargic regime and may result in FIR pulses with a duration (bandwidth) which is shorter (greater) than that of the pump (Akhmanov, 1971; DeTemple, 1979; Lemley and Nurmikko, 1979). A test of this will require the development of new correlation techniques for the detection of ultrashort FIR pulses. As a crude indication of the potential of this approach, CO_2 mode-locked pulses of 80–800 psec and single pulses of 30 psec have been generated (Abrams, 1975; Kwok, 1977; Corkum, 1978).

VI. Summary

The progress made since the first demonstrations in 1973 of kilowatts in the FIR has been steady, impressive, and complimentary of the approach. Using the oscillator–amplifier configuration, FIR powers in the range of

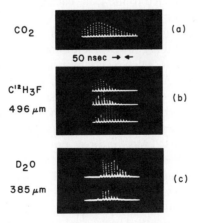

FIG. 28. Time behavior of a mode-locked TEA laser (a) and of a synchronously pumped FIR laser, (b) and (c). with 1-Torr fill gas. FIR pulses of <1 nsec were generated with peak powers in the range of 1–10 kW (Lee, 1979).

$\frac{1}{2}$–1 MW on a single mode with efficiencies between 10 and 20% of the Manley–Rowe limit have been achieved. Even higher power and efficiency are available in the superfluorescent mode with some sacrifice in spectral purity. The identification of two-wave interactions, such as AC Stark shifts and stimulated Raman emission, has contributed to our fundamental understanding of the nature and complexity of the FIR generation processes.

Because of the complexity, a detailed theory of the overall FIR evolution in the Raman and SF modes has not been demonstrated. Such a theory is prerequisite to a complete understanding of the identification of the appropriate molecular parameters and pumping characteristics needed for further development of these sources, and for the extensions of the basic approach to other molecules and pumping schemes. The results of the three-wave interaction can be taken to indicate that an even further degree of complexity is present, recalling the cascades in Figs. 1, 5, 6, and 7.

Regarding the sparsity of strongly emitting molecules and lines, strong cw emitters such as CH_3OH emit rather weakly on a pulsed basis (Mathieu, and Izatt, 1977) as do many other species (Plant et al., 1974; Semet and Luhmann, 1977); ground state absorption is suspected although the details are unknown (DeTemple and Lawton, 1978). The bent A_2B molecular class appears very attractive by analogy with D_2O and may not have the equivalent K bottleneck present in the CH_3X class, suggesting longer pulse operation. Other pumping approaches include overtone bands (Fetterman et al., 1974), isotopic CO_2 lines (Freed et al., 1974), CO_2, sequence bands (Weiss et al., 1977; Feldman et al., 1978; Danielewicz, 1978), and multiphoton pumping, all of which will increase the available number of pumping lines and candidate molecules.

Tunability is closely related to the gain–bandwidth, which will be dominated by either the frequency spread in a K-multiplet or the pump bandwidth, both being in the GHz range. Consequently, an isolated line pumped by a single mode will have a tuning range of a few linewidths or about 100 MHz, which may be extended to a few GHz by using a tunable CO_2 pump in conjunction with the Raman effect. Alternately, the tunable Raman emission from HCl observed over a 60-GHz range suggests that the greatest tunability will occur in systems in which the pump has the greatest tunability, i.e., dye laser systems. The dye laser approach would have more appeal and a greater efficiency if the dye laser itself was the direct, pump, which would require that the molecule absorb in the visible. Other approaches suggested would be to invert Rydberg states which are Stark tunable or selectively excited with a tunable pump (Grischkowsky et al., 1977; Chang et al., 1978).

Approaches towards ultrashort pulse generation are less well defined. Superradiance has been identified and partially characterized but the details and practicalities of steady-state superradiance are unknowns, perhaps requiring unrealistically long path lengths. Stimulated backward Raman emission is present but, again, may require long path lengths to be efficient and to enter the short pulse regime. The use of a mode locked pump, CO_2, or even dye laser, which is relatively unexplored, offers the shortest pumping pulses and may be the most compact of all the approaches although not necessarily resulting in the most intense FIR signals.

In view of the youthfulness of the optical pumping approach, the demonstrated performance, and the new physical phenomena inherent to the spectral regime, one may anticipate continued contributions to the field of quantum electronics, quantum optics, and nonlinear optics for some time to come.

ACKNOWLEDGMENTS

It is a pleasure to acknowledge the many contributions made to this chapter by individuals at the Aerospace Corporation, Bell Telephone Laboratory, Culham Laboratory, Harry Diamond Laboratory, MIT Lincoln Laboratory, MIT National Magnet Laboratory, Redstone Arsenal, University of California at Los Angeles, University of Strathclyde, and students and colleagues, particularly P. D. Coleman, at the University of Illinois. A special thanks is due to Mrs. J. Smith for her effort in typing this and other manuscripts.

REFERENCES

Abrams, R. L., and Wood, O. R. (1975). *Appl Phys. Lett.* **19**, 518–520.
Akhmanov, S. A., Drabovich, K. N., Sukhorukov, A. P., and Chirkin, A. S. (1971). *Sov. Phys. JETP* **32**, 266–273.
Akhmanov, S. A., D'yakov, Yu. E., and Pavlov, L. I. (1974). *Sov. Phys.-JETP* **39**, 249–256.
Allen, L., and Eberly, J. H. (1975). "Optical Resonance and Two-Level Atoms." Wiley, New York.
Allen, L., Dodel, G. and Magyar, G. (1979). *Optics Commun.* **28**, 383–388.
Arimondo, E., Glorieux, P., and Oka, T. (1978). *Phys. Rev. A* **17**, 1375–1393.
Autler, S. H., and Townes, C. H. (1955). *Phys. Rev.* **100**, 703–722.
Beck, R., Englisch, W., and Gürs, K. (1976). "Tables of Laser Lines in Gases and Vapors," Springer-Verlag, Berlin and New York.
Bischel, W. K., Kelly, P. J., and Rhodes, C. K. (1975). *Phys. Rev. Lett.* **34**, 300–303.
Bluyssen, H. J. A., McIntosh, R. E., van Ettegen, A. F., and Wyder, P. (1975). *IEEE J. Quantum Electron.* **QE-11**, 341–348.
Bonifacio, R., and Lugiato, L. A. (1975). *Phys. Rev. A* **11**, 1507–1521.
Bonifacio, R., Hopf, F. A., Meystre, P., and Scully, M. O. (1975). *Phys. Rev. A* **12**, 2568–2573.
Bowden, C. M., Ehrlich, J., Howgate, D. W., Lehnigh, S. H., Rosenberger, A. T., and DeTemple, T. A. (1977). *Bull. Am. Phys. Soc.* **22**, 1269.
Brewer, R. G., and Hahn, E. L. (1975). *Phys. Rev. A* **2**, 1641–1649.
Brewer, R. G., and Shoemaker, R. L. (1971). *Phys. Rev. Lett.* **27**, 631–634.
Brown, F., Silver, E., Chase, C. E., Button, K. J., and Lax, B. (1972). *IEEE J. Quantum Electron.* **QE-8**, 499–500.
Brown, F., Horman, S. R. Palevsky, A., and Button, K. J. (1973). *Opt. Commun.* **9**, 28–30.
Brown, F., Kronheim, S., Silver, E. (1974). *Appl. Phys. Lett.* **25**, 394–396.

Brown, F., Hislop, P. D., and Kronheim, S. R. (1976). *Appl. Phys. Lett.* **28**, 654–656.
Brown, F., Hislop, P. D., and Tarpinian, J. O. (1977). *IEEE J. Quantum Electron.* **QE-13**, 445–446.
Broyer, M., Gouedard, G., Lehmann, J. C., and Nigné, J. (1976). *Adv. At. Mol. Phys.* **12**, 165–213.
Carman, R. L., Shimizu, F., Wang, C. S., and Bloembergen, N. (1970). *Phys. Rev. A* **12**, 60–72.
Casperson, L. W., and Yariv, A. (1972). *IEEE J. Quantum Electron.* **QE-8**, 80–85.
Chan, C. K., and Sari, S. O. (1974). *Appl. Phys. Lett.* **25**, 403–406.
Chang, T. Y. (1974). *IEEE Trans. Microwave Theory Tech.* **MTT-22**, 983–988.
Chang, T. Y. (1977). *IEEE J. Quantum Electron.* **QE-13**, 937–943.
Chang, T. Y., and Bridges, T. J. (1970). *Opt. Commun.* **1**, 423–425.
Chang, T. Y., Damen, T. C., McGee, J. D., Nguyen, V. T., and Bridges, T. J. (1978). *Appl. Phys. Lett.* **32**, 633–635.
Cohn, R. R., Fuse, T., Button, K. J., Lax, B. and Drozdowicz, Z. (1975). *Appl. Phys. Lett.* **27**, 280–282.
Corkum, P. B., Alcock, A. J., Rollin, D. F., and Morrison, H. D. (1978). *Appl. Phys. Lett.* **32**, 27–29.
Cotter, D., Hannan, D. C., Tuttlebee, W. H. W., and Yuratich, M. A. (1977). *Opt. Commun.* **22**, 190–194.
Danielewicz, E. J., and Keilmann, F. (1979). *IEEE J. Quantum Electron.* **QE-15**, 8–11.
Danielewicz, E. J., and Weiss, C. O. (1978). *Optics Comm.* **27**, 98–100.
Dangoisse, D., Willemot, E., Deldalle, A. and Bellet, J. (1979). *Optics Commun.* **28**, 111–116.
deMartino, A., Frey, R., and Pradere, F. (1978). *Optics Comm.* **27**, 262–266.
Deroche, J. C. (1978). *J. Mol. Spectrosc.* **69**, 19–24.
Deroche, J. C., and Betrencourt-Stirneman, C. (1976). *Molec. Phys.* **32**, 921–930.
DeTemple, T. A. (1979). *Proc. Int. Conf. Lasers*, 1978, 104–110.
DeTemple, T. A., and Danielewicz, E. J. (1976). *IEEE J. Quantum Electron.* **QE-12**, 40–47.
DeTemple, T. A., and Lawton, S. A. (1978). *IEEE J. Quantum Electron.* **QE-14**, 762–768.
DeTemple, T. A., and Nurmikko, A. V. (1971). *Opt. Commun.* **4**, 231–233.
DeTemple, T. A., Plant, T. K., and Coleman, P. D. (1973). *Appl. Phys. Lett.* **22**, 644–646.
Dicke, R. H. (1954). *Phys. Rev.* **93**, 99–110.
Dodel, G., and Magyar, G. (1978). *Appl. Phys. Lett.* **32**, 44–46.
Drozdowicz, Z., Temkin, R. J., Button, K. J., and Cohn, D. R. (1976). *Appl. Phys. Lett.* **28**, 328–330. ᵛ
Drozdowicz, Z., Woskoboinikow, P., Isobe, K., Cohn, D. R., Temkin, R. J., Button, K. J., and Waldman, J. (1977). *IEEE J. Quantum Electron.* **QE-13**, 413–417.
Drozdowicz, Z., Lax, B., and Temkin, R. J. (1978). *Appl. Phys. Lett* **33**, 154–156.
Earl, B. L., Isolani, P. C., and Ronn, A. M. (1976). *Chem. Phys. Lett.* **39**, 95–97.
Ehrlich, J. J., Bowden, C. M., Howgate, D. W., Rosenberger, A. T., and DeTemple, T. A., (1978). *In* "Coherence and Quantum Optics IV (1978)" (L. Mandel and E. Wolf, eds.), pp. 923–937, Plenum Press, NY.
Evans, D. E., Sharp, L. E., James, B. W., and Peebles, W. A. (1975). *Appl. Phys. Lett.* **26**, 630–632.
Evans, D. E., Peebles, W. A., Sharp, L. E., and Taylor, G. (1976). *Opt. Commun.* **18**, 479–484.
Evans, D. E., Sharp, L. E., Peebles, W. A., and Taylor, G. (1977a). *IEEE J. Quantum Electron.* **QE-13**, 54–58.
Evans, D. E., Guinee, R. A., Huckridge, D. A., and Taylor, G. (1977b). *Opt. Commun.* **22**, 337–342.
Ewanizky, T. F., Rohde, R. S. and Bayha, W. P. (1979). *IEEE J. Quantum. Electron.* **QE-15**, in press.

Fano, U. (1957). *Rev. Mod. Phys.* **29**, 74–93.

Feld, M. S., and Javan, A. (1969). *Phys. Rev.* **177**, 540–562.

Feldman, B. J., and Feld, M. S. (1972). *Phys. Rev. A* **5**, 899–918.

Feldman, B. J., Fisher, R. A., Pollock, C. R., Simmons, S. W., Tencovier, R. H. (1978). *Opt. Lett.* **2**, 16–18.

Fenillade, C. (1976). *Chem. Phys. Lett.* **41**, 529–534.

Fenillade, C., and Bottecher, C. (1977). *Chem. Phys. Lett.* **52**, 603–605.

Fetterman, H. R., Scholssberg, H. R., and Waldman, J. (1972). *Opt. Commun.* **6**, 156–159.

Fetterman, H. R., Schlossberg, H. R., and Parker, C. D. (1974). *IEEE J. Quantum Electron.* **QE-10**, 740–741.

Fetterman, H. R., Tannenwald, P. E., Parker, C. D., Melngailis, J., Williamson, R. C., Woskoboinikow, P., Praddaude, H. C., and Mulligan, W. J. (1979). *Appl. Phys. Lett.* **34**, 123–125.

Freed, C., Spears, D. L., and O'Donnell, R. G. (1974). *In* "Laser Spectroscopy" (R. G. Brewer and A. Mooradian, eds.), pp. 171–192. Plenum Press, New York.

Frenkel, L., Marantz, H., and Sullivan, T. (1971). *Phys. Rev. A* **3**, 1640–1651.

Flusberg, A., Mossberg, T., and Hartmann, S. R. (1976). *Phys. Lett.* **58A**, 373–375.

Freund, S. M., Duxbury, G., Romheld, M., Tiedje, J. T., and Oka, T. (1974). *J. Mol. Spectrosc.* **52**, 38–57.

Frey, R., Pradere, F., and Ducuing, J. (1977). *Opt. Commun.* **23**, 65–68.

Gallagher, J. J., Blue, M. D., Bean, B., and Perkowitz, S. (1977). *Infrared Phys.* **17**, 43–55.

Gerry, E. T. (1966). *Appl. Phys. Lett.* **7**, 6–8.

Gibbs, H. M. (1977). *In* "Cooperative Effects in Matter and Radiation" (C. M. Bowden, D. W. Howgate, and R. Robl, eds.), pp. 61–78. Plenum Press, New York.

Gondhalekar, A., Heckenberg, N. R., and Holzhouer, E. (1975). *IEEE J. Quantum Electron.* **QE-11**, 103–108.

Graner, G. (1975). *Opt. Commun.* **14**, 67–69.

Grischkowsky, D. R., Lankard, J. R., and Sorokin, P. P. (1977). *IEEE J. Quantum Electron* **QE-13**, 392–395.

Gross, M., Fabre, C., Pillet, R., and Haroche, S. (1976). *Phys. Rev. Lett.* **36**, 1035–1038.

Gullberg, K., Hartman, B., and Kleman, B. (1973). *Phys. Scripta* **8**, 177–182.

Hacker, M. P., Drozdowicz, Z., Cohn, D. R., Isobe, K., and Temkin, R. J. (1976). *Phys. Lett.* **57A**, 328–330.

Hänsch, T., and Toschek, P. (1970). *Z. Phys.* **235**, 213–244.

Henningsen, J. O. (1977). *IEEE J. Quantum Electron.* **QE-13**, 435–441.

Henningsen, J. O., and Jensen, H. G. (1975). *IEEE J. Quantum Electron.* **QE-11**, 248–252.

Heppner, J., and Weiss, C. O. (1977). *Opt. Commun.* **21**, 324–326.

Heppner, J., Weiss, C. O., and Plainchamp, P. (1977). *Opt. Commun.* **23**, 381–384.

Heritage, J. P., and Jain, R. K. (1978). *Appl. Phys. Lett.* **32**, 101–103.

Herman, I. P., MacGillivray, J. C., Skribanowitz, N., and Feld, M. S. (1974). *In* "Laser Spectroscopy" (R. G. Brewer and A. Mooradian, eds.), pp. 379–412. Plenum Press, New York.

Hodges, D. T., and Hartwick, T. S. (1973). *Appl. Phys. Lett.* **23**, 252–253.

Hodges, D. T., and Tucker, J. R. (1975). *Appl. Phys. Lett.* **27**, 667–669.

Hopf, F., Meystre, P., and McLaughlin, D. W. (1976). *Phys. Rev. A* **13**, 777–783.

Hutchinson, D. P. (1977). *Proc. Soc. Photo-Opt. Inst. Eng.* **105**, 80–83.

Hutchinson, D. P., and Vander Sluis, K. L. (1977). *Appl. Opt.* **16**, 293–294.

Izatt, J. R., Budhiraju, C. J., and Matheiu, P. (1977). *IEEE J. Quantum Electron.* **QE-13**, 396–397.

Jacobs, R. R., Prosnitz, D., Bischel, W. K., and Rhodes, C. K. (1976). *Appl. Phys. Lett.* **29**, 710–712.

Javan, A. (1957). *Phys. Rev.* **107**, 1579–1589.

Javan, A., and Szöke, A. (1965). *Phys. Rev.* **137**, A536–A549.

Johns, J. W. C., McKeller, A. R. W., Oka, T., Römheld, M. (1975). *J. Chem. Phys.* **62**, 1488–1496.

Kachen, G. I., and Lowdermilk, W. H. (1977). *Phys. Rev. A* **16**, 1657–1664.

Keilmann, F., Sheffield, R. L., Leite, J. R. R., Feld, M. S., and Javan, A. (1975). *Appl. Phys. Lett.* **26**, 19–22.

Kumar, S. R., Tansey, R. J., and Waldman, J. (1977). *IEEE J. Quantum Electron.* **QE-13**, 30–32.

Kwok, H. S., and Yablonovitch, E. (1975). *Appl. Phys. Lett.* **27**, 583–585.

Kwok, H. S., and Yablonovitch, E. (1977). *Appl. Phys. Lett.* **30**, 158–160.

Lachambre, J. L., Lavigne, P., Otis, G., and Noël, M. (1976). *IEEE J. Quantum Elec.* **QE-12**, 756–764.

Leap, J. W., Kim, K. J., Malk, E. G., and Coleman, P. D. (1978). *IEEE MTT-S Int. Microwave Symp., Ottawa, Canada, 1978*, Paper C6.1.

Lee, S. H., Petuchowski, S. J., Rosenberger, A. T., and DeTemple, T. A. (1979). *Optics. Lett.* **4**, 6–9.

Leite, J. R. R., Sheffield, R. L., Ducloy, M., Sharma, R. D., Feld, M. S. (1976). *Phys. Rev. A* **14**, 1151–1168.

Leite, J. R. R., Ducloy, M., Sanchez, A., Seligson, D., and Feld, M. S. (1977a). *Phys. Rev. Lett.* **39**, 1465–1469.

Leite, J. R. R., Ducloy, M., Sanchez, A., Seligson, D., and Feld, M. S. (1977b). *Phys. Rev. Lett.* **39**, 1469–1472.

Lemley, W., and Nurmikko, A. V. (1979). Unpublished.

Lin, C. L., and Shaw, J. H. (1977). *J. Mol. Spectrosc.* **66**, 441–447.

Lipton, K. S., and Nicholson, J. P. (1977). *IEEE J. Quantum Electron.* **QE-13**, 811–812.

Lipton, K. S., and Nicholson, J. P. (1978). *Opt. Commun.* **24**, 321–326.

Lipton, K. S., Nicholson, J. P., and Illingworth, R. (1977). *Opt. Commun.* **21**, 42–45.

Loudon, R. (1973). "The Quantum Theory of Light," Chapters 7, 8, 10, 12. Oxford Univ. Press, London and New York.

MacGillivray, J. C., and Feld, M. S. (1976). *Phys. Rev. A* **14**, 1169–1189.

MacGillivray, J. C., and Feld, M. S. (1977a). *Appl. Phys. Lett.* **31**, 74–76.

MacGillivray, J. C., and Feld, M. S. (1977b). *In* "Cooperative Effects in Matter and Radiation" (C. M. Bowden, D. W. Howgate, and R. Robl, eds.), pp. 1–14, Plenum Press, New York.

MacGurk, J. C., Schmaltz, T. G., and Flygare, W. H. (1974). *Adv. Chem. Phys.* **25**, 1–28.

Maier, M., Kaiser, W., and Giordmaine, J. A. (1969). *Phys. Rev.* **177**, 580–599.

Malk, E., Niesen, J., and Coleman, P. D. (1978). *IEEE J. Quantum Electron.* **QE-14**, 544–550.

Mathieu, P., and Izatt, J. R. (1977). *IEEE J. Quantum Electron.* **QE-13**, 465–468.

Murphy, J. S., and Boggs, J. E. (1967). *J. Chem. Phys.* **47**, 4152–4158.

Murray, J. R., Goldhar, J., and Szöke, A. (1978). *Appl. Phys. Lett.* **32**, 551–553.

Nicholson, J. P., and Lipton, K. S. (1977). *Appl. Phys. Lett.* **31**, 430–432.

Oka, T. (1973). *Adv. At. Mol. Phys.* **9**, 127–206.

Okada, J., Ikeda, K., and Matsuoka, M. (1978). *Optics Comm.* **26**, 189–192.

Panock, R. L., and Temkin, R. J. (1977). *IEEE J. Quantum Electron.* **QE-13**, 425–434.

Pantell, R. H., and Puthoff, H. E. (1969). "Fundamentals of Quantum Electronics," Chapters 1–3. Wiley, New York.

Petuchowski, S. J., Oberstar, J. D., and DeTemple, T. A. (1979). Unpublished.

Petuchowski, S. J., Rosenberger, A. T., and DeTemple, T. A. (1977). *IEEE J. Quantum Electron.* **QE-13**, 476–481.

Pichamuthu, J. P., and Sinha, U. N. (1978). *Opt. Commun.* **24**, 195–198.

Plant, T. K., and DeTemple, T. A. (1976). *J. Appl. Phys.* **47**, 3042–3044.

Plant, T. K., Newman, L. A., Danielewicz, E. J., DeTemple, T. A., and Coleman, P. D. (1974). *IEEE Trans. Microwave Theory Tech.* **MTT-22**, 988–900.

Radford, H. E., Peterson, F. R., Jennings, D. A., and Mucher, J. A. (1977). *IEEE J. Quantum Electron.* **QE-13**, 92–94.

Rehler, N. E., and Eberly, J. H. (1971). *Phys. Rev. A* **3**, 1735–1751.

Rosenberger, A. T., Petuchowski, S. J., and DeTemple, T. A. (1977). *In* "Cooperative Effects in Matter and Radiation" (C. M. Bowden, D. W. Howgate, and H. R. Robl, eds.), pp. 15–35. Plenum Press, New York.

Rosenberger, A. T., Petuchowski, S. J., and DeTemple, T. A. (1978) in "Coherence and Quantum Optics IV (1978)" (L. Mandel and E. Wolf, eds.), pp. 555–565, Plenum Press, NY.

Rosenbluh, M., Temkin, R. J., and Button, K. S. (1976). *Appl. Opt.* **15**, 2635–2644.

Sargent, M., Scully, M. O., and Lamb, W. E. Jr., (1974). "Laser Physics" Chapter VII. Addison-Wesley, Reading, Massachusetts.

Sattler, J. P., and Simonis, G. P. (1977). *IEEE J. Quantum Electron.* **QE-13**, 461–465.

Schalow, A. L., and Townes, C. A. (1955). "Microwave Spectroscopy," Chapters 3, 4. McGraw Hill, New York.

Schmidt, J., Berman, P. R., and Brewer, R. G. (1973). *Phys. Rev. Lett.* **31**, 1103–1106.

Scott, A. C., Chu, F. Y. F., and McLaughlin, D. W. (1973). *Proc. IEEE* **61**, 1443–1483.

Scully, M. O., Louisell, W. H., and McKnight, W. B. (1975). *Opt. Commun.* **9**, 246–248.

Seligson, D., Ducloy, M., Leite, J. R. R., Sanchez, A., and Feld, M. S. (1977). *IEEE J. Quantum Electron.* **QE-13**, 468–472.

Semet, A., and Luhmann, N. C. Jr., (1976). *Appl. Phys. Lett.* **28**, 659–662.

Semet, A., and Luhmann, N. C. Jr., (1977). *Proc. Soc. Photo-Opt. Inst. Eng.* **105**, 83–92.

Sharp, L. E., Peebles, W. A., James, B. W., and Evans, D. E. (1975). *Opt. Commun.* **14**, 215–218.

Shoemaker, R. L., Stenholm, S., and Brewer, R. G. (1974). *Phys. Rev. A* **10**, 2037–2054.

Skribanowitz, N., Kelly, M. J., and Feld, M. S. (1972a). *Phys. Rev. A* **6**, 2302–2311.

Skribanowitz, N., Herman, I. P., Osgood, R. M. Jr., Feld, M. S., and Javan, A. (1972b). *Appl. Phys. Lett.* **20**, 428–431.

Skribanowitz, N., Herman, I. P., MacGillivray, J. C., and Feld, M. S. (1973). *Phys. Rev. Lett.* **30**, 309–312.

Smith, P. W. (1972). *Proc. IEEE.* **60**, 442–440.

Stroud, C. R. Jr., Eberly, J. H., Lama, W. L., and Mandel, L. (1972). *Phys. Rev. A* **5**, 1094–1104.

Strumia, F., and Tonelli, M. (1975). *Proc. Summer School Fisica Atomica an Moleculare, L'Aguila, Italy.*

Takami, M. (1976). *Jpn. J. Appl. Phys.* **15**, 1063–1071.

Temkin, R. J. (1977). *IEEE J. Quantum Electron.* **QE-13**, 450–454.

Temkin, R. J., and Cohn, D. R. (1976). *Opt. Commun.* **16**, 213–217.

Tucker, J. R. (1974). *IEEE Trans. Microwave Theory Tech.* **MTT-22**, 1117.

Tucker, J. R. (1976). *Opt. Commun.* **16**, 209–212.

Ueda, Y., and Shimoda, K. (1975). *In* "Lecture Notes in Physics-Laser Spectroscopy" (S. Haroche, J. C. Pebay-Peyroula, T. W. Hänsch, and S. E. Harris, eds.), Vol. 43, pp. 186–197. Springer-Verlag, Berlin and New York.

Ulrich, R., Bridges, T. J., and Pollack, M. A. (1970). *Appl. Opt.* **11**, 2511–2516.

Vrehen, Q. H. F. (1977). *In* "Cooperative Effects in Matter and Radiation" (C. M. Bowden, D. W. Howgate, and R. Robl, eds.), pp. 79–100. Plenum Press, New York.

Vrehen, Q. H. F., and Hikspoors, H. M. T. (1977). *Opt. Commun.* **21**, 127–131.

Vrehen, Q. H. F., Hikspoors, H. M. T., and Gibbs, H. M. (1977). *Phys. Rev. Lett.* **38**, 764–767.

Weber, B. A., Simonis, G. J., and Kulpa, S. M. (1978). *Optics Lett.* **3**, 229–231.

Weiss, C. O. (1976). *IEEE J. Quantum Electron.* **QE-12**, 580–584.

Weiss, J. A., and Goldberg, L. S. (1972). *IEEE J. Quantum Electron.* **QE-8**, 757–758.

Weiss, C. O., Grinda, M., and Siemsen, K. (1977). *IEEE J. Quantum Electron.* **QE-13**, 892.

Weitz, E., and Flynn, G. W. (1973). *J. Chem. Phys.* **58**, 2781–2793.

Wiggins, J. D., Drozdowicz, Z., and Temkin, R. J. (1978). *IEEE J. Quantum Electron.* **QE-14**, 23–30.

Worchesky, T. L., Ritter, K. J., Sattler, J. P., and Riessler, W. A. (1978). *Opt. Lett.* **2**, 70–71.

Woskoboinikow, P., Drozdowicz, Z., Isobe, K., Cohn, D. R., and Temkin, R. J. (1976). *Phys. Lett.* **59A**, 264–266.

Woskoboinikow, P., Mulligan, W., and Cohn, D. R. (1977). *Bull. Am. Phys. Soc.* **22**, 1175.

Woskoboinikow, P., Praddande, H. C., Mulligan, W. J., and Lax, B. (1979). *J. App. Phys.*, **50**, 1125–1127.

Yamanaka, M. (1976). *Rev. Laser Eng.* (*Japan*) **3**, 253–294.

Yamanaka, M., Homma, Y., Tanaka, A., Takada, M., Tanimoto, A., and Yoshinaga, H. (1974). *Jpn. J. Appl. Phys.* **13**, 843–850.

CHAPTER 4

Backward Wave Oscillators

G. Kantorowicz   and   P. Palluel

I. Electron Tubes for Millimeter or Far-Infrared Wavelengths

Electron tubes were the first available sources for generation and amplification of coherent microwaves at the centimetric wavelengths; they were able to operate very soon at shorter wavelengths. This was accomplished by the backward wave oscillator, whose operation has reached the millimeter and even submillimeter wavelengths. Speaking in terms of frequency, the progress towards the higher frequencies, beginning near 1 GHz, has been extended to near 1200 GHz.

In a linear-beam microwave tube, which transforms dc energy into rf energy, a beam of electrons moving in vacuum interacts with a circuit supporting the generated or amplified electromagnetic energy. The tubes that will be described here operate at voltages below 10 kV. They are able to deliver power levels very much higher than those that can be obtained with electrons moving in a semiconductor. Even though greater electron current densities may be created in the latter, the charge velocity is limited by the interaction with the crystal lattice. Backward wave tubes are also able to offer electronic tunability in a simpler way than quantum generators. Tubes using relativistic effects are beyond the scope of this chapter and are described in Chapter 1 in this volume; they are able to generate very large powers at millimeter wavelengths but need either an ultrafast beam, in the megavolt range, or a magnetic field so intense that it relies upon cryogenic technics. Such tubes are also not easily tunable.

185

Although the different types of conventional microwave tubes have certain general features in common, they differ by the nature or arrangement of their rf circuits: resonant cavities as in the klystron, or periodic slow-wave structure as in the traveling-wave tube (TWT) and backward wave oscillator (BWO). In most cases the transverse dimensions of the circuit elements are shorter than the wavelength and become so small in the lowest wavelength range that critical fabrication problems are encountered, which eventually limit the frequency attainable.

Crossed-field tubes are impractical at short wavelengths because they need a transverse electric field that becomes prohibitive. For the klystron, reliable construction of cavities and tuning systems becomes delicate, while only narrow bandwidths can be covered. Moreover, the gain of amplifiers is drastically reduced at very high frequencies so that oscillators are preferable in all respects.

These trends explain the major place taken in the millimeter and submillimeter ranges by the BWO, which also has the great advantages of being continuously tunable electrically, and being able to cover broad bandwidths. This tunability, which makes the frequency a function of the beam voltage, could seem to preclude the use of the BWO when a stable frequency is needed. However it has been shown that these tubes can be easily phase- or frequency-locked to an external stable reference source of lower frequency.

Because of their small transverse dimensions, the traveling-wave structures will be preferably vane circuits or gratings, which are easier to make than the conventional interdigital comb lines used in the centimeter range, disk loaded waveguides, or coupled-cavity circuits. A better regularity of construction may be achieved with the vane circuit, allowing the tube to work at very low wavelengths; nevertheless the propagation losses still increase with frequency and it will be shown that the circuit losses limit both the achievable frequency and power.

While the beam voltage is not critically limited at short wavelengths, the beam current has on the contrary some limitations that are functions of practical or physical limits, such as the beam radius, or thickness in the case of a sheet beam, the cathode density, gun convergence, and magnetic focusing field. It is even found that at very short wavelengths the predominant effect of the thermal velocity of cathode emission may lead to having a beam current proportional to λ^4.

The maximum achievable power decreases because of the above limitations on the dc beam power, whereas the efficiency is also reduced by the predominant effect of large rf losses of various origins. The power–frequency dependence has been plotted for various tubes in the millimeter and submillimeter ranges (Fig. 1) for continuous wave or pulsed operation. On the same plot, the tubes have been compared to solid-state devices and to the

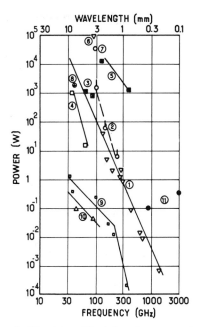

FIG. 1. Output power of millimetric and far-infrared sources vs. frequency. Tubes: (1) BWO cw; (2) EIO pulse; (3) TWT cw solenoid; (4) TWT cw PPM; (5) gyrotron cw; (6) BWO pulse; (7) magnetron pulse; and (8) klystron cw. Solid State: (9) IMPATT diodes; and (10) TEO. Lasers: (11) cw FIR lasers.

lines of some lasers. The results given here are the best points obtained in the laboratory, independent of the bandwidth; commercially available tubes with guaranteed specifications are at a somewhat lower level. Also given for comparison are some cw results for relativistic tubes such as the gyrotrons, which can deliver very high power levels.

While at low frequencies the power–frequency dependence varies according to a $1/f^2$ law for tubes or for solid-state devices, in the millimeter range this dependence very soon becomes a $1/f^4$ law or even $1/f^{4.5}$. This effect is due to limitations in beam current as well as to a decrease in the power generation efficiency. There are at least two decades between the power levels of tubes and solid-state devices.

Some other approaches have been attempted to overcome the critical problem of small transverse dimensions at short wavelengths, involving the use of overdimensioned circuits. For example, frequency multiplier klystrons —amplifiers and oscillators—have demonstrated interesting performances in some cases, which, however, are too limited in frequency coverage. In devices such as the orotron and the ledatron, an effect similar to the BWO interaction is produced along a corrugated or grating circuit surrounded by an oversized open resonator. Such devices may cover very large frequency ranges but require simultaneous mechanical and electrical tuning.

This chapter is mainly devoted to the description and performance of the BWO or Carcinotron.[1] This tube, which has been able to cover almost continuously the range of millimeter and submillimeter wavelengths from 10 to 0.25 mm, is also very useful due to its electronic tunability over a wide range. Other related devices, such as orotrons or ledatrons, will also be considered briefly.

In Section II, the usual device configuration is given, and after a description of the principle of operation and design parameters of the BWO, it is compared to related devices.

In Section III, the problems arising in the millimeter and submillimeter ranges are analyzed, and some examples of BWO characteristics are given.

II. The Backward Wave Oscillator: Principle and Design

A full description of the BWO has already been covered by numerous papers (Heffner, 1954; Grow and Gunderson, 1970; Guénard, 1975); in this section we will try to give to the readers of this series, who are much more familiar with quantum generators than with conventional electron tubes, a qualitative aspect of some features of the BWO that are useful to understand and to assess the performances in the millimeter and submillimeter ranges.

Before the description of the principles of BWO operation and the problems raised in the millimeter and submillimeter ranges, let us first give a brief survey of the history of these tubes.

The BWO was invented in 1951[2] in two versions, one using crossed fields (M-type Carcinotron or CM), the other using a linear beam (O-type Carcinotron or CO). The first models covered an octave bandwidth in the range 30 cm–8 mm, and delivered 100 mW to 1 W cw.

The first results of millimetric wave generation by linear Carcinotrons were presented as early as 1953. The range was extended to 1.5 mm in 1957. The "one millimeter frontier" was crossed in 1960, extended rapidly downwards step by step, until the junction with pulsed molecular generators was made in 1965 with a tube operated down to 0.345 mm. The lowest wavelength ever mentioned with an electron tube was around 0.25 mm in 1969.

---

[1] Carcinotron is the trade name for BWO's manufactured by the Electron Tube Division of THOMSON–CSF.

[2] French patent 1,035,379; British patent 699,893; US patent 2,880,335.

A pulsed experimental model operated at 100-kW peak, 300-W average at 94 GHz (Sedin *et al.*, 1963) should also be cited, and for comparison a 900-W cw TWT at the same frequency (Heney, 1970).

A. DESCRIPTION OF OPERATION OF THE BACKWARD WAVE OSCILLATOR

Consider an electron beam moving through a slow-wave structure along which a wave is propagating. Strong interaction occurs when the phase velocity of the wave is close to the electron velocity. The longitudinal component of the slow-wave electric field modulates the electron velocity and density, and a part of the beam kinetic energy is consequently exchanged with the wave electromagnetic energy.

The BWO design of preferred configuration will require a choice for the various elements of the tube, which are the slow-wave structure, electron gun, magnetic field guiding the beam through the circuit, and spent electron beam collector. These various parts are shown schematically in Fig. 2. We shall give in this section some topics related to the gun, the circuit, and the backward wave interaction.

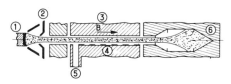

FIG. 2. Schematic of a BWO: (1) electron emitter; (2) anode; (3) focusing magnetic field; (4) slow-wave circuit; (5) rf output; and (6) electron collector.

1. *The Electron Gun Configuration*

The electron beam is initiated at a cathode, the radius of which must be larger than the beam radius in order to take into account the limited available current density of electron emitters. The beam is then concentrated and focused in the circuit entry by a multielectrode electron gun, which must perform well over a rather wide range of voltages corresponding to the tuning bandwidth of the BWO. The beam is guided along the circuit by a magnetic field, and the cathode is partly shielded from this field in such a way that the magnetic flux through the beam is approximately constant.

These conditions result in some limitations for the gun convergence. If a ratio of 1000 has been demonstrated in some cases, such a gun, besides a risk of poor accuracy and bad alignment, will provide a beam with a wide velocity spread. This is due to a gain in transverse energy, which is approximately proportional to the magnetic field, originating in the thermal velocity spread of the emission at the cathode. Therefore a practical limit on gun convergence is set at 100.

2. Backward Wave Circuits

The slow-wave structures which are needed to support an rf electric field with a longitudinal component are periodic in the direction of the beam and behave like microwave filters with passbands and stopbands. Due to the periodicity of the geometry, the fields are identical from cell to cell except for a constant phase shift ϕ which is said to be the phase shift per cell of the structure. This phase shift, purely real in the passbands of a lossless structure, varies with frequency. According to Floquet's theorem, the electric field can be described, at an angular frequency ω, by a sum of "space harmonics":

$$E(z, t) = \sum_{n=-\infty}^{+\infty} E_n e^{j(\omega t - k_n z)} \qquad (1)$$

where the propagation constant k_n of each harmonic is expressed as

$$k_n = (\phi + 2n\pi)/p \qquad (0 < \phi < 2\pi) \qquad (2)$$

p being the pitch of the circuit and n an integer.

In the Brillouin diagram (k, ω), the dispersion is represented as in Fig. 3a for a circuit having a forward fundamental $(n = 0)$. According to the choice of n, the phase velocity of a particular space harmonic

$$v_n = \omega/k_n \qquad (3)$$

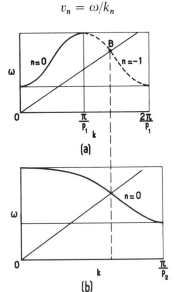

FIG. 3.   Dispersion of the slow-wave circuit and synchronism condition for backward wave interaction: (a) forward fundamental synchronism is at point B, intersection of the $n = -1$ space harmonic with the line of slope v_e; and (b) backward $(n = 0)$ fundamental.

has or has not the same sign as the group velocity

$$v_g = d\omega/dk_n = p\, d\omega/d\phi. \tag{4}$$

This last expression is independent of n, so that the group velocity is the same for all space harmonics. The group velocity may be shown to be also the energy velocity.

A periodic structure can thus support both forward and backward space harmonics, which are not modes of the field and cannot exist independently, even if a beam can be coupled to only one of them. According to the type of circuit, the fundamental is forward and the first negative space harmonic ($n = -1$) is backward (Fig. 3a); or the fundamental is backward, the first positive space harmonic ($n = 1$) being forward (Fig. 3b).

As the magnitude of the space harmonics decreases rapidly when the value of n is large, the interaction can be significant only with the fundamental or the first space harmonic. Among the structures having a backward fundamental space harmonic, the interdigitated structure (Fig. 4a) and the coupled-cavity circuit (Fig. 4b) can be cited. The former leads to dimensions too small for operation at millimeter wavelengths; the latter is bulkier but more difficult to manufacture with good precision.

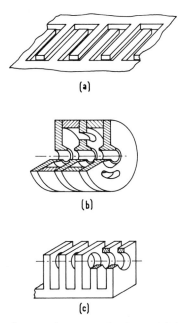

(a)

(b)

(c)

FIG. 4.   Configuration of some slow-wave structures: (a) interdigitated structure; (b) coupled-cavity circuit (both circuits have a backward fundamental space harmonic); and (c) vane-type slow-wave structure with a forward fundamental space harmonic.

A representative and widely used circuit for operation with the first space harmonic is the vane or grating circuit (Fig. 4c), which can be machined to good precision without undue difficulty.

With vanes of pitch p and height h, the approximate expression of the first backward space harmonic dispersion is given by

$$k = (2\pi/p) - (\omega/c)[1 + \tfrac{1}{4}tg^2(\omega h/c)]^{1/2} \tag{5}$$

valid far from the π-mode corresponding to $k = \pi/p$.

3. Backward Wave Interaction and Frequency Tunability

Assuming a harmonic field is propagating on the circuit, if one considers the rf electric field as seen by N different electrons entering the circuit at equidistant time intervals T/N during the rf period T (Fig. 5a), the first half of them will see, for example, an accelerating field while the other half will see a decelerating field. Thus if the position of the N electrons vs time after they have entered the structure is plotted (Fig. 5b), the accelerated electrons will bunch together with the electrons which entered the structure during the last part of the preceding period, the bunch being located around the electron entering at time $t = 0$, if the entering electrons have exactly the same velocity as the wave. In this case there will be an equal number of electrons accelerated or decelerated, and the exchanged energy will be null on the average.

If the wave travels with a phase velocity slower than the unmodulated beam velocity, the center of the bunch is located inside the decelerating field region, making the energy exchange effective, the beam being slowed down

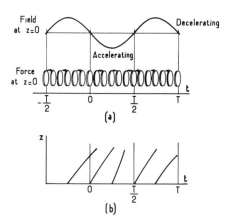

FIG. 5. Principle of electron bunching by a traveling electric field: (a) forces exerted on the electrons vs their time of entrance in the electric field; and (b) line of flight of the electrons interacting with the traveling wave.

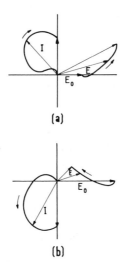

(a)

(b)

FIG. 6. Phasor diagram of traveling wave interaction. It represents the variation of the rf beam current I and that of the circuit electric field E in a frame of coordinate rotating with E_0, the cold circuit electric field, with the phase velocity of the wave, before coupling to the beam: (a) forward wave interaction; and (b) backward wave interaction.

on the average and the field amplitude being modified as it travels along the circuit (Pierce, 1950). The beam density modulation is seen from the circuit as an rf current modulation which has a phase lag of nearly $\pi/2$ behind the traveling wave electric field, at the first stage of the interaction.

Up to this point, it has not been specified whether the wave is a forward wave or a backward one. In the case of a forward interaction, the group velocity of the wave and the beam velocity are codirected. The current induces upon the wave a positive phase increment and the wave is slowed down. It gains energy according to an exponential law, while the current phase increases, at least in the first part of interaction (Fig. 6a).

When the wave is a backward wave, the group velocity of the wave and the beam velocity are contradirected, the phase increment induced by the current on the wave is first negative and the wave velocity increases. After that it loses energy while the phase of the current increases (Fig. 6b). The wave magnitude decreases; the field variation is not exponential but sinusoidal (Louisell, 1960).

As the rf energy is injected at the circuit end opposite to the beam entrance, the rf field increases from the beam output to input, while the rf current modulation increases in the opposite direction. This can be looked at as an internal reaction, as the coupling of the beam to the contradirected wave through a backward space harmonic creates an amplification. For a sufficient length of interaction, the gain may become infinite, and an oscillation will start and develop along the circuit.

The beam is represented in the (k, ω) diagram by a straight line of slope v_e, and synchronism is obtained at point B of Fig. 3a. The actual operating point is not far from point B in millimeter tubes. When the oscillation starts, the beam and wave velocities are related by

$$v_e/v = 1 + (\omega_q/\omega). \tag{6}$$

ω_q is the reduced plasma frequency, which usually in millimetric tubes is very low as compared to the angular frequency ω; the interaction is thus only slightly out of the synchronism condition (Beck, 1958).

The phase velocity of a backward wave being frequency dependent by nature, different oscillation frequencies are allowed when the point B is moved along the dispersion curve by changing the beam voltage, thus the slope v_e of the beam line. A broad tuning bandwidth can be obtained if the group velocity varies slowly vs frequency; under these conditions, moreover, the starting oscillation current has a slow variation with the frequency.

4.   Comparison with Related Oscillators—Ledatron and Orotron

These devices were proposed with the aim of overcoming the practical problems inherent to the small dimensions of electron beams and circuit element at very short wavelengths. They were presented as being suitable for operation down to 0.1 mm. However most of the results presented until recently have been in the millimeter range, or devoted to clarifying the problems of operating modes (Rusin and Bogomolov, 1969; Mizuno et al., 1970). Such devices are briefly considered here because of their relationship with BWO's.

The names "ledatron" or "orotron" refer to two kinds of operation of the same device, which utilizes the synchronous interaction between a flat beam of large cross section and a grating surrounded by an oversized open resonator (Fig. 7). Operation takes place at a beam transit angle θ per period of the grating near π or 2π. The orotron operation is similar to an extended interaction klystron, that of a ledatron to a BWO, with the important difference that the interacting rf field is built essentially by the resonator feedback.

In both cases the operating frequency is selected by mechanically tuning the resonator, while the synchronism condition at the optimum requires simultaneous voltage tuning. Owing to a slight variation of the electronic susceptance with the voltage, a narrow smooth electrical tuning band is possible.

The mechanical tuning range of the oversized resonator is quite large, particularly in the ledatron. As such a resonator has a large number of

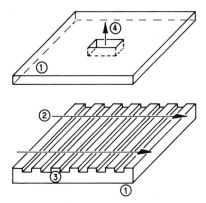

FIG. 7.   Schematic of the ledatron: (1) mirrors of the open resonator, (2) ribbon electron beam; (3) grooves insuring a periodic stationary electric field interacting with the beam; and (4) rf output.

resonant modes, several frequencies may be excited at the same voltage. As shown by Mizuno *et al.* (1974), the main mode ranges (TEM$_{00}$, TEM$_{01}$, TEM$_{02}$) overlap, and the starting currents are not very different. Therefore careful adjustments are necessary for the operation on the desired mode, while the stabilization procedure might be disturbed by frequency ambiguities and mode jumps.

Though the beam densities are much lower than for BWO's of comparable circuit length, the beam currents are found to be considerably higher. In fact, the feedback mechanism seems unable to overcome the loss of coupling that results from both the large beam cross section and small fraction of the resonator energy converted into the synchronous component.

It is felt that further work on the ledatron and orotron for submillimeter operation, which seems necessary in order to obtain a better appreciation of their feasibility, will show that the optimal design of cw models in this range will bear more similarity to that of the BWO.

B.   DESIGN PARAMETERS

The small-signal theory of oscillation starting was analyzed very early (Kompfner and Williams, 1953). Large-signal operation was described only approximately, however, allowing rough estimations of the rf efficiency and power saturation, which in many cases were found to be in fair agreement with the test results.

In the submillimeter-wave range, exact knowledge of the true design parameters is often questionable because of many causes of uncertainty in their determination, so that accurate large signal evaluations are still not of great importance. The most recent efforts were devoted instead to estimating the effects of the magnitude of some major design parameters on the tube performance, namely on the expectable power–frequency characteristics and the achievable tuning range.

1. Basic Parameters

Only a few main parameters need to be considered:

(a) the beam voltage V_0 and current I_0,
(b) the beam transfer impedance

$$Z_B = 2(V_0/I_0)(\omega_q/\omega),   \tag{7}$$

where ω_q is the angular plasma frequency,

(c) the circuit-to-beam coupling impedance Z_c,
(d) the circuit length L expressed as the number

$$N = (1/2\pi)(\omega L/v)   \tag{8}$$

of delayed wavelengths at the beam velocity, and

(e) the loss of the circuit α expressed in decibels per delayed wavelength.

2. Starting Conditions

The classical small-signal theory in the case of negligible circuit losses leads to the following oscillation-starting condition (referred to the subscript st)

$$N \geq \tfrac{1}{2}(Z_B/Z_C)_{st}^{1/2}   \tag{9}$$

When circuit losses are taken into account, a new parameter ψ should be defined (Guénard, 1975) as

$$\sin \psi_{(st)} = (\alpha/54.5)(Z_B/Z_C)_{st}^{1/2}   \tag{10}$$

and the starting condition becomes

$$N \geq \tfrac{1}{2}\{(Z_B/Z_C)^{1/2}[(1 + 2\psi/\pi)/\cos \psi]\}_{st}   \tag{11}$$

Most of the quantities present in such relations are also frequency dependent so that these relations also give the frequency range where oscillations occur, i.e., the tuning range of the tube. Equation (10) sets up an absolute limit for oscillation: $\sin \psi < 1$, such that the required circuit length would be infinite at the limit, when for example α approaches a critical value depending on both circuit and beam characteristics:

$$\alpha < 54.5(Z_C/Z_B)^{1/2}.   \tag{12}$$

The role of losses on the power generation is illustrated by Fig. 8, which gives the relative variations of the rf voltage V and rf beam current I, along the circuit, for various values of ψ. When the attenuation is negligible

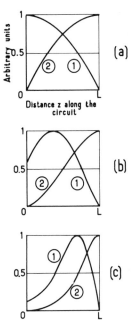

FIG. 8.   Distribution of the RF electric field (1) and beam current (2) vs the abscissa on the slow-wave circuit for various loss factors. (a) No losses: the electric field is maximum at the output $z = 0$; (b) moderate losses: the electric field is maximum near the output; and (c) large losses: the electric field is maximum near the collector end of the circuit $z = L$.

($\psi \simeq 0$), V and I are maximum at the circuit ends, respectively at $z = 0$ and $z = L$ (Fig. 8a). At increasing ψ, the maximum rf voltage occurs nearer the circuit end ($z = L$) while the output voltage at $z = 0$ decreases more and more (Fig. 8b,c). The rf current is still maximum at $z = L$, but the current modulation builds up very slowly along the circuit.

For a given length of the circuit, the oscillation will start for the value of the loss factor ψ_{st} given as a function of the total present circuit loss $\mathscr{L}$ by the relation:

$$\mathscr{L}_{dB} = 27.25\{[1 + (2\psi/\pi)]tg\psi\}_{st} \tag{13}$$

which, through Eq. (11), allows defining the starting current I_{st} from the knowledge of other parameters such as the beam voltage V_0 and the circuit parameters α and Z_C.

3.   RF Efficiency and Output Power

The rf-modulated current I_m rises with a further increase of the beam current above I_{st}. The rf power transferred to the circuit saturates when the ratio of I_m to I_0 is maximum:

$$P_B = \tfrac{1}{2} Z_B I_m^2 = I_0 V_0 [(I_m/I_0)^2(\omega_q/\omega)]. \tag{14}$$

A fraction of this power is absorbed by the circuit losses; it has been shown (Convert *et al.*, 1959) that it is equivalent to a loss $\mathscr{L}/2$ inserted in the output circuit. Then the approximate expressions for efficiency and output power are

$$\eta \simeq (I_m/I_0)^2(\omega_q/\omega)e^{-\mathscr{L}/8.68} \tag{15}$$

and

$$P \simeq \tfrac{1}{2}(I_m/I_0)^2(\omega_q/\omega)I_0 V_0 e^{-\mathscr{L}/8.68}. \tag{16}$$

4. *Figure of Merit*

Using some other parameters, such as the wavelength λ, the normalized angular beam radius γ_b (in the range of unity), and the beam current density J_0, Eq. (15) can be expressed as

$$1.75 \times 10^5 \, P/(J_0^{3/2} V_0^{7/4} \lambda^3) = e^{-\mathscr{L}/8.68}(I_m/I_0)^2(\gamma_b)^2 = M, \tag{17}$$

where the units are W, A/cm^2, kV, and mm.

M may be regarded as a "merit factor" depending upon the effective achievement of several dimensionless characteristics, and therefore helping to compare or judge the quality of tubes having various power performances. Examples will be shown in Section III.B.1.

5. *Main Trends in BWO Design*

Many guidelines may be followed in designing BWO's. However, performances in power and tunability are the predominant criteria, and we will now attempt to briefly justify the corresponding choices.

In early developments the preference was given to designs allowing the use of the highest beam power compatible with the available focusing magnetic fields, that is, also with the shortest possible length of the rf circuit, which consequently was used in its more dispersive region near the π-mode where the coupling impedance is maximum. These trends resulted in models of relatively high power, operating however over narrow tuning bandwidths. A dispersive circuit being more lossy, this approach, which is still appropriate for high-power tubes, risks failure at very short wavelengths because high current can no longer be provided.

Using a less dispersive circuit leads to decrease of both the coupling impedance Z_c and the circuit loss α, which are approximately proportional to the ratio v_ϕ/v_g between the phase and group velocities, at the benefit of a lower value of the term $\alpha/Z_c^{1/2}$ present in ψ. A parameter ψ of smaller value and therefore less variable with the wavelength is favorable for designing broad-band tunable models. This reduction of ψ also moves the design limitations towards shorter wavelengths, which finally includes those arising from the decrease of the allowable beam current.

C. Spectral Purity and Noise

The BWO is a voltage tunable oscillator, whose voltage tuning rate is directly related to the propagation characteristics of the circuit. The oscillation starts at a frequency where the wave propagating on the circuit is synchronous with the slow space-charge wave of the beam, the velocity of which is not very different from the beam velocity in millimeter wave tubes. Inherently the BWO is more sensitive than other oscillators to external fluctuations. Nevertheless its ability to be phase- or frequency-locked has been demonstrated, leading to successful operation as a heterodyne local oscillator.

1. *Frequency Stability*

The frequency–voltage sensitivity, obtained by differentiating the oscillation start condition with respect to the frequency, is given by the relation

$$\Delta f/f = \tfrac{1}{2}[1/(1 + |v_\phi/v_g|)](\Delta V_0/V_0) \tag{18}$$

The factor $|v_\phi/v_g|$ that characterizes the circuit dispersion is lower for wide-band than for medium-band tubes. For example, a tube tunable over a 20% band, at 0.8 mm, has a frequency sensitivity varying from 25 to 10 MHz/V across the tuning range, while a 10%-bandwidth tube presents a frequency sensitivity varying between 13 and 6 MHz/V.

The residual modulation of the voltage supply will generate a noise spectrum according to this voltage sensitivity. But stabilization has been demonstrated in the submillimeter range by phase-locking the BWO to a lower frequency stable generator; some examples of this ability are given in Volume 3 of this book.

The oscillation frequency is also sensitive to the beam current. This sensitivity is called "frequency pushing." The current fluctuations at low frequencies are mainly due to the anode voltage supply, and the frequency sensitivity to the anode voltage is given by the relation

$$|\Delta f/f| = \tfrac{3}{4}[\omega_q/\omega/(1 + |v_\phi/v_g|)](\Delta V_a/V_a). \tag{19}$$

This sensitivity as compared to the cathode voltage sensitivity is reduced by the ratio ω_q/ω, which is of the order of a few times 10^{-2}.

2. *Noise*

Until recently very little work has been done on BWO noise, except for the suppression of secondary emission effects and ionic instabilities. It was found experimentally for several X-band BWO's (Mackenzie *et al.*, 1965) that at 30 MHz from the carrier, the carrier–noise ratio ranged from 93 to 127 dB per MHz bandwidth, the two noise sidebands being unsymmetrical. Measurements of noise on submillimeter-wave BWO's performed recently (de Graauw et al., 1978) have shown that a signal–noise ratio of approximately 120 dB/MHz could be expected in this wavelength range. In heterodyne detection using a BWO as a local oscillator, this figure corresponds to a noise temperature added by the oscillator of only 1000–3000K. More details on these experiments will be given in Volume 3 of this series.

III. Millimeter Wave and Far-Infrared BWO's

In Section II the general features of the BWO have been described. This section is devoted to a survey of the problems related to millimeter and submillimeter wavelengths, allowing an identification of the reasons of the successful operation at short wavelengths.

A. PROBLEMS RELATED TO THE MILLIMETER AND FAR-INFRARED RANGES

Scaling the dimensions of an electron tube proportionally to the wavelength (Warnecke and Guénard, 1951) keeps the gun perveance constant; the circuit dimensions vary as the wavelength and the square root of the voltage, the electronic efficiency is constant, the output power varies as the 5/2 power of the voltage; moreover, the fractional bandwidth is unmodified. But as far as millimeter or submillimeter wavelengths are concerned, some limitations appear in the scaling factors, caused, for example, by the available cathode current density or by the breakdown electric field. Other factors that are not taken into account by the scaling laws become nonnegligible or even of primary importance, such as ohmic losses, irregularities of the circuit, or thermal velocity spread of the emitted electrons. All these limitations will now be reviewed.

1. The Circuit Losses

As described in Section II.B.3 the efficiency and the available rf power decrease at short wavelengths because of the increasing effect of large rf losses, which comes from the fact that they increase with the frequency. The BWO slow-wave structures are all-metal circuits. The skin depth at the conductor surface is reduced proportionally to the square root of the frequency. It is of the order of 0.1 μm at 1-mm wavelength for copper. The roughness of the surface comes to play a significant role and the treatment of the surface will greatly affect the circuit loss. For example, a surface roughness of one-half or one skin depth increases the loss by a factor of 20 or 60%, respectively (Convert and Yeou, 1964).

Ohmic losses of the circuit are one of the sources of detrimental effects on the oscillator. Also, the velocity spread of the electrons due to the thermal velocity distribution of the emission process at the cathode, as well as the random imperfections of the circuit, can be described by an equivalent circuit loss playing the same role as the ohmic losses. Some aspects of these effects are considered below.

2. The Dimensions of the Circuit

As the transverse dimensions of the circuit elements become smaller at short wavelengths, the simplest construction, in spite of possible lower interaction, should be preferred, in order to avoid additional losses due to manufacturing irregularities.

For example, a slot-coupled cavity circuit (Gross, 1973) is manufactured by assembling and welding under pressure several hundreds of thin punched disks of various profiles and orientations and with a thickness of less than 100 μm for operation below a wavelength of 1 mm. A mechanical precision of 1 μm for all dimensions is required at several steps of the fabrication, which can hardly be achieved by multiple-element assembling techniques.

A structure that better fits the millimeter-wave requirements is the vane-type circuit, manufactured by milling slots in a block of copper. The vanes are shorter than $\lambda/4$, and their thickness is less than one-tenth of a wavelength. The same mechanical accuracy of 1 μm is easier to obtain with this one-part circuit. The open structure allows optical checking of the circuit regularity and even makes it possible to correct at least some irregularities.

The random manufacturing irregularities of the slow-wave structure may be considered as equivalent to a phase velocity spread or to random reflections, and their effect, analogous to the beam velocity spread, is taken into account as an addition to the circuit losses (Convert et al., 1959; Convert and Yeou, 1964; Pease, 1961). It has been shown that this effect is less important with the shorter vanes of a low-dispersion circuit.

3. The Circuit Heat Dissipation

The rf efficiency of millimeter and submillimeter wave tubes is low; moreover, as a consequence of uncontrollable inaccuracies in the tube construction or positioning in the magnet, the percentage of the beam power transmission to the cooled collector may sometimes be rather poor. In practice the rf circuit should be designed to support the full beam interception without excessive heating; on the other hand, convenient heat shields are provided against direct beam bombardment at the circuit entrance.

As an example, for a BWO working at 320–390 GHz, the vanes have the following dimensions: 0.17-mm height and 0.2 × 0.05-mm cross section. Their thermal resistance is approximately 35°C/W between the vane top grazed by the electron beam and the cooled base; thus the power intercepted by any vane must be kept below ~ 10 W for safe operation. From measurements made under normal operating conditions, less than an average of 2 W per vane is dissipated in such tubes at a total beam power of about 300 W.

Experience has shown that the heat-dissipation capability of an rf circuit of a given type is approximately proportional to the wavelength. However such limitations are only encountered in very high-power tubes at relatively low frequencies. As seen from the above example, a safe margin still exists for several hundred watt electron beams in tubes operating at wavelengths under one millimeter. Moreover the available beam power in the submillimeter range is more severely restricted by other limitations, as shown in Sections II.B.4 and 6. It is therefore expected than even at the shortest wavelengths envisaged, the circuit dissipation will not be a critical factor concerning reliability and life, provided sufficient care is taken in tube fabrication and throughout the cycles of processing and operation.

4. The Electron Beam

The electron beam presents the most difficult problems in the design of millimeter and submillimeter-wave BWO's because of the severe requirements imposed by the small dimensions of the interaction space, particularly with respect to the high values of beam-current density and magnetic focusing field required.

Axially symmetric systems are the most suitable for achieving such conditions. For a beam radius b and a wave number γ, a first condition is that γb be not too much greater than 1. From the definition of the beam density, the beam current expressed as a function of the gun convergence S and cathode emission density J_c by the relation

$$I = 0.31 \times 10^{-6} S J_c (\gamma b)^2 V \lambda^2 \tag{20}$$

shows, for sufficient beam current at short wavelengths, the need of large $S J_c$ products. Both of these parameters encounter some limitations of a technological or even physical character.

The minimum magnetic field B_B for guiding a cylindrical electron beam corresponds to a zero magnetic flux through the cathode and is called the Brillouin field. A confined flow is usually preferred in order to improve the beam stiffness and reduce scalloping and beam interception; this requires a magnetic field B_C at the cathode and a main field B greater than B_B:

$$B = B_B / \sqrt{1 - K^2}, \tag{21}$$

where

$$K = S B_c / B \tag{22}$$

is the beam confinement ratio. A confined flow would normally be achieved only with a very strong field, a situation which can only be approached. A value of 2–2.5 times the Brillouin field is usually applied, corresponding to a magnetic flux through the cathode on the order of 90% of the flux through the beam and to a main field B generally in excess of 1 T (10 kG) below 1-mm wavelength.

5. The Velocity Dispersion

In highly convergent electron guns, the thermal velocity spread of the electrons emitted at the cathode of temperature T produces a transversal motion resulting in an increase in the beam radius. To maintain the gun convergence, the magnetic field must be increased by a further factor (Herrmann, 1958)

$$B = B_B(1 + \theta)^{1/2}/(1 - K^2)^{1/2} \tag{23}$$

where

$$(1 + \theta)^{1/2} = 1 + \frac{2S}{(\gamma b)^2} \frac{kT}{eV} \frac{\omega^2}{\omega_q^2} = 1 + \frac{1.85 \times 10^{-2}}{(J_c/T)(\gamma b)^2 V^{1/2} \lambda^2}. \tag{24}$$

The thermal velocity spread in the electron beam also produces random phase fluctuations, which disturb the synchronism conditions between the electrons and the wave. The effect of the thermal velocity distribution has been found to be equivalent to an increase of the beam plasma frequency (Parzen, 1952):

$$\omega_p^2(T) = \omega_p^2 + S(kT/2eV)\omega^2 = \omega_p^2[1 + S(kT/2eV)(\omega/\omega_p)^2], \tag{25}$$

which corresponds to an increase in the beam space-charge impedance Z_B, and therefore to an increase in the loss factor ψ, which is equivalent to a higher circuit loss α.

6. Cathode Emission Density

Another important problem raised by the technology of submillimeter wave tubes concerns the life of the cathode, which must be able to operate at space-charge electron emission densities of 10–20 A/cm², or even more, if possible, at short submillimeter wavelengths. This problem may become fundamental because in many cases the required emission density will be obtained only at a cathode temperature exceeding the values usually recommended for long life. For example, a BWO delivering a few watts at 300 GHz has a cathode loading of 15 A/cm², needing operation near 1250°C.

The cathodes of millimetric and submillimetric BWO's are usually made from a porous tungsten matrix impregnated with barium and calcium aluminates. At the operating temperature, some active material, essentially barium, is produced and migrates to the cathode surface, where it constitutes the emissive layer of low thermionic work function. The evaporation of barium from the emissive layer is compensated for totally or partially by barium dispensed from the inside of the cathode to its surface.

A large reserve of barium can be provided for a long life, but the barium content in the tungsten plug decreases with the operating time, thus reducing the replenishment rate progressively. These mechanisms are diversely accelerated by the high temperature needed for high emission density at the expense of cathode life. Moreover, temperatures as high as previously mentioned also pose problems concerning the cathode heater life.

Detailed investigations of the cathode mechanisms, analysis of practical results on operating tubes, and accumulated life tests performed to satisfy the present demand for very reliable and long life TWT's for space communications now allow prediction of the possible trade off between the emission and life requirements. For example, it is found that both the saturated electron emission and barium consumption are exponentially related to the temperature by the relation (Shroff and Palluel, 1978)

$$t_T \sim 1/(J_T)^n, \tag{26}$$

allowing some predictions about the cathode life t_T at the temperature required for the saturated density J_T, compatible within a safety margin with the needed operating density. Such a relation is well verified in practice, as shown in Fig. 9, which is a plot of measured lifetimes vs rated current density for different kinds of microwave tubes. This plot includes experimental

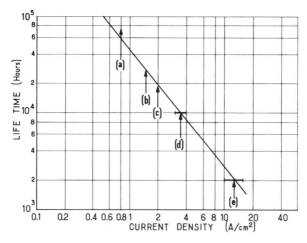

FIG. 9. Lifetime of impregnated tungsten cathodes vs the operating current density: (a) space TWT; (b, c, d) communication TWT; and (e) BWO.

results showing two or three thousand hours of life for a series of 1 mm Carcinotrons.

Also keeping in mind the interest in the highest possible cathode density for the best performance at short submillimeter wavelengths, it is thought that in the near future wide use could be made of new cathode structures, which are presently being investigated with the aim of increasing the emission at lower temperatures. Promising results, such as 80,000 h at 980°C and 10-A/cm² saturation with osmium-coated cathodes and many thousand hours at 1050°C and 100-A/cm² saturation with barium scandate cathodes, have recently been mentioned (van Ostrom, 1978).

7. Limitations at Very Short Wavelengths

To summarize, the available beam current decreases radically at very short wavelengths due to combined limitations related to the gun convergence, the magnetic field, and the cathode electron emission. If the beam voltage, magnetic field, and cathode emission densities are considered to be the main parameters, beam current and gun convergence may be written

$$I_0 = 0.143 \times 10^{-12}(\gamma b)^2[(1 - K^2)/(1 + \theta)]B^2 V_0^{3/2}\lambda^2 \tag{27}$$

$$S = 0.46 \times 10^{-6}[(1 - K^2)/(1 + \theta)]B^2 V_0^{1/2}/J_c. \tag{28}$$

As θ, which varies as λ^{-2}, may become significantly larger than 1, dependences on the wavelength may occur according to the relations:

$$I_0 \sim \lambda^4, \qquad S \sim \lambda^2, \qquad \text{and} \qquad \sin \psi \sim 1/\lambda^{3/2}. \tag{29}$$

Relationships of this kind will be found below a wavelength of around 0.3 mm, depending in fact upon the feasibility of high values of cathode density, magnetic field, and beam voltage.

B. BWO PERFORMANCES

1. Series of Millimeter and Far-Infrared BWO Models

Early models have been fabricated for the short millimeter and far-infrared ranges with moderate tuning bandwidth (Table I). More recently, two series of broadband models for the millimeter-wave (Gross, 1971, 1973) and the submillimeter-wave ranges (Golant et al., 1969) have been briefly described. Their performances are presented in Tables II and III.

A summary of these results is given in Figs. 10–11 where the output power and the efficiency of the different models, respectively, are compared. The output power follows the rule:

$$P \sim \lambda^{5/2} - \lambda^3, \tag{30}$$

TABLE I

Characteristics of Medium Bandwidth BWO's[a]

Wavelength (mm)	4	2	1	0.5	0.4	0.35
Maximum output power (mW)	38×10^3	8×10^3	1.5×10^3	15	9	0.25
Relative tuning bandwidth (%)	—	—	6	13	5	—
Maximum voltage (kV)	6	6	10	9.5	8	10
Beam current (mA)	65	45	30	35	35	45
Magnetic field (T)			0.8–0.95			
Maximum power consumption (W)	390	270	300	330	280	450
Maximum efficiency (%)	11	4	0.5	4×10^{-3}	2×10^{-3}	6×10^{-5}

[a] Vane-type circuit (Convert and Yeou, 1964).

TABLE II

Characteristics of Wideband BWO's[a]

Wavelength (mm)	11–7	7.5–5	5.3–3.3	3.5–2.7
Maximum power (mW)	300	100	35	20
Relative tuning bandwidth (%)	35	30	35	25
Maximum voltage (kV)	2.6	2.6	2.4	2.2
Beam current (mA)		12–15		
Magnetic field (T)		0.25		
Maximum power consumption (W)		300–350		
Maximum efficiency (%)	1.25	0.4	0.2	0.1

[a] Coupled-cavity circuit (Gross, 1971).

TABLE III

Characteristics of Wideband BWO's[a]

Wavelength (mm)	1.7–1.1	1.2–0.8	0.81–0.56	0.57–0.42	0.42–0.32	0.325–0.24
Maximum output power (mW)	75	—	—	—	—	0.8
Relative tuning bandwidth (%)	43	40	36	30	27	30
Maximum voltage (kV)	4	4			6.5	6.5
Beam current (mA)			50			
Magnetic field (T)	0.6	0.7/0.9			1/1.2	1/1.2
Maximum power consumption (W)		200			325	
Maximum efficiency (%)	$4 \; 10^{-2}$	—	—	—	—	$2.5 \; 10^{-4}$

[a] Golant et al., 1969.

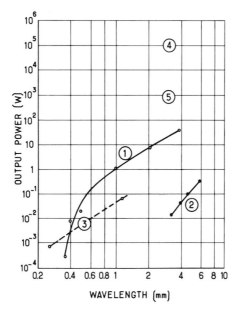

FIG. 10. Typical output of various models of BWO: (1, 2, 3) models of Tables I (Convert and Yeou, 1964), II (Gross, 1971), and III (Golant *et al.*, 1969), respectively; (4) pulse BWO (Sedin *et al.*, 1963); and (5) cw TWT for comparison (Heney, 1970).

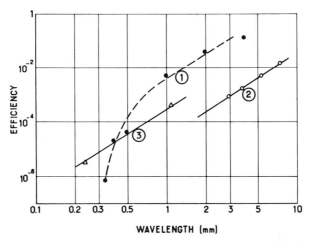

FIG. 11. Efficiency of various models of BWO: (1) models of Table I; (2) models of Table II; and (3) models of Table III.

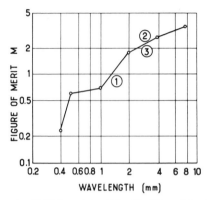

FIG. 12.   Figure of merit of BWO models of Fig. 10: (1) models of Table I; (2) pulse BWO at 94 GHz; and (3) cw TWT at 94 GHz.

except for the models of Table 1 for λ below 0.5 mm. A similar rule applies for the efficiencies. It must be noted that the models of Table 1 exhibit an output power greater than the others, by at least an order of magnitude. This can be explained by their relatively smaller tuning bandwidths. In fact, a broader bandwidth involves a lower circuit dispersion, a lower value of the coupling impedance Z_c, and in some cases a longer circuit. All these factors act in the same way to increase the loss factor ψ and reduce the output power. But at short submillimetric wavelengths the circuit losses α increase more rapidly for the dispersive circuit than for the wideband circuit, explaining the limit in frequency, which is lower for the smaller bandwidth tubes than for the large bandwidth models.

In each series of models (Tables I and III), the beam currents are kept almost constant by progressive modifications of either parameter, the strong limitation at very short wavelengths being in the limits of feasibility of high voltage, cathode-emission densities, magnetic fields, and power dissipation of the circuit. Values as high as 10 kV, 10 A/cm², 1 T or more were already present in these models and can be overcome only at the price of important efforts in tube technology.

Whereas the power performances of the tubes mentioned are radically different, the merit factors that can be calculated from published data, as shown by the plot of Fig. 12, are coherent and demonstrate the homogeneity of the designs of the corresponding models, along with reasonable absolute values of the M factor at least above 1 or 0.5 mm.

2.   Examples of Tube Characteristics

Some operating characteristics of typical BWO's are given below for an illustration of the evaluation and limits detailed in the preceding sections.

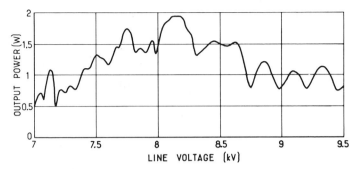

FIG. 13. Output power characteristics vs tuning voltage of a 6% bandwidth high power BWO in the 280 GHz range.

a. 280 *GHz Medium Bandwidth BWO.*   This tube, developed at THOMSON–CSF, has a tuning bandwidth of $\pm 3\%$ centered around 280 GHz. The voltage tuning range is 7–9.5 kV.

Figure 13 shows the typical output power measured vs the tuning voltage. The collector current is 6 mA for a cathode current of 17 mA. The output power varies between 1 and 2 W over most of the bandwidth. The power characteristic has a fluctuating structure, caused by reflections at both ends of the circuit.

The frequency characteristic vs the tuning voltage is given in Fig. 14. The frequency sensitivity to the voltage varies in the bandwidth from 6 to 15 MHz/V. These rather low values, which are typical of dispersive delay circuits, correspond to a medium bandwidth, which has its counterpart in the relatively high value of the output power.

b. 360 *GHz Extended Bandwidth BWO.*   This model was developed under an ESA contract (Bonnefoy *et al.*, 1977) in order to review the present possibility of manufacturing a low power consumption BWO, broadband

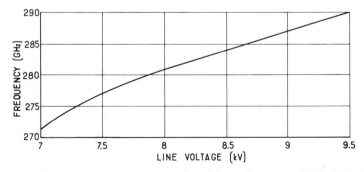

FIG. 14. Frequency characteristics vs tuning of the medium power BWO of Fig. 13.

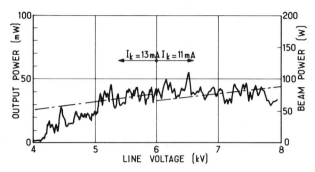

FIG. 15.   Output power characteristics vs tuning of a 20% bandwidth medium power BWO in the 360 GHz range.

tunable continuously around 360 GHz, and to investigate the qualities of the tube as a stable local oscillator for low noise heterodyne detection.

Figures 15 and 16 show the variations of the frequency, rf output power, and beam power consumption vs the applied voltage. A continuous spectrum coverage 0.94–0.74-mm wavelength is measured over the voltage range 4–9.5 kV. The rf output is greater than 10 mW over the frequency range 360 GHz ± 10%. The total bandwidth is swept with a half-magnitude change in the beam current, pointing out the possibility of reducing the power consumption at the high voltage end of the spectrum. The maximum consumption of less than 100 W is a rather interesting figure for a tube working in this frequency range. The power ripples correspond to a power–frequency slope

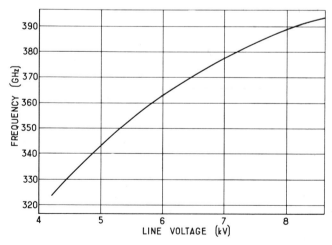

FIG. 16.   Frequency characteristics vs tuning voltage of the high power BWO of Fig. 15.

well below 0.01 dB/MHz as in the best matched TWT for communications. The voltage to frequency conversion ranges from 10 to 30 MHz/V.

Studies concerning frequency stabilization, noise characteristics, and operation performance will be described in Volume 3.

IV. Conclusion

The generation of coherent waves in the millimeter and far-infrared spectral ranges is feasible with the backward wave oscillator, a conventional electron tube which does not differ in principle from tubes able to work in the microwave region of the electromagnetic spectrum.

The BWO brings to the short wavelengths its specificity of being electrically tunable, under some conditions over a broad bandwidth. The BWO has proven to be useful as a local oscillator in various heterodyne detection experiments in the far infrared, where its low noise characteristics, added to its tunability, allow the use of receivers with a lower noise temperature. The BWO may also be used as a power generator in applications where wideband tuning is not required. It can deliver up to a few watts of output power at short millimetric wavelengths. This is far below the capability of relativistic generators, but the BWO does not require use of sophisticated technology.

Some improvements may be made to the present state of the art, using new technological developments. Up to now one drawback was the reduced lifetime due to the high current loading of the cathode, requiring cathode operation at high temperature. This will be overcome by the use of new emitters capable of high current densities at lower temperatures. Development of rare-earth magnetic materials, with high energy density, should allow reducing the mass and dimensions of the permanent magnets, and would even make possible an increase in the achievable output power.

REFERENCES

Beck, A. H. W. (1958). "Space Charge Waves." Pergamon, Oxford.
Bonnefoy, R., Kantorowicz, G., and Palluel, P. (1977). *Proc. Eur. Microwave Conf., 7th,* Copenhagen.
Convert, G., and Yeou, T. (1964). *In* " Millimeter and Submillimeter Waves " (F. A. Benson, ed.), Chapter 4, pp. 57–78. Illiffe Books, London.
Convert, G., Yeou, T., and Pasty, B. (1959). *Symp. mm-Waves,* Brooklyn Poly. Press, Brooklyn. New York, p. 313.
de Graauw, Th., Anderegg, M., Fitton, B., Bonnefoy, R., and Gustincic, J. J. (1978). *Int. Conf. Submm. Waves, 3rd, Guildford University of Surrey.*
Golant, M. B., Elekceenko, Z. T., Korotkova, Z. S., Lunkind, L. A., Negerev, A. A., Petrova, O. P., Rebrova, T. B., and Saveleva, V. S. (1969). *Prib. Tekh. Eksp.* **3**, 231–237.
Gross, F. (1971). *Microwave J.* **14**, 55–57.
Gross, F. (1973). *Proc. Eur. Microwave Conf., 3rd, Brussels.*
Grow, R. W., and Gunderson, D. R. (1970). *IEEE Trans. Electron. Dev.* **ED-17**, 1032–1039.

Guénard, P. (1975). THOMSON-CSF Internal Memorandum.

Heffner, H. (1954). *Proc. IRE* **42**, 930–937.

Heney, J. F. (1970). *Int. Conf. M.O.G.A., 8th, Amsterdam.*

Herrmann, G. (1958). *J. Appl. Phys.* **29**, 127–136.

Kompfner, R., and Williams, N. T. (1953). *Proc. IRE* **41**, 1602.

Louisell, V. H. (1960). "Coupled Modes and Parametric Electronics." Wiley, New York.

Mackenzie, L. A., Mosher, C. H., and Dalman, G. G. (1965). *Congr. Tubes Hyperfréquence, 5th, Paris, 1964* pp. 23–26.

Mizuno, K., Ono, S., Shibata, Y. (1970). *Proc. Symp. Submm. Waves, Brooklyn Poly. Press* pp. 115–134.

Mizuno, K., Kuje, N., and Ono, S. (1974). *Proc. Eur. Microwave Conf., 4th, Montreux* pp. 46–50.

Parzen, G. (1952). *J. Appl. Phys.* **23**, 394–406.

Pease, M. C. (1961). *In* "Crossed-Field Microwave Devices" (E. Okress, ed.), Vol. 1, Chapter 2.5. Academic Press, New York.

Pierce, J. R. (1950). "Traveling Wave Tubes." Van Nostrand-Reinhold, Princeton, Jersey.

Rusin, F. S., and Bogomolov, G. D. (1969). *Proc. IEEE Lett.* **57**, 720–722.

Sedin, J. W., Purnell, M., and Slocum, K. W. (1963). Technical Rep. RTD-TDR-63-4174-AFAL.

Shroff, A. M., and Palluel, P. (1978). Tri-Service Cathode Workshop, Naval Research Laboratory, Washington, D.C.

van Ostrum, A. (1978). Tri-Service Cathode Workshop, Naval Research Laboratory, Washington, D.C.

Warnecke, R. R., and Guénard, P. (1951). "Les tubes à modulation de vitesse," Chapter XX. Gauthier-Villars, Paris.

CHAPTER 5

The Ledatron*

K. Mizuno   *and*   *S. Ono*

I. Introduction

Many approaches to generating tunable submillimeter waves are being investigated and developed (Martin and Mizuno, 1976). Among them there are free electron beam devices. These are helical beam devices such as the gyrotron (Flyagin *et al.*, 1977), periodic-circuit devices such as the Carcinotron, and relativistic-electron beam devices (Granatstein and Sprangle, 1977). Although this spectrum range is now also obtained with far-infrared lasers, solid state generators like IMPATT diodes (Ohmori *et al.*, 1977; Mizuno *et al.*, 1978), and harmonic generators, the electron beam device will be unsurpassed as a tunable high-power source. The Ledatron is a member of the periodic-circuit devices, where a bunched electron beam interacts with an electromagnetic field which is carried by the circuit.

A high-frequency limit of coherent interaction between an electron beam and a resonant cavity has been analyzed from the quantum-mechanical point of view (Senitzky, 1954). The quantization of the field and the wave

---

* In Greek mythology, Leda was the mother of the twins, Castor and Pollux, who were believed to be the children of Zeus, who visited her in the form of a swan, and "tron" is a familiar ending for electron tubes. The title Ledatron, therefore, stands for an electron tube having two mode oscillations.

properties of the electrons modifies both the amplitude and the phase of the interaction. The high-frequency limit of around 3 THz ($\lambda = 0.1$ mm) has been estimated. Another restriction on the coherent interaction considered has been on the density of electron beam needed to obtain sufficient bunching action (Mizuno and Ono, 1975). The Debye length of the electron beam was used as a criterion for the electron beam to operate in a collective manner. The high-frequency limit for an electron beam of available density is again 3 THz. The present day short wavelength limit for the periodic-circuit device is around 0.2 mm for bwo's (backward wave oscillators).

The conventional period-circuit devices such as bwo's, magnetrons, klystrons, twt's (traveling wave tubes), etc., need a smaller interaction circuit structure with higher operating frequencies. Consequently, the generation of short millimeter or submillimeter high-power waves will become very difficult, because of a decrease in the permissible heat dissipation power and fabrication difficulties encountered in longitudinal field structures, etc. To avoid these difficulties, the Ledatron uses the Fabry–Perot interferometer, the familiar physical optics device which has proven its utility in laser research, as a practical resonator for the generation of submillimeter waves by electron beam techniques. The Fabry–Perot resonator has dimensions several orders larger than the operating wavelength; its reasonable size in the submillimeter region provides practical means of circumventing the above difficulties at short wavelengths. Two different mode interactions, the Fabry–Perot mode and the surface-wave mode exist in the Ledatron (Mizuno *et al.*, 1973). Both of these interactions result in oscillations of electromagnetic waves. In the Fabry–Perot mode an electron beam interacts with a standing wave in the Fabry–Perot resonator. On the other hand, the surface-wave mode involves the interaction between an electron beam and a backward wave contained in the surface wave guided by the grating. With the same structure of the grating, oscillation wavelength of the Fabry–Perot mode is shorter than that of the surface-wave mode at a constant electron accelerating voltage. These two modes, however, can be separated by proper selection of the mirror spacing of the Fabry–Perot resonator.

In the USSR, open resonator technologies has been systematically studied (Hibben, 1969), and an electron tube called Orotron similar to the Ledatron was proposed independently by us and has been studied in the Institute of Physical Problems in Moscow (Rusin and Bogomolov, 1969; Bogomolov *et al.*, 1970). In this chapter, however, we shall describe studies which have been done mainly in Sendai.

We shall consider in the next section the operational principles of the Fabry–Perot mode and the surface-wave mode. In Section III theoretical and experimental studies on the Fabry–Perot mode will be presented. Section IV presents operational characteristics of the surface-wave mode only briefly, because this mode is essentially the same as that in bwo's which were considered in Chapter 4 of this volume. Section V consists of summary and conclusion.

II. Operational Principle

A. FABRY–PEROT MODE OPERATION

The fundamental configuration of this tube is shown in Fig. 1. A reflection grating is used as one of the mirrors of the Fabry–Perot resonator. If the pitch of the grating is shorter than that of the operating wavelength (Mizuno *et al.*, 1970), then no lateral diffraction maxima corresponding to unwanted losses are possible at the reflection of an electromagnetic wave. When the spacing between the mirrors is adjusted properly, a resonance will occur for

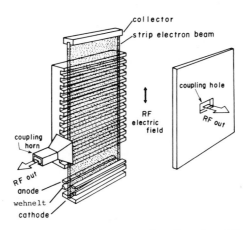

FIG. 1. Perspective view of the fundamental configuration of the Ledatron.

the wave. If the polarity of the resonant electric field is perpendicular to the grooves, the amplitude of electric fields just in front of the grating will be given as shown in Fig. 2, i.e., the fields will be only on the grooves and not on the ridges. Such an electric field distribution makes a space harmonic interaction possible. For simplicity, we assume that the amplitude of electric field is constant on all of the grooves and zero on the ridges. Hence the electric field along the y axis is given by the following equation:

$$E_y = \sum_{m=-\infty}^{\infty} A_m e^{j(\omega t - \beta_m y)} \tag{1}$$

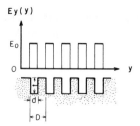

FIG. 2.   Idealized distribution of the electric field intensity in the region just in front of the grooved mirror. This distribution is expressed by a Fourier series, consisting of many space harmonics with various phase velocities.

where

$$A_m = (d/D)E_0 \qquad\qquad (m = 0)$$
$$= (2E_0/m\pi)\sin[(md/D)\pi] \qquad (m \neq 0), \qquad (2)$$

$$\beta_m = 2m/D, \qquad\qquad (3)$$

and where d is the groove width, D the groove pitch, and ω the angular frequency. Therefore, although the electromagnetic energy actually does not propagate in the y direction, there are many space harmonic waves propagating in this direction, and the phase velocity of the mth harmonic wave v_{pm} is given by the equation $v_{pm} = \omega/\beta_m$. When an electron passes near the grating at a velocity close to the phase velocity of one of the space harmonics produced along the grating, an effective interaction takes place between the electron and the resonant electromagnetic field.

The Fabry–Perot mode interaction may be considered as Smith–Purcell radiation (Smith and Purcell, 1953) with internal feedback by the resonator.

B.   SURFACE-WAVE MODE OPERATION

The electromagnetic characteristics of corrugated structures have been studied by many authors. The general theory indicates that the surface will act as a guide for electromagnetic waves, and the guided surface wave consists of an infinite number of space harmonic waves. Therefore, when an electron beam passes near the surface at a velocity close to the phase velocity of one of the backward space harmonics, an effective interaction may take place between the electron and the wave, resulting in backward wave oscillation.

The method of field matching, taken at the surface of the grating, gives the determinantal equation as follows:

$$\sum_n \frac{\sin \frac{1}{2}\beta_n d}{\beta_n \sqrt{\beta_n^2 - k^2}} = \frac{D}{2k}\cot kt \qquad (4)$$

where t is the groove depth, $k = \omega/c$, and $\beta = \beta_0 + 2\pi n/D$. The phase velocity of the nth space harmonic is given by $v_{pn} = \omega/\beta_n$.

Since there is some leakage energy along the grating, the oscillating wave can be enhanced by putting a reflector in suitable position, thereby forming a Fabry–Perot resonator together with the grating. Since the leakage energy is small, the reflector does not play an essential role in this interaction. In our experiments, the oscillation wavelengths was found to be almost completely determined by the groove dimension of the grating and the electron velocity, and not by the reflector position. Therefore, this interaction is completely different from that of the Fabry–Perot mode, where the oscillation frequency can be tuned by changing the mirror spacing.

III. Fabry–Perot Mode Operation

A. Grooved Mirror (Grating)

1. *Design*

In the Ledatron, we utilize the periodic electric field distribution in the region just in front of the grooved mirror to obtain the interaction of an electron with the resonant field. The groove structure must be designed to obtain maximum coupling between the electron and the field. The design consideration consists of two steps:

   (i)   to maximize the harmonic content with which the electron beam interacts in comparison with other harmonics, and
   (ii) to maximize the field intensity just in front of each groove.

   a. *Groove Width to Pitch.* An amplitude of each harmonic wave is expressed in Eq. (2). When the mth harmonic is chosen to interact with the electron beam, the following condition has to be satisfied in order to maximize the amplitude of the harmonic:

$$md_{\text{eff}}/D = \tfrac{1}{2}, \tag{5}$$

where d_{eff} is the effective width of the groove with respect to the field distribution, and will be described in the next section.

   b. *Groove Depth.* For each groove structure optimization, we investigate (Mizuno et al., 1970) a reflection of a plane electromagnetic wave which, having only E_y and H_x as its components, enters perpendicularly a surface of the grooved mirror. From the periodic nature of the structure, we can consider this as a TM_{0n} mode problem in a parallel line strip with an interval D. The model for this analysis is shown in Fig. 3. The definition of Regions (I) and (II) and the coordinate axes used in this analysis are also shown in Fig. 3.

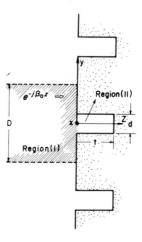

FIG. 3.   Grooved mirror structure.

As usual, we describe the incident plane wave by the function $\exp(-j\beta_0 z)$, where $\beta_0 = 2\pi/\lambda_0$. Moreover, the following conditions are taken into account:

(i)   The electric field at $z = 0$ is represented by the eigenfunction in Region (I).

(ii)   Only the mode that is symmetric about $y = 0$ is excited.

We get the electric field expression for Region (I) as follows:

$$E_y^I(y, z) = -2j \sin(\beta_0 z) + \sum_{n=0}^{\infty} A_{2n} \cos\left(\frac{2n\pi}{D} y\right) e^{j\beta_{2n} z}, \qquad (6)$$

where

$$\beta_{2n} = -j\sqrt{(2n\pi/D)^2 - (2\pi/\lambda)^2} = -j(2\pi/\lambda)\sqrt{(n\lambda/D)^2 - 1}. \qquad (7)$$

Equation (7) holds under the condition for operation of the Fabry–Perot resonator $D < \lambda$ (λ is the operating wavelength).

The magnetic field expression in Region (I) is similarly given by

$$H_x^I(y, z) = -\frac{2\omega\varepsilon}{\beta_0} \cos(\beta_0 z) + \sum_{n=0}^{\infty} \frac{\omega\varepsilon}{\beta_{2n}} A_{2n} \cos\left(\frac{2n\pi}{D} y\right) e^{j\beta_{2n} z}. \qquad (8)$$

In Region (II), noting that the line is shortened at $z = t$, the fields are given by

$$E_y^{II}(y, z) = \sum_{s=0}^{\infty} B_{2s} \cos\left(\frac{2s\pi}{d} y\right) \frac{\sin[\zeta_{2s}(z - t)]}{-\sin(\zeta_{2s}t)}, \tag{9}$$

$$\zeta_{2s} = j\sqrt{(2s\pi/d)^2 - (2\pi/\lambda)^2} = j(2\pi/\lambda)\sqrt{(\lambda s/d)^2 - 1}, \tag{10}$$

$$H_x^{II}(y, z) = \sum_{s=0}^{\infty} j\omega\varepsilon B_{2s} \cos\left(\frac{2s\pi}{d} y\right) \frac{\cos[\zeta_{2s}(z - t)]}{\zeta_{2s} \sin(\zeta_{2s}t)}. \tag{11}$$

For fulfilling the boundary conditions at $z = 0$, the undetermined coefficients A_{2n} and B_{2s} must now satisfy two relations which follow from the continuity of E_y

$$\sum_{n=0}^{\infty} A_{2n} \cos\left(\frac{2n\pi}{D} y\right) = \sum_{s=0}^{\infty} B_{2s} \cos\left(\frac{2s\pi}{d} y\right), \tag{12}$$

and the continuity of H_x

$$\sum_{s=0}^{\infty} jB_{2s} \cos\left(\frac{2s\pi}{d} y\right) \frac{\cos(\zeta_{2s}t)}{\zeta_{2s} \sin(\zeta_{2s}t)}$$

$$= -\frac{2}{\beta_0} + \sum_{n=0}^{\infty} \frac{A_{2n}}{\beta_{2n}} \cos\left(\frac{2n\pi}{D} y\right) \qquad \left(|y| < \frac{d}{2}\right). \tag{13}$$

Solving Eq. (12), we have

$$A_{2n} = \frac{4}{D} \int_0^{D/2} B_{2s} \cos\left(\frac{2s\pi}{D} y\right) \cos\left(\frac{2n\pi}{D} y\right) dy$$

$$= \frac{4}{D} \int_0^{d/2} B_{2s} \cos\left(\frac{2s\pi}{D} y\right) \cos\left(\frac{2n\pi}{D} y\right) dy$$

$$= \sum_{s=0}^{\infty} B_{2s} b(n, s) \qquad (n \neq 0)$$

$$= B_0 b(0, 0) \qquad (n = 0) \tag{14}$$

where

$$b(n, s) = \frac{\sin[\pi(s/d + n/D)d]}{\pi(s/d + n/D)D} + \frac{\sin[\pi(s/d - n/D)d]}{\pi(s/d - n/D)D},$$

and

$$b(0, 0) = d/D.$$

Similarly we obtain

$$\frac{d}{D} B_0 \left[j \frac{\cos(\zeta_0 t)}{\zeta_0 \sin(\zeta_0 t)} D - \frac{d}{\beta_0} \right] - \sum_{n=1}^{\infty} \frac{A_{2n}}{\beta_{2n}} \frac{D}{n\pi} \sin\left(\frac{n\pi d}{D}\right) = -\frac{2}{\beta_0} d \qquad (s = 0),$$

(15)

$$j \frac{\cos(\zeta_{2s} t)}{\zeta_2 \sin(\zeta_{2s} t)} B_{2s} \frac{d}{2} = \sum_{n=0}^{\infty} A_{2n} a(n, s) \qquad (s \neq 0),$$

where

$$a(n, s) = \left\{ \frac{\sin[\pi(n/D + s/d)d]}{2\pi(n/D + s/d)} + \frac{\sin[\pi(n/D - s/d)d]}{2\pi(n/D - s/d)} \right\} \left(\frac{1}{\beta_{2n}}\right)$$

Substituting Eq. (14) into Eq. (15), we arrive at the relationship

$$\begin{vmatrix} C_{00} & C_{01} & \cdots & C_{07} & \cdots \\ C_{10} & C_{11} & \cdots & C_{17} & \cdots \\ \vdots & & & & \\ C_{70} & C_{71} & \cdots & C_{77} & \cdots \end{vmatrix} \begin{vmatrix} B_0 \\ B_2 \\ \vdots \end{vmatrix} = \begin{vmatrix} -(2/\beta_0)d \\ 0 \\ \vdots \end{vmatrix}$$

(16)

where

$$C_{i,l} = -\sum_{k=0}^{\infty} a(k, i)b(k, l),$$

$$C_{i,i} = j \frac{d}{2} \frac{\cos(\zeta_{2i} t)}{\zeta_2 \sin(\zeta_{2i} t)} - \sum_{k=0}^{\infty} a(k, i)b(k, i) \qquad (i = l \neq 0),$$

$$C_{0l} = -\sum_{k=1}^{\infty} a(k, 0)b(k, l),$$

$$C_{0,0} = j \frac{d}{\zeta_0} \frac{\cos(\zeta_{0t})}{\sin(\zeta_{0} t)} - \frac{d^2}{D\beta_0} - \sum_{k=1}^{\infty} a(k, 0)b(k, 0) \qquad (i = l = 0).$$

The simultaneous equations (Eqs. 16) have been solved to find the field intensity on the groove. Figure 4 shows the optimum dimensions of the groove for establishing the maximum field at $z = 0$ (mirror surface), namely, the antiresonance condition of the groove. Figure 5 shows the field intensity changes at various point of z for $D = 0.6\lambda$, $d = 0.2\lambda$, and $t = 0.2\lambda$. The figure is plotted in the region from $y = 0$ to $y = D/2$ because of the symmetry about $y = 0$. It is found that in the region from $z = 0$ to $z = -0.1\lambda$, the space harmonic interaction with electron beams is feasible. It must be noted that the effective width of the groove considering the field intensity, d_{eff}, is larger than the actual width, being about 1.5 times the actual width.

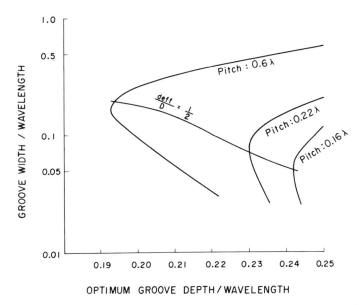

FIG. 4.   Calculated relations between the groove width and the optimum groove depth.

c. *Reflection Coefficient of the Grating.* From Eq. (6) the reflection coefficient Γ of the grating is given by

$$\Gamma = e^{j\theta} = A_0 - 1. \tag{17}$$

The mirror spacing of the Fabry–Perot resonator with the grating of reflection coefficient Γ is

$$l_{sp} = (\lambda/4\pi)\theta + (\tfrac{1}{2}n - \tfrac{1}{4})\lambda, \tag{18}$$

where n is the order of the resonance. Figure 6 shows the comparison (Mizuno *et al.*, 1974) of the mirror spacing between experimental results and the theoretical ones from Eq. (18). The experimental results show the oscillation wavelength as a function of the mirror spacings for a 65-GHz Ledatron. Excellent agreement between the theory and the measurements suggests that the theory is valid and that the contribution of the electronic susceptance to the reflection coefficient of the grating is negligibly small.

2.   *Machining*

The grooves on the gratings (30 × 30 mm) have been cut with a milling machine (Mizuno *et al.*, 1969). The distance between the grooves, the smoothness of their inner faces, and their width and depth are highly critical. Cu–Si–Ni (98–1–1) alloy has been chosen for the material of the grating for machining easiness. The milling machine used has an optically indexed

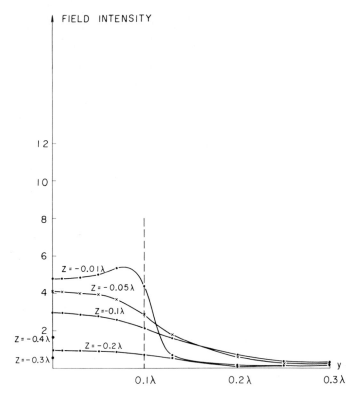

FIG. 5. Calculated field intensity on the groove for several values of z ($D = 0.60\lambda$, $d = 0.20\lambda$, and $t = 0.20\lambda$).

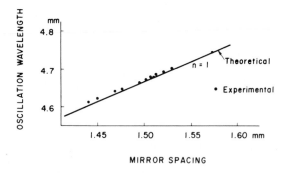

FIG. 6. Oscillation wavelength vs. mirror spacing. The solid line is a theoretical result. Electron accelerating voltage is 11.6 kV ($D = 999 \, \mu m$, $d = 316 \, \mu m$, and $f = 990 \, \mu m$).

precision scale and is carefully adjusted. For example, the machine is permitted to idle in order to expand all parts fully before the operation is started. The milling cutter is a 30-mm diameter cemented carbide (tungsten carbide) circular saw. Tolerances of ~ 1 μm for the dimensions of grooves have been maintained.

Grooved mirror structures considered suitable for operating down to 0.5-mm wavelength have been constructed by using a tungsten carbide milling cutter, 50 μm thick.

B. START-OSCILLATION CONDITION

1. Electronic Conductance

The Ledatron is very similar in concept to a floating-drift-tube klystron with the exception that extended interaction replaces the single-gap coupling to the beam, and is in the monotron oscillator family since only a single resonator with internal feedback is used. Formerly, Hechtel (1958) and Fujisawa (1964), and Wessel-Berg (1957) analyzed the interaction of this kind in kinematic theory and space charge theory, respectively. In actual operating conditions, the reduced plasma wavelength is much larger than the interaction length, and we assume that the electronic transit angle on a groove is small, so that we can reduce the effect of the angle on induced current to consideration of the beam coupling coefficient only. Therefore let us analyze the interaction according to the kinematic bunching theory in cascaded interaction.

The notation used in this analysis is the following: $\overline{V}_b$ is the dc beam voltage, $\overline{I}_b$ the dc beam current, $\overline{G}_b = \overline{I}_b/\overline{V}_b$ the dc beam conductance, $\overline{\theta}$ the dc electronic transit angle between two adjacent grooves, I_{cn} the induced current at the nth groove, β_0 the beam coupling coefficient on a groove, V_{0n} the rf voltage on the nth groove, and N_0 the total number of grooves.

Consider a small-signal operation, so that the terms of V_{0n} greater than the second order can be neglected. Then we can get the total effect of interaction grooves by superposing the modulation effect of each groove. A current induced at the second groove, resulting from the bunching effect of electron beams, velocities of which are modulated at the first groove, is given by

$$I_{c2} = \tfrac{1}{2}\beta_0^2 \overline{G}_b V_{01} j\overline{\theta} e^{-j\overline{\theta}}. \tag{19}$$

The induced current at the third groove is given by the sum of both the effects of the first groove and the second one.

$$I_{c3} = \tfrac{1}{2}\beta_0^2 \overline{G}_b(V_{01} j2\overline{\theta} e^{-j2\overline{\theta}} + V_{02} j\overline{\theta} e^{-j\overline{\theta}}). \tag{20}$$

I_{cn} is similarly obtained. Consequently the electronic admittance of the nth groove g_{en} is given by

$$g_{en} = I_{cn}/V_{0n} = \tfrac{1}{2}\beta_0^2(\bar{G}_b/V_{0n})V_{01}j(n-1)\bar{\theta}e^{-j(n-1)\bar{\theta}} + \cdots + V_{0,n-1}j\bar{\theta}e^{-j\bar{\theta}},$$

$$= \tfrac{1}{2}\beta_0^2(\bar{G}_b/V_{0n})\sum_{k=1}^{n-1} V_{0k}\,j(n-k)\bar{\theta}e^{-j(n-k)\bar{\theta}}. \tag{21}$$

The total electronic admittance $G_t^{(e)}$ is given by

$$G_t^{(e)} = \sum_{n=2}^{N_0} g_{en}$$

$$= \tfrac{1}{2}\beta_0^2\bar{G}_b \sum_{i=2}^{N_0} V_{0i}^{-1} \sum_{k=1}^{i-1} V_{0k}\,j(i-k)\bar{\theta}e^{-j(i-k)\bar{\theta}}$$

$$= \tfrac{1}{2}\beta_0^2\,\bar{G}_b\,f(\bar{\theta}, N_0), \tag{22}$$

where $f(\bar{\theta}, N_0)$ is a normalized electronic admittance.

Figures 7a, 7b, and 7c show the calculated normalized electronic conductance as a function of the dc transit angle of an electron between two adjacent grooves in the case of TEM_{00}, TEM_{01}, and TEM_{02} resonance of the Fabry–Perot mode (Fox and Li, 1961), respectively, because a Fabry–Perot resonator is a multimode system. The field distribution patterns were

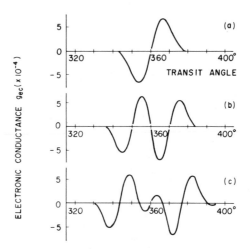

FIG. 7. Calculated normalized electronic conductance vs. dc transit angle: (a), (b), and (c) are for TEM_{00}, TEM_{01}, and TEM_{02} modes, respectively. The field distribution patterns were assumed to be sinusoidal. (Total number of grooves $N_0 = 30$; $D = 990$ μm, $d = 316$ μm, and $t = 990$ μm.)

assumed to be sinusoidal along the grating. It is worth noting that in the TEM_{02} interaction, the transit angle giving negative conductance is larger than that in the others, resulting in a reduction of electron accelerating voltage.

The maximum negative conductance is obtained at

$$\bar{\theta} = \bar{\theta}_{0p} \simeq 2m\pi - \pi/N_0 \qquad (TEM_{00} \text{ resonance})$$

and

$$\bar{\theta} = \bar{\theta}_{0p} \simeq 2m\pi - 2\pi/N_0 \qquad (TEM_{01} \text{ resonance})$$

where $m(1, 2, 3, \ldots)$ is the order of the space harmonic wave with which the electron beam interacts.

2. Load Conductance

In order to obtain the load conductance of this interaction, we begin by calculating the power loss in the planar Fabry–Perot resonator exclusive of the effects of grooves and output coupler. The electromagnetic fields of the TEM_{00q} mode in the resonator are expressed as

$$E = -jE_0 \sin(kz)F(x, y),$$
$$H = E_0\sqrt{\varepsilon_0/\mu_0} \cos(kz)F(x, y). \tag{23}$$

where $F(x, y)$ is the normalized field distribution function. The power dissipation due to ohmic loss is calculated from

$$P_r = \frac{1}{2\delta\sigma} \iint_{\substack{\text{mirror} \\ \text{surface}}} |H|^2_{z=0} \, dx \, dy, \tag{24}$$

where δ is the skin depth for the operating frequency and the cavity material, and σ the conductivity of the metal. On the other hand, the power loss due to diffraction per transit of an electromagnetic wave was obtained by Fox and Li (1961) as

$$\delta_d = 0.189 N_F^{-1.43} \tag{25}$$

in the case of strip mirrors and a TEM_{00} mode, where N_F is the Fresnel number for the mirror configuration. A spherical mirror configuration is desirable for reduced diffraction loss. The power loss in reflection is given by

$$\delta_r = 2\delta(\omega/c). \tag{26}$$

Then the total cavity loss, including ohmic and diffraction losses, is

$$P_t = P_r(1 + \delta_d/\delta_r). \tag{27}$$

The 35 GHz model experiment (Mizuno et al., 1970) has been performed to compare the Q-value of the TEM_{00q} mode resonance for the Fabry–Perot resonator with a grooved mirror to that for the resonator with flat mirrors. The experimental result pointed out that the power loss increased about seven times by cutting grooves on the mirror. As for TEM_{00q} mode resonance, this relation may hold in any frequency region for the resonator with the dimensions proportional to operating wavelengths. Therefore, the total cavity loss of the Fabry–Perot resonator with a grooved mirror is given by

$$P_{L,g} = 7P_t. \tag{28}$$

The load conductance is given by

$$G_0 = P_{L,g}/V_t^2, \tag{29}$$

where $V_t = \sum_i^{No} |V_{0i}|$.

Considering the effect of an output coupler, we have the total shunt conductance of the loaded cavity

$$G_t^{(l)} = (Q_0/Q_1)G_0, \tag{30}$$

where Q_0 and Q_1 are the unloaded and loaded Q values, respectively. If we suppose $Q_1 = \frac{1}{2}Q_0$ (the maximum output condition), we get

$$G_t^{(l)} = 2G_0. \tag{31}$$

3.   Starting Current

The preceding analyses show that the oscillation can occur because the total shunt conductance $G_t^{(e)} + G_t^{(l)}$ can be negative. The start-oscillation condition is expressed as follows:

$$G_t^{(e)} + G_t^{(l)} = 0. \tag{32}$$

With this condition the starting current is given by

$$(\bar{I}_b)_{start} = \bar{V}_b G_t^{(l)}/\tfrac{1}{2}\beta_0^2 \, f(\bar{\theta}, N_0). \tag{33}$$

The starting current of the Ledatron is relatively high in comparison with other submillimeter wave tubes such as Carcinotrons, mainly because of the small coupling impedance of the Ledatron.

Figure 8a shows experimentally obtained starting currents as a function of oscillation wavelengths for a 70-GHz band tube. The dimension of the grating surface used in this tube is 30 × 30 mm, and there are thirty grooves with a pitch of 990 μm. The width and depth of the groove are 316 and 990 μm, respectively. A convergent electron gun produced a ribbon-like electron beam (20 × 0.3 mm), and a magnetic field of about 4000 G was necessary for proper beam focusing. The experimental tube was operated under pulsed conditions (2 μs, 50 Hz). The experimental values of the starting current are

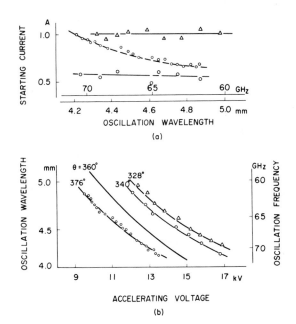

FIG. 8. (a) Starting current vs. oscillation frequency ($D = 990$ μm, $d = 316$ μm, and $t = 990$ μm). (b) Oscillation frequency vs. electron accelerating voltage. The oscillation due to the TEM_{02} interaction has been observed at the transit angle of $\sim 376°$. Legend: O, TEM_{00}; $\triangle$, TEM_{01}; $\circ$, TEM_{02}.

larger than the theoretical ones calculated from Eq. (33) by about a factor of 5. This discrepancy would be due to electron beam laminarity, current density distribution in the beam, effect of coupling hole for the output, etc.

4. Oscillation Mode

Oscillation has been observed in the transit angle at around 340, 328, and 376°, corresponding to the TEM_{00}, TEM_{01}, and TEM_{02} interaction, respectively (Fig. 8b). The TEM_{02} interaction was observed also at the transit angle of 325°. The experimental results can be explained by the theoretical results shown in Fig. 7, where several assumptions have been used to simplify the analysis. Of greater importance is the fact that an electron accelerating voltage can be reduced when the TEM_{02} interaction is specified. The reduction of the voltage is of great practical advantage, especially in the generation of submillimeter waves, where a high accelerating voltage is usually needed.

Figure 9 shows experimentally obtained field distributions in the resonator when the tube is in an oscillation state. In the figure, the field intensity is expressed as a function of the groove number along the direction of the

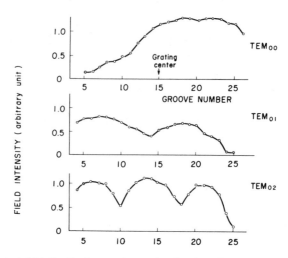

FIG. 9. Typical field distribution on the grating for the tube in an oscillating state. The center of the grating is at the fifteenth groove. The total number of grooves is thirty.

electron beam. The center of the grating is at the fifteenth groove. When a mode-selective configuration (Kawakami and Nishida, 1971) of the Fabry–Perot mirrors is used in the Ledatron, a single-mode interaction can be obtained. The field distributions shown in Fig. 9 were measured by moving a waveguide horn along the grating side (Fig. 1). The horn can be used as an output port of the Ledatron (Mizuno and Ono, 1973). The coupling coefficient of the resonator is varied by changing the position of the horn and the number of grooves which can be seen by the horn. This method greatly simplifies the tube configuration, especially when sealed-off tubes are manufactured, since the coupling system can be separated from a drift device which is used to vary the mirror spacing. The other method of coupling was through a hole in a diaphragm placed in the center of the smooth mirror of the Fabry–Perot resonator (Fig. 1). Almost the same output power has been obtained with both methods.

C. FREQUENCY TUNING CHARACTERISTICS

1. Mechanical and Electronic Tuning

An example of the mechanical and electronic tuning ranges of the Fabry–Perot mode in the 70-GHz Ledatron are shown in Figs. 10 and 11, respectively. In the tube a 30 × 30-mm grating and a concave smooth mirror with a radius of curvature of 70 mm are used as the Fabry–Perot mirrors. The mechanical and electronic tunings are done by changing the mirror spacing and the electron accelerating voltage, respectively. The oscillation occurs in

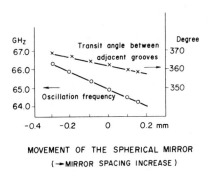

MOVEMENT OF THE SPHERICAL MIRROR
(→MIRROR SPACING INCREASE)

FIG. 10. Mechanical tuning range of the Fabry–Perot mode (accelerating voltage $V_0 = 11.6$ kV).

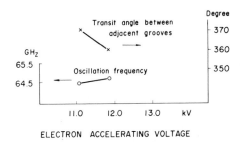

ELECTRON ACCELERATING VOLTAGE

FIG. 11. Electronic tuning of the Fabry–Perot mode at a constant mirror spacing.

the range from 358 to 369° of the transit angle between adjacent grooves. The mechanical tuning range of this tube is normally 4%, at fixed voltage. The electronic tuning range is some 0.4%, and the modulation sensitivity 0.35 MHz/volt. The frequency change as a function of the operating current— or frequency pushing—has not been observed in the Fabry–Perot mode. The output power as a function of the mechanically-tuned wavelength at a fixed voltage is shown in Fig. 12.

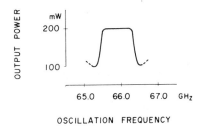

OSCILLATION FREQUENCY

FIG. 12. Typical characteristic of output power vs. oscillation wavelength for the Fabry–Perot mode (accelerating voltage $V_0 = 12.5$ kV; operational current $I_0 = 1.0$ A).

2. Frequency Tuning Range

Figure 13 shows a typical example of overall frequency tuning characteristics of the 70-GHz band Ledatron. In the Fabry–Perot mode, frequency tuning is obtained with both mechanical and electronic tunings. The overall tuning range of 42% has been obtained in this tube. It is worth noting that compared with the surface-wave mode described in Section IV, the Fabry–Perot mode needs lower beam accelerating voltages when operating at the same wavelength, because in the Fabry–Perot mode operation, the cutoff is specified. These two modes can be separated by proper selection of the mirror spacing of the Fabry–Perot resonator.

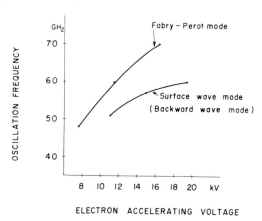

FIG. 13. Typical example of frequency tuning of a 70 GHz Ledatron. [Grating #11 (30 × 30 mm), $D = 990$ μm, $d = 316$ μm, and $t = 990$ μm.]

The wide frequency tuning range and small power variation over this range are considered excellent features of the Fabry–Perot mode, which make it especially desirable as a spectroscopic source and as a local oscillator for heterodyne detection.

IV. Surface-Wave Mode Operation

The surface-wave mode in the Ledatron is essentially the same operation as that in the backward wave oscillator, which was described in detail in Chapter 4. Therefore in this chapter, we will describe only briefly some operational characteristics of the surface-wave mode.

A. Frequency Tuning Characteristic

In Fig. 13, the tuning characteristic is shown for the tube with the same grating as that for the 70-GHz Fabry–Perot mode. The mirror spacing was adjusted so as not to give oscillations in the Fabry–Perot mode. The tuning characteristic agrees well with the theoretical value calculated from Eq. (14). The output power vs. electron accelerating voltage has a fine grain component with power fluctuations up to 10 dB, which are caused by multiple reflections from small discontinuities within the tube. The average output power of this tube is 100 mW. The beam current is 200 mA to 1.2 A. Output power was found to be almost a linear function of the beam current. The frequency pushing figure of 2.0–1.5 MHz/mA was observed in this tube.

B. START-OSCILLATION CONDITION

The starting currents of the surface-wave mode have been measured for tubes with various grating structures, and were found to agree with theoretical values. The typical characteristic of the starting current vs oscillation frequency is shown in Fig. 14. The dimensions of the grating groove in the tube used are 664 μm in the pitch, 318 μm in the width, and 690 μm in the depth. The theoretically obtained curve, according to the following equation (Heffner, 1954, Johnson, 1955), taking no account of space charge and losses, is also shown.

$$(I_0)_{start} = 32\pi^3 (0.314)^3 V_0 / K(\beta_e L)^3 \tag{34}$$

where V_0 is the dc beam voltage, K the coupling impedance for the particular space harmonic, β_e the propagation constant of electrons, and L the total active length of the circuit. The coupling impedance was calculated in the same way as for the backward wave oscillator with the vane structure circuit.

It has been observed that the mirror spacing changed the starting current by 14%, but did not affect the oscillation frequency by more than 0.1% when the spacing was larger than one oscillation wavelength.

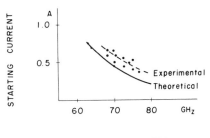

FIG. 14. Starting current vs. oscillation wavelength for the surface-wave mode. [Grating #12 (30 × 30 mm), $D = 664$ μm, $d = 318$ μm, and $t = 690$ μm.]

V.   Summary and Conclusion

The Ledatron, an electron tube with a Fabry–Perot resonator and a grating, has been discussed. Two different modes of operation, the Fabry–Perot mode and the surface-wave mode, have been described. The Fabry–Perot mode needs a lower electron accelerating voltage than the surface-wave mode to obtain the same wavelength of oscillation. The small variation of the power-vs-frequency characteristic in the Fabry–Perot mode is also an excellent feature of the tube. Using both modes, a wide tuning range of about 1 octave is obtained with the tube. The output power is 10 mW to 1 W in the region of short millimeter and long submillimeter waves.

The Ledatron is still in an early stage of development, and further investigations should be focused on the increase in coupling impedance of the grating between the electron beam and the field in order to reduce its operational current. A study of an electron gun producing a ribbonlike beam with high electron density would also be needed.

The position of the Ledatron among submillimeter sources, however, is not definite at present. Solid state sources, especially IMPATT diodes, seem to be promising tunable sources of long submillimeter waves with reasonable power. High-power submillimeter waves have been produced by the gyrotron with high efficiency. Production of 1.5 kW of cw power at $\lambda = 900 \ \mu m$ has been demonstrated. Relativistic electron beam devices are still in the stage of basic research, but 1 MW of peak power was measured at $\lambda = 400 \ \mu m$ from a 50-ns pulsed intense beam.

Probably the Ledatron will find its position as a submillimeter tunable source with more than 50 mW of cw output power and wide tuning range of 30 %.

ACKNOWLEDGMENTS

The material covered in this chapter is the result of cooperative studies with many colleagues of the vacuum electronics group at Tohoku University, They include Messrs. N. Taguchi, N. Kuji, H. Yonezawa, M. Ohuchi, and K. Sagae. At the early stage of investigation, this group was led by Professor Y. Shibata. Many other people assisted these members during the course of studies. The authors would like to express their sincere appreciation for their cooperation and contributions. Special thanks are also due to Professor D. H. Martin of Queen Mary College (London) for his interest and encouragement, and to Professor H. L. Hartnagel of University of Newcastle Upon Tyne for his comments on the manuscript.

REFERENCES

Bogomolov, G. D., Rusin, F. S., and Kushch, V. S. (1970). *Radio Eng. Electron. Phys.* **15**, 730–731.
Flyagin, V. A., Gaponov, A. V., Petelin, M. I., and Yulpatov, V. K. (1977). *IEEE Trans. Microwave Theory Tech.* **MTT-25**, 514–521.

Fox, A. G., and Li, T. (1961). *Bell Syst. Tech. J.* **40**, 453–488.

Fujisawa, K. (1964). *IEEE Trans. Electron Devices* **ED-11**, 381–391.

Granatstein, V. L., and Sprangle, P. (1977). *IEEE Trans. Microwave Theory Tech.* **MTT-25**, 545–550.

Hechtel, R. (1958). *Telefunken Röhre* **35**, 5.

Heffner, H. (1954). *Proc. IRE* **42**, 930–937.

Hibben, S. G. (1969). *Microwave J.* **12**, No. 11, 59–65.

Johnson, H. R. (1955). *Proc. IRE* **43**, 684–697.

Kawakami, S., and Nishida, S. (1971). *IEEE Trans. Microwave Theory Tech.* **MTT-19**, 403–406.

Martin, D. H., and Mizuno, K. (1976). *Adv. Phys.* **25**, 211–246.

Mizuno, K., and Ono, S. (1973). *Proc. Eur. Microwave Conf., Brussels* C13.5.

Mizuno, K., and Ono, S. (1975). *J. Appl. Phys.* **46**, 1849.

Mizuno, K., Ohuchi, M., Ono, S., Shibata, Y., and Kamiryo, K. (1969). *Rec. Elec. Commun. Eng. Conversat. (Tohoku Univ.)* **37**, 24–29.

Mizuno, K., Ono, S., and Shibata, Y. (1970). *Proc. Symp. Submillimeter Waves* pp. 115–134. Polytechnic Press, New York.

Mizuno, K., Ono, S., and Shibata, Y. (1973). *IEEE Trans. Electron Devices* **ED-20**, 749–752.

Mizuno, K., Kuji, N., and Ono, S. (1974). *Proc. Eur. Microwave Conf., Brussels* C13.5.

Mizuno, K., Ohmori, M., Miyazawa, K., Morimoto, M., Kodaira, S., and Ono, S. (1978). *Infrared Phys.* **18**, 401–403.

Ohmori, M., Ishibashi, T., and Ono, S. (1977). *IEEE Trans. Electron Devices* **ED-24**, 1323–1329.

Rusin, F. S., and Bogomolov, G. D. (1969). *Proc. IEEE* **57**, 720–722.

Senitzky, I. R. (1954). *J. Appl. Phys.* **46**, 904–911.

Smith, S. J., and Purcell, E. M. (1953). *Phys. Rev.* **92**, 1069.

Wessel-Berg, T. (1957). Rep. No. 376, Stanford Microwave Lab., Palo Alto, California.

CHAPTER 6

Infrared and Submillimeter-Wave Waveguides

F. K. Kneubühl   and   E. Affolter

I. Introduction

The waveguides actually used in infrared and submillimeter-wave laser systems originate in the waveguiding of microwaves and in the rather crude channeling of light and infrared in metal light pipes. Laser resonators with oversized dielectric or metallic waveguides were first suggested by Marcatili and Schmeltzer (1964). In recent years waveguide resonators have earned full recognition by laser scientists and engineers. This review gives an account

235

of the development, properties and application of waveguides feasible for infrared and submillimeter-wave lasers, with special emphasis on hollow guides.

A. MICROWAVE WAVEGUIDES

Metal waveguides are characteristic for microwave techniques (e.g., Borgnis and Papas, 1958; Ramo et al., 1965). Best known are the hollow rectangular guide, hollow circular guide, and parallel plane guide. The waves or modes propagating in these guides are usually classified into the following types:

1. Transverse electromagnetic waves (TEM) that contain neither electric nor magnetic field in the direction of propagation. They are the usual transmission line waves along a multiconductor guide. Closed hollow metal guides do not support TEM waves.
2. Transverse magnetic waves (TM or E) that contain an electric field but no magnetic field in the direction of propagation.
3. Transverse electric waves (TE or H) that contain a magnetic field but no electric field in the direction of propagation.

These definitions hold for a lossless dielectric (e.g., vacuum) in the guide and for walls of infinite electrical conductivity. For low-loss dielectrics (e.g., air) in the guide and walls of a real metal (e.g., copper) the waves also have exceedingly small components of the forbidden field in the direction of propagation.

The metal waveguides used for microwaves are capable of guiding submillimeter or infrared waves if they possess modes that show decreasing attenuation with increasing frequency. This attenuation is caused by the loss of the dielectric in the guide and by the surface resistivity R_S of the metal walls. Since we may assume that the guide is under vacuum or filled with a laser-active gas, we have to consider the effects of the surface resistivity on the propagation properties of the guides. The surface resistivity R_S is determined by the normal skin effect. Therefore, it obeys the relation $R_S = (\sigma\delta)^{-1}$, where σ is the electrical conductivity and δ the penetration depth. In the microwave and infrared region of the electromagnetic spectrum the frequency and wavelength dependence of the surface resistivity of the common metals at room temperature is approximately

$$R_S = (\omega\mu\mu_0/2\sigma)^{1/2} = K\lambda^{-1/2} \tag{1}$$

where μ indicates the magnetic permeability of the metal. Thus, R_S increases with decreasing vacuum wavelength λ. In Table I we show $R_S(\lambda)$ for metals used as guide walls.

TABLE I

SURFACE RESISTIVITY R_S AS APPROXIMATE
FUNCTION OF THE VACUUM WAVELENGTH λ
(in μm)

Metal	R_S (ohm)
Silver	$4.36\lambda^{-1/2}$
Copper	$4.52\lambda^{-1/2}$
Aluminum	$5.65\lambda^{-1/2}$
Brass (example)	$8.7\lambda^{-1/2}$
Solder (example)	$13.4\lambda^{-1/2}$

In spite of the fact that the surface resistivity of the metal walls increases with increasing frequency, there are modes of certain microwave guides with decreasing attenuation. Unfortunately, there are no modes of the hollow rectangular guide which show this effect.

For the parallel-plane guide (shown in Fig. 8) the attenuation α of the TE_{n0} waves is

$$\alpha(TE_{n0}) = (R_S/2Zb)(\lambda/\lambda_c)^2[1 - (\lambda/\lambda_c)^2]^{-1/2} \tag{2}$$

where Z is the wave impedance of the lossless dielectric, b the separation of the planes, and $\lambda_c = 2b/n$ the cutoff wavelength. This expression demonstrates the continuous decrease of the attenuation with decreasing wavelength.

The attenuation α of the TE_{0m} modes of the hollow circular guide shows a similar behavior. It obeys the equation

$$\alpha(TE_{0m}) = (R_S/Zr_0)(\lambda/\lambda_c)^2[1 - (\lambda/\lambda_c)^2]^{-1/2} \tag{3}$$

where r_0 represents the inner radius of the guide and $\lambda_c = 2\pi r_0/u_{0m}$ the cutoff wavelength. u_{0m} is the mth zero of the Bessel function $J_1(z)$. Figure 7 shows the field distribution of the TE_{01} mode in a hollow circular metal guide. The TE_{0m} modes yield a decreasing attenuation for a circular metal tube of fixed diameter as the frequency is increased. Because of this property, these waves have received a good deal of attention for possible long-distance propagation of millimeter waves. Attenuations as low as 2 dB/mile were attained (Miller, 1954). The major problem in the use of the TE_{0m} modes arises because they are not the modes of lowest cutoff frequency. In addition, the deformation of the circular cross section of a metal guide to an ellipse changes its mode structure and the unique frequency characteristics of the losses of the TE_{0m} modes. Unfortunately, there exists no mode in a metal guide with elliptic cross section whose loss decreases with increasing frequency (Chu, 1938).

If the TE_{01} mode is above cutoff, there will be at least the four modes: TE_{11}, TM_{01}, TE_{21}, and TM_{11} propagating. This multimode character has two serious disadvantages. First, it requires a method of exciting the desired mode with sufficient purity. Second, one must guard against coupling from the desired TE_{0m} mode to undesired modes by irregularities in the guide shape or direction. The undesired modes cause higher losses as well as distortion of the signal by different phase velocities. For these reasons special mode filters have been devised.

From these considerations we conclude that, in principle, it is possible to extend the microwave techniques of metallic waveguides into the submilli-meter-wave and infrared region of the spectrum. However, restrictions with respect to the type of modes and difficulties with mode excitation and un-desired modes demonstrate that this does not give the proper solution of the problem.

The use of a hollow dielectric tube as a microwave guide was proposed by Unger (1954) on the basis of a detailed calculation of its modes. He found that for a sufficiently large dielectric constant ε_a of the wall material, the wave in the dielectric tube with optimum microwave propagation characteristics is the so-called HE_{11} mode. The comparison with the TE_{01} mode of the hollow circular metal guide shows that for the same attenuation characteristics, the HE_{11} mode of the dielectric tube requires a considerably smaller diameter in spite of the large diameter of its electromagnetic field. In addition, the HE_{11} mode of the dielectric tube causes less signal distortions than the TE_{01} mode of the metal tube.

The hollow dielectric guide has found very restricted applications in micro-wave techniques. Furthermore, attempts by Steffen and Kneubühl (1967) to apply the above theory to submillimeter-wave waveguides for laser reson-ators were not successful. For this problem they had to rely upon their own calculations (Steffen and Kneubühl, 1968a) and, later, on the basic theory by Marcatili and Schmeltzer (1964).

B. Metal Light Pipes

The first waveguides to channel infrared radiation were metal light pipes, which have been used as a simple means to transmit radiation from an exit slit of a spectrometer or an interferometer to a solid sample and a detector in a cryostat, under certain circumstances surrounded by an electromagnet. In this and other applications, the use of light pipes instead of mirror and lens optics results in increased economy and simplicity of the apparatus.

In some cases the light pipe allows experiments to be performed where conventional optics are not convenient on account of aperture restrictions. For submillimeter waves, where energy is at a premium, Ohlmann et al. (1958) calculated the fractional transmission of radiation through metal light pipes and compared it with experimental data. From a consideration of the reflection and transmission of nonskew infrared rays, they derived the equation

$$T = \tfrac{1}{2}e^{-2q} + (1 - e^{-q/2F^2})F^2/q \simeq \tfrac{1}{2}(1 + e^{-2q}) - q/8F^2, \qquad (4)$$

where

$$q = (2\varepsilon_0 \omega\rho)^{1/2}L/d = 0.18(\rho/\lambda)^{1/2}L/d$$

for the fractional energy transmission T of the metal light pipe. T is written in terms of the f/number of the incoming radiation F, the resistivity of the metal ρ, the length of the pipe L, its diameter d, the angular frequency ω, and the wavelength λ. The first term of the equation corresponds to the radiation with its polarization parallel to the plane of incidence on the metal wall, the second term to the radiation with perpendicular polarization. The former is much more rapidly attenuated than the latter and its attenuation is independent of the f/number in this approximation.

Figure 1 shows the transmission of brass, copper, aluminum, and silvered glass light pipes as observed by Ohlmann *et al.* (1958). For comparison, the calculated transmission for a resistivity of $\rho = 18$ μohm cm and for wavelengths $\lambda = 70$ and 140 μm is indicated. This figure demonstrates that whenever $(L/d)(\rho/\lambda)^{1/2}$ can be made small enough, the transmission is sufficient for the substitution of conventional optics by metal light pipes.

Metal light pipes can also direct radiation around corners. However, if a pipe is merely bent, only about 50% of the radiation is transmitted around the bend, even for a large radius. Ohlmann *et al.* (1958) found that a right angle turn may be manufactured that transmits 95% of the radiation by joining two pipes perpedicularly, and placing a polished plane metal mirror across the joint so that radiation from one pipe is reflected down the other.

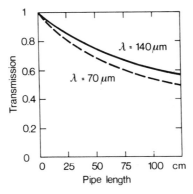

FIG. 1. Transmission T of a metal pipe with bulk resistivity $\rho = 18$ μohm cm as a function of the pipe length L, for input radiation f/1.5 and vacuum wavelengths 70 and 140 μm [according to Ohlmann *et al.* (1958)].

C. WAVEGUIDE LASERS

In 1964 Marcatili and Schmeltzer determined the field configurations and the propagation characteristics of the modes in hollow circular waveguides made of dielectric material or metal. At the end of their paper they suggested the use of the hollow dielectric waveguides both to guide the laser radiation and to confine the gas discharge. They proposed the construction of waveguide lasers working at wavelengths between 0.4 and 4 μm.

The first experimental evidence of waveguiding inside a gas laser resonator was reported three years later, when Schwaller et al. (1967) observed that the resonator mode spectrum and losses of the low Fresnel number submillimeter-wave laser working with HCN and ICN at wavelengths between 311 and 774 μm were in essential disagreement with the results of the conventional resonator theory by Bergstein and Schachter (1964). This observation was subsequently interpreted by Steffen and Kneubühl (1968a,b) to mean that the waveguiding properties of the glass laser tube dominated over those of the resonator mirrors. Therefore, they calculated the mode spectrum and mode losses of a hollow dielectric tube terminated at each end by a perfectly conducting flat. By laser resonator interferometry they found good agreement of this theory with experiment. This method relies on the fact that doppler and pressure broadening of the low-pressure HCN and ICN emissions are smaller than the mode separation of a submillimeter-wave laser resonator.

The laser resonators considered by Steffen and Kneubühl (1968a,b) were characterized by low Fresnel numbers

$$N = a^2/\lambda L \leq 1, \tag{5}$$

where λ is the wavelength, a the radius, and L the length of the resonator. Standard resonators of optical lasers have Fresnel numbers of order of magnitude 50.

According to conventional diffraction theory the radiation loss per pass of an optical resonator is considerable for low Fresnel numbers. The only means to reduce the large diffraction losses of resonators with low Fresnel numbers are waveguiding effects.

While the properties of the standard optical high Fresnel-number resonators are derived from scalar diffraction theory, an understanding of waveguiding in laser resonators requires approximate or precise solutions of Maxwell's equations. For the dielectric tube this was done by Marcatili and Schmeltzer (1964) as well as by Steffen and Kneubühl (1968a,b). They found in agreement that for a reasonable dielectric tube resonator the radius to wavelength ratio should approximately fulfill the condition

$$a/\lambda \simeq 50. \tag{6}$$

Taking into account that standard optical resonators fail to work for $N \lesssim 1$ according to Eq. (5), waveguide resonators are feasible for a length to radius ratio $L/a \gtrsim 50$.

Waveguide lasers differ from conventional lasers by the mark that the radiation in the resonator is guided and does not obey the rules of free space propagation. Therefore, spatial mode distributions, frequency spectra, losses, and stability conditions of waveguide lasers cannot be derived from conventional resonator theories (e.g., Bergstein and Schachter, 1964; Kogelnik and Li, 1966). On the basis of the above definition, the concept waveguide laser can be used for a large variety of devices that work with gaseous, liquid, and solid media.

Explicit reference to the initial considerations by Marcatili and Sch-meltzer (1964) was made in literature, when Smith (1971) reported on the first successful operation of a He–Ne optical waveguide gas laser. This achievement gave a remarkable impetus to the rapid development of a large variety of waveguide gas lasers in the infrared and in the submillimeter-wave spectral region. These lasers will be discussed in Section V. A general survey on waveguide lasers was pbulished by Degnan (1976), whereas the review by Yamanaka (1977) is restricted to optically pumped submillimeter-wave lasers. The relevant waveguide structures used in various types of sub-millimeter-wave gas lasers were recently discussed by Kneubühl (1977).

Laser-pumped oscillation in narrow capillary liquid dye lasers has been reported by Zeidler (1971) and by Wang (1974a). Dye laser materials have also been impregnated in gelatinous films and thin film solutions in order to test distributed feedback structures (Chang et al., 1974).

The guiding media in these structures had refractive indices smaller than that of the active region which allowed true guided wave propagation. Burlamacchi and co-workers (1973, 1973a,b, 1974a,b) have operated a number of flashlamp-pumped lasers based on waveguide propagation in liquid solutions of Rhodamine 6G dye. Gains sufficient for superradiance have been achieved in pyrex capillaries with a length of only a few centi-meters and a bore diameter as small as 60 μm (Wang, 1974a).

Solid-state waveguide lasers, such as heterostructure junction lasers and optically pumped thin film lasers, play an important role in integrated optics. They are the subject of recent reviews by Kogelnik (1975), Chang et al. (1974), Taylor and Yariv (1974), Panish (1975), and a book edited by Tamir (1975).

II. Leaky Modes in Hollow Waveguides

A closed hollow metal waveguide with perfectly conducting walls and filled with a lossless medium represents a closed region in space where electromagnetic energy can neither leave nor enter. This ideal waveguide can guide electromagnetic energy in the form of characteristic modes correspond-ing to discrete pure positive imaginary or pure positive real values of the propagation constant γ. These modes form a complete orthonormal set. Therefore, any physically realizable field in this waveguide can, at any particular frequency, be resolved into a discrete spectrum of these modes (Marcuwitz, 1949; Ramo et al., 1965).

For an imperfect waveguide, the losses in the medium filling the guide cause only a change in the propagation coefficients and the other param-eters of the modes, but otherwise do not affect the nature of the solution for the ideal waveguide. However, if the walls of the waveguide are imperfect, energy can leave the system, which then becomes an open system. In open systems difficulties arise because in such cases the spectrum of characteristic modes must then be supplemented by a properly selected continuous spec-trum (Marcuwitz, 1956; Shevchenko, 1971; Marcuse, 1974). For imperfect waveguides with lossy walls this difficulty can be circumvented by introducing the concept of surface impedance (Karbowiak, 1955; Ramo et al., 1965; Kurokawa, 1969).

Here, the walls of the waveguide structure are assumed to represent a finite impedance to the waves, in order to simulate the energy leaving the guide. Under this assumption and provided the surface impedance is small, a new set of slightly different modes and characteristic propagation constants can be derived. This artifice works in a large group of problems, but it does not help in the correct formulation of the physical situation. Moreover, it is of no avail with waves which are guided in intrinsically open structures such as hollow dielectric or surface waveguides.

Hollow dielectric (Marcatili and Schmeltzer, 1964; Steffen and Kneubühl, 1968a) and (see Fig. 9) metal–dielectric (Adam and Kneubühl, 1975) wave-guides in infrared and submillimeter-wave lasers represent open structures as well as the Fabry–Perot resonators (Weinstein, 1964, 1969) used in optical lasers. Due to the refraction of the guided waves at the dielectric walls, these waveguides are continuously losing power by radiation. Open structures possess a continuous spectrum which is interpreted as a radiation field with its origin at the source of the field. The rigorous solution gives the radiation field with a continuous spectrum, and no indication of the guided waves with a discrete spectrum, which contradicts experimental evidence. This paradox can be resolved by the introduction of the concept of leaky modes (Marcuwitz, 1956) or quasimodes (Karbowiak, 1964).

For lossless dielectric walls, a characteristic of the quasimodes in hollow dielectric (Marcatili and Schmeltzer, 1964; Steffen and Kneubühl, 1968a; Degnan, 1976) and metal–dielectric (Adam and Kneubühl, 1975) wave-guides is the exponential increase of the electromagnetic field in transverse direction away from the core (Karbowiak, 1964; Marcuse, 1974; Adam and

Kneubühl, 1975). The losses in real dielectric walls play an important part as far as the nature of the guided waves is concerned (Adam and Kneubühl, 1975). When the dissipation in the dielectric walls just balances the power leaking out of the guide, the transverse exponential increase of the field will vanish. For losses beyond that critical value, the fields of the guided waves decrease exponentially in the transverse direction, and a description in terms of proper modes once again becomes possible. The resolution of the field into quasimodes, however, remains valid (Karbowiak, 1964). The influence of the losses in the dielectric walls on the nature of the guided waves in a rectangular metal–dielectric waveguide was studied in some detail by Adam and Kneubühl (1975). They found the following conditions for an exponential decrease of the fields in the transverse direction:

$$\varepsilon'' > \begin{cases} (\lambda/2\pi a)[\varepsilon'^2/(\varepsilon' - 1)^{1/2}] & \text{for the LSM modes,} \\ (\lambda/\pi a)(\varepsilon' - 1)^{1/2} & \text{for the LSE modes,} \end{cases} \qquad (7)$$

where the loss in the dielectric wall is represented by the imaginary part ε'' of the relative dielectric constant $\varepsilon = \varepsilon' + i\varepsilon''$, and $2a$ is the distance between the dielectric walls. For a diameter to wavelength ratio $a/\lambda \geq 100$, these conditions are fulfilled by all dielectrics in the submillimeter-wave region.

III. Hollow Waveguides with Constant Cross Section

Most hollow waveguide resonators incorporated in electrically and optically excited waveguide lasers (Degnan, 1976; Yamanaka, 1977; Kneubühl, 1977) exhibit a constant cross section along the laser axis. The calculation of the fields and the losses of the modes in these waveguides is simple compared with the theory of waveguide structures with cross sections which vary in the longitudinal direction. An important class of guides with varying cross section is the entirety of periodic structures used in distributed feedback (DFB) lasers.

The losses δ of modes in hollow waveguides with constant cross section obey scaling laws (Kneubühl, 1977) which permit the comparison of losses δ per length L of modes in waveguides of identical structure but with different diameters $2a$ and operated at different wavelengths λ. Losses of modes of hollow dielectric and hollow ideal conductor–dielectric waveguides are described by the equation

$$\delta \simeq \lambda^2 a^{-3} L F_1[\varepsilon_a(\lambda), \varepsilon_i(\lambda), \text{mode}] \qquad (8)$$

where $\varepsilon_a(\lambda)$ and $\varepsilon_i(\lambda)$ indicate the relative dielectric constants of the dielectric wall and hollow core. For gas lasers $\varepsilon_i(\lambda) = 1$ in the first approximation. The effect of the temporal variation of $\varepsilon_i(\lambda)$ of the plasma in a pulsed HCN gas laser on the resonance conditions and the pulse shapes of the modes was studied experimentally by Steffen and Kneubühl (1968b).

For the modes in hollow circular metal waveguides at frequencies far from cutoff, the scaling laws for losses caused by the walls are

$$\delta \simeq \lambda^{3/2} a^{-3} L F_2 (\text{metal, mode}) \qquad \text{for } TE_{0m} \text{ modes}$$
$$\delta \simeq \lambda^{-1/2} a^{-1} L F_3 (\text{metal, mode}) \qquad \text{for } TM_{nm} \text{ modes} \qquad (9)$$

in accordance with Eq. (3) in Section I.A. The attenuation of the $TE_{n \neq 0, m}$ and TM_{nm} modes increases with frequency, whereas the losses of the TE_{0m} modes decrease. Therefore, hollow metal tubes with circular cross sections can be operated at submillimeter and infrared wavelengths only in the TE_{0m} modes.

The wall loss of modes in metal waveguides can be calculated on the basis of the surface impedance approximation discussed in Section II. Under these circumstances the wall loss is essentially determined by the quotient R_S/Z, where R_S and Z indicate the surface resistance of the metal wall and wave impedance of the core. For simplicity we consider a waveguide core with $Z = Z_0 = (\mu_0/\varepsilon_0)^{1/2}$ and $\varepsilon = \mu = 1$.

At microwave and submillimeter-wave frequencies the dielectric constant ε and refractive index n of a metal are described by the Hagen–Rubens relation (Mott and Jones, 1936)

$$\varepsilon \simeq i\sigma/\omega\varepsilon_0, \qquad n \simeq (1 + i)(\sigma/2\omega\varepsilon_0)^{1/2}. \qquad (10)$$

On the basis of this approximation we can derive Eq. (1) for the surface resistance R_S and the ratio

$$R_S/Z_0 = (\varepsilon_0 \omega/2\sigma)^{1/2} = (\pi/Z_0 \sigma\lambda)^{1/2}. \qquad (11)$$

In the infrared, the Hagen–Rubens relation for metals is not valid. However, we still may assume that ε fulfills the condition

$$\varepsilon = n^2 = \varepsilon' + i\varepsilon'' \qquad \text{with} \qquad \varepsilon'' \gg 1. \qquad (12)$$

As examples we mention the refractive indices at the vacuum wavelength $\lambda = 10.6 \ \mu m$ for freshly evaporated aluminum, $n = 20.5 + i58.6$ (Beattie, 1955), and for freshly evaporated copper, $n = 14.2 + i64.5$ (Lenham and Treherne, 1966). In this case the surface impedance concept leads to the relation

$$R_S/Z_0 = \text{Re}(\varepsilon^{-1/2}) = \text{Re}(n^{-1}). \qquad (13)$$

The waveguiding by different walls in various guides is illustrated in Fig. 2. Here, the losses in decibels of the lowest modes in a number of structures with identical length $L = 1$ m are shown for the wavelength $\lambda = 337 \ \mu m$ as a

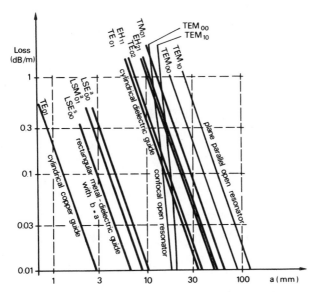

FIG. 2.   Loss δ per unit length $L = 1$ m for modes of typical waveguides and resonators as a function of the characteristic dimension or radius a in the range 1–100 mm for electromagnetic waves with the vacuum wavelength $\lambda = 337$ μm [according to Kneubühl (1977)].

function of the radius a. The guiding effect obviously increases in the sequence: resonators without walls, dielectric walls, walls composed of a metal and a dielectric, metallic walls. This can be easily checked in Fig. 2 by looking at the characteristic radius a of the various structures for a fixed mode loss $\delta = 1$ dB per unit length $L = 1$ m.

In the following sections we shall consider the modes of the most important waveguide structures with constant cross section in more detail.

A.   DIELECTRIC HOLLOW CIRCULAR GUIDES

The modes of dielectric hollow circular waveguides with diameters $2r_0$ considerably larger than the vacuum wavelength λ were first derived by Marcatili and Schmeltzer (1964), later by Steffen and Kneubühl (1968a) and by Degnan (1973). Steffen and Kneubühl used the designation "tube modes," whereas Degnan rederived the field components of the modes in the earlier field notation of Snitzer (1961), which is more common among scientists working on optical dielectric waveguides. The designation by Marcatili and Schmeltzer (1964) is now established in the laser literature (Degnan, 1976).

The modes in the dielectric hollow circular guide can be split into three categories: transverse electric TE_{0m}, transverse magnetic TM_{0m}, and hybrid

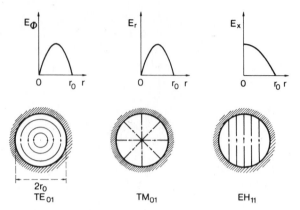

FIG. 3.   Characteristic modes of the dielectric hollow circular waveguide.

EH_{nm} modes. Figure 3 shows the most important mode of each category: the TE_{01}, TM_{01}, and EH_{11} modes. In a first approximation the components of the electric fields are represented by Bessel functions and the characteristic parameters u_{nm}:

(i)   for TE_{0m}: $m \geq 1$, $J_1(u_{0m}) = 0$:

$$E_\phi = E_0 J_1(u_{0m}r/r_0)e^{i(\gamma_{0m}z - \omega t)},$$
$$E_r = 0,$$
$$E_z = 0. \tag{14a}$$

(ii)   for TM_{0m}: $m \geq 1$, $J_1(u_{0m}) = 0$:

$$E_\phi = 0,$$
$$E_r = E_0 J_1(u_{0m}r/r_0)e^{i(\gamma_{0m}z - \omega t)},$$
$$E_z = O(\lambda/r_0). \tag{14b}$$

(iii)   for EH_{nm}: $n \neq 0$, $m \geq 1$, $J_{n-1}(u_{nm}) = 0$:

$$E_\phi = E_0 J_{n-1}(u_{nm}r/r_0) \cos(n\phi)e^{i(\gamma_{nm}z - \omega t)},$$
$$E_r = E_0 J_{n-1}(u_{nm}r/r_0) \sin(n\phi)e^{i(\gamma_{nm}z - \omega t)},$$
$$E_z = 0 \, (\lambda/r_0). \tag{14c}$$

The propagation constants $\gamma_{nm} = \beta_{nm} + i\alpha_{nm}$ are given by the expression

$$(\lambda/2\pi)\gamma_{nm} = 1 - \tfrac{1}{2}(u_{nm}\lambda/2\pi r_0)^2\{1 - i(\lambda/\pi r_0)\varepsilon_M[\varepsilon_i(\varepsilon_a - \varepsilon_i)]^{-1/2}\} \tag{15}$$

where

$$\varepsilon_M = \begin{cases} \varepsilon_i & \text{for } TE_{0m} \text{ modes} \\ \varepsilon_a & \text{for } TM_{0m} \text{ modes} \\ \tfrac{1}{2}(\varepsilon_a + \varepsilon_i) & \text{for } EH_{nm} \text{ modes.} \end{cases}$$

$\lambda_{nm} = 2\pi/\beta_{nm}$ represents the guide wavelength and α_{nm} the attenuation. For real ε_i and ε_a, the mode with the lowest attenuation is TE_{01} for $\varepsilon_a > 4.08\varepsilon_i$ and EH_{11} for $\varepsilon_a < 4.08\varepsilon_i$. The significance of the imaginary part of ε_a for the nature of the modes was discussed in Section II. In the above approximation the propagation constants of the hybrid $EH_{n+2,m}$ and $EH_{-|n|,m}$ modes are identical for a fixed value of $n \geq 2$. Hence, these modes are degenerate (Marcatili and Schmeltzer, 1964; Degnan, 1973, 1976). The increased attenuation of the modes in bent dielectric hollow cylindrical guides has also been investigated by Marcatili and Schmeltzer (1964).

B. DIELECTRIC HOLLOW RECTANGULAR GUIDES

The propagation of an electromagnetic wave in a dielectric rod with rectangular cross section surrounded by dielectrics of smaller indices of refraction was first investigated by Marcatili (1969). Later, Krammer (1976) modified this calculation for the hollow rectangular dielectric guide shown in Fig. 4.

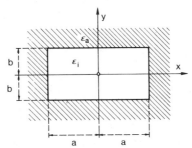

FIG. 4.   Geometry and parameters of a dielectric hollow rectangular waveguide.

The hybrid modes of this rectangular guide are denoted by $E^y H^x_{mn}$ and $E^x H^y_{mn}$, where the superscripts indicate the major polarization of the transverse components of the electric and magnetic fields. We may restrict our considerations to the $E^y H^x_{mn}$ modes since the corresponding equations for the $E^x H^y_{mn}$ modes can be derived by rotating the coordinate system by $\pi/2$ and interchanging the transverse dimensions a and b. For the square guide with $a = b$, the $E^y H^x_{mn}$ mode is degenerate with the $E^x H^y_{mn}$ mode. In a first approximation, for $\lambda \ll a,b$, the hybrid modes approach TEMlike solutions. The electric field components are then given by

$$E_y = E_0 \sin[\pi m(a + x)/2a] \sin[\pi n(b + y)/2b]e^{i(\gamma_{mn}z - \omega t)}$$
$$E_x = 0, \qquad E_z = O(\lambda/b). \tag{16}$$

Figure 5 presents the electric field distribution of the $E^y H^x_{11}$ mode. The complex propagation constants γ_{mn} of the $E^y H^x_{mn}$ modes are given by the

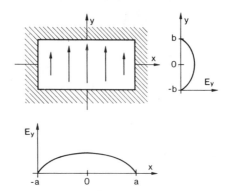

FIG. 5. $E^y H^x_{11}$ mode of a dielectric hollow rectangular waveguide.

following equation:

$$(\lambda/2\pi)\gamma_{mn} = 1 - \tfrac{1}{2}(m\lambda/4a)^2\{1 - i(\lambda/\pi a)\varepsilon_i[(\varepsilon_a - \varepsilon_i)\varepsilon_i]^{-1/2}\}$$
$$- \tfrac{1}{2}(n\lambda/4b)^2\{1 - i(\lambda/\pi b)\varepsilon_a[(\varepsilon_a - \varepsilon_i)\varepsilon_i]^{-1/2}\}. \qquad (17)$$

For $a > b$ and real ε_i and ε_a, the $E^y H^x_{11}$ mode exhibits the least attenuation. For a rectangular guide in which all walls are made of a dense dielectric, no low-loss mode exists because of the brewster's angle problem. This is in contrast to a hollow circular guide made of a dense dielectric with $\varepsilon_a \gg \varepsilon_i$. Such a structure supports low-loss TE modes, which have a small attenuation because their E-fields are tangential to the walls.

A more detailed calculation of field configurations and propagation constants of hollow rectangular dielectric waveguides was performed by Laakmann and Steier (1976) including guides with two dissimilar wall materials. For a hollow dielectric square guide with $2a = 1$ mm and a vacuum wavelength $\lambda = 10.6$ μm, they predict an attenuation of 0.140 dB/m for SiO_2 walls and 0.032 dB/m for BeO walls.

C.   DIELECTRIC HOLLOW SLAB GUIDES

Hollow slab or dielectric planar guides have been studied by Burke (1970), Marcuse (1972, 1974), and Krammer (1976). There are two types of modes in these guides, the transverse electric TE_{m-1} and transverse magnetic TM_{n-1} modes. For the slab guide shown in Fig. 6 the approximation $\lambda \ll a$ yields the following electric field equations:

(i)   for $TE_{m-1}, m \geq 1$:

$$E_y = E_0 \sin[\pi m(a + x)/2a]e^{i(\gamma_{m-1}z - \omega t)}$$
$$E_x = 0, \qquad\qquad\qquad\qquad\qquad (18a)$$
$$E_z = 0.$$

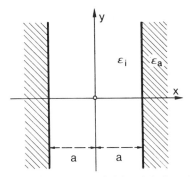

FIG. 6. Geometry and parameters of dielectric hollow slab waveguide

(ii)   for TM_{n-1}; $n \geq 1$:

$$E_x = E_0 \sin[\pi m(a + x)/2a]e^{i(\gamma_{n-1}z - \omega t)}$$
$$E_y = 0, \tag{18b}$$
$$E_z = 0 \ (\lambda/a).$$

For the propagation constants $\gamma = \beta + i\alpha$ we find

$$(\lambda/2\pi)\gamma(TE_{m-1}) = 1 - \tfrac{1}{2}(m\lambda/4a)^2\{1 - i(\lambda/\pi a)\varepsilon_i[(\varepsilon_a - \varepsilon_i)\varepsilon_i]^{-1/2}\}$$
$$(\lambda/2\pi)\gamma(TM_{n-1}) = 1 - \tfrac{1}{2}(n\lambda/4a)^2\{1 - i(\lambda/\pi a)\varepsilon_a[(\varepsilon_a - \varepsilon_i)\varepsilon_i]^{-1/2}\}. \tag{19}$$

For real ε_i and ε_a the TE_0 mode suffers the lowest loss.

D.   METAL HOLLOW CIRCULAR GUIDES

At submillimeter-wave and infrared frequencies only the TE_{0m} modes of the hollow circular metal tubes are capable of guiding electromagnetic energy with low loss. Figure 7 shows the fields of the TE_{01} mode.

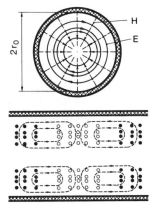

FIG. 7.   Electric (full lines) and magnetic (broken lines) fields of the TE_{01} mode of the circular metal waveguide.

For large diameters $2r_0 \gg \lambda$ of the guide we can approximate the loss

$$\alpha(TE_{0m}) = (R_S/Z_0)(u'_{0m}/2\pi)^2 \lambda^2 r_0^{-3} \tag{20}$$

where u'_{0m} is defined by $J'_0(u'_{0m}) = 0$. This equation represents the concept of surface impedance. For submillimeter waves R_S/Z_0 is approximated by Eq. (11), and for the infrared by Eq. (13).

E. METAL PLANAR WAVEGUIDES

In the parallel plane guide shown in Fig. 8 the TE modes exhibit low attenuation, whereas the TM modes are subjected to high losses. For short wavelengths ($\lambda \ll b$), the losses of the TE_{n0} modes are described by the surface impedance approximation

$$\alpha(TE_{n0}) = (R_S/Z_0)n^2\lambda^2(2b)^{-3} \tag{21}$$

in accordance with Eq. (2). Again R_S/Z_0 is given by Eq. (11) for submillimeter waves and by Eq. (13) for the infrared.

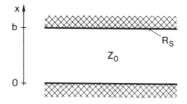

FIG. 8.   Geometry and parameters of the metal planar waveguide.

F. METAL–DIELECTRIC HOLLOW RECTANGULAR GUIDES

Hollow rectangular waveguides with combined metal and dielectric walls have been applied successfully for the construction of transversely excited HCN (Adam *et al.*, 1973) and CO_2 (Smith *et al.*, 1976) waveguide lasers, as well as for the observation of the Stark effect in optically pumped farinfrared lasers (Tobin and Jensen, 1976, 1977). Fields, losses, and resonance conditions for waveguide resonators with two walls made of an ideal conductor and two dielectric walls as shown in Fig. 9 were calculated by Adam and Kneubühl (1975). It can be demonstrated (Argence and Kahan, 1964) that for waveguides with inhomogeneous cross sections it is not possible to separate the field components into the well known E- and H-type modes. However, their modes can be represented by linear combinations of modes of these two classes. According to Collin (1960) the modes of the metal–dielectric guide exhibit longitudinal components of the electric and the magnetic field. They split into two classes, which are characterized by the vanishing of the fields

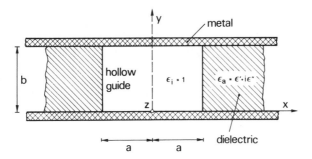

FIG. 9. Geometry and parameters of the metal–dielectric hollow rectangular waveguide.

perpendicular to the dielectric walls. Modes with vanishing electric field ($E_x = 0$) normal to the dielectric interface are called longitudinal section electric (LSE) modes, those with vanishing magnetic field ($H_x = 0$) normal to the dielectric interface are named longitudinal section magnetic (LSM) modes. In addition, a distinction is made between symmetric and anti-symmetric modes.

In oversized guides with $a,b \gg \lambda$, the E_y and H_z components of the LSM modes and the E_z and H_y components of the LSE modes are small. If they are neglected the LSM and the LSE modes convert into the TM and the TE modes.

Assuming a core with $\varepsilon_i = 1$ and taking an ideal conductor for the metal walls, the propagation constants γ of the modes in the guide shown in Fig. 9 can be approximated (Adam and Kneubühl, 1975) by

$$\frac{\lambda}{2\pi}\gamma = \left\{1 + \frac{i}{\pi}\left(\frac{\lambda}{2a}\right)^3 (\varepsilon_a - 1)^{-1/2} \begin{cases} \varepsilon_a(n+1)^2 & \text{for } \text{LSM}^s_{n,\,m>0} \text{ modes} \\ \varepsilon_a(n+\tfrac{1}{2})^2 & \text{for } \text{LSM}^a_{n,\,m>0} \text{ modes} \\ (n+\tfrac{1}{2})^2 & \text{for } \text{LSE}^s_{n,\,m} \text{ modes} \\ (n+1)^2 & \text{for } \text{LSE}^a_{n,\,m} \text{ modes.} \end{cases} \right. \tag{22}$$

Calculations and measurements of the losses of approximate TE and TM modes in metal–dielectric hollow rectangular waveguides with real metal walls have been performed by Nishihara et al. (1974), Garmire et al. (1976a,b), Garmire (1976) and, Tobin and Jensen (1976). On the basis of Eqs. (2), (19), (21), and (22) we can derive approximations of the losses of these modes:

$$\alpha(\text{TE}_{n-1,\,m}) = (R_S/Z_0)2/b + (n^2\lambda^2/16a^3)\varepsilon_i[(\varepsilon_a - \varepsilon_i)\varepsilon_i]^{-1/2}$$
$$\alpha(\text{TM}_{m-1,\,n}) = (R_S/Z_0)n^2\lambda^2/(2b)^3 + (m^2\lambda^2/16a^3)\varepsilon_a[(\varepsilon_a - \varepsilon_i)\varepsilon_i]^{-1/2}. \tag{23}$$

Again R_S/Z_0 is given by Eq. (11) for submillimeter waves and by Eq. (13) for the infrared. Equations (23) are in good agreement with the formulas presented by the authors mentioned above.

G. H-Guides

Batt et al. (1974) proposed a modified microwave H-guide as a low-loss guide in far-infrared (FIR) waveguide lasers in addition to the hollow planar, rectangular, and circular structures. This guide uses the low-loss modes between parallel conducting planes, with a thin dielectric film stretched between the planes to concentrate the propagated energy in the required direction. Propagation in the nth order mode is possible if the separation b of the metal planes is larger than $n\lambda/2$. The dielectric thickness determines the extension of the fields away from the dielectric film. The characteristics of the guide have been investigated at 3 cm, 8 mm, and 0.337 mm using a resonance technique. The measurements at 0.337 mm were performed with an HCN laser as source, and the observed loss was as low as 8 dB/m.

H. Curved Hollow Guides

The influence of the curvature of bent hollow circular waveguides on the attenuation constants of the different modes was discussed already by Marcatili and Schmeltzer (1964) in their relevant paper on hollow metallic and dielectric waveguides. For the lowest-loss TE_{0m} modes in a curved hollow circular metal guide of inner radius r_0 they found the following dependence of the attenuation constant on the radius R of curvature:

$$\alpha_{0m}(R) \simeq \alpha_{0m}(\infty)[1 + \tfrac{4}{3}(2\pi r_0/\lambda u_{0m})^4(r_0/R)^2] \qquad (24)$$

where $\alpha_{0m}(\infty)$ indicates the attenuation constant in the straight guide and u_{0m} the mth root of the equation $J_1(u_{0m}) = 0$. More complicated expressions occur for the R dependence of the attenuation constants of the other modes in the hollow circular metal guide and hollow circular dielectric guides.

Curved planar waveguides with ideal dielectric walls were studied by Marcatili (1969) and by Heilblum and Harris (1975), whereas those with real dielectric or metal walls were investigated in detail by Garmire et al. (1976a,b, 1977), Garmire (1976), and Krammer (1977). In agreement with Krammer (1977) we can write the attenuation constants of the TE modes in curved hollow planar metal waveguides with a large radius R of curvature as follows:

$$\alpha(R) = \alpha(\infty) + (R_S/Z_0)R^{-1} \qquad (25)$$

where R_S/Z_0 is given by Eq. (11) in the submillimeter-wave region and, according to Eq. (13), equals $\mathrm{Re}(n^{-1}) = \mathrm{Re}(\varepsilon^{-1/2})$ in the infrared.

IV. Periodic Waveguide Structures

In principle, it should be possible to build an optically pumped or electrically excited distributed feedback (DFB) gas laser (for definition see Section V) with the aid of a hollow periodic waveguide as resonator. The first proposal of a DFB gas laser using a dielectric tube as resonator was made by Marcuse (1972). Later, Miles and Grow (1976) reported the first evidence for the successful operation of a DFB CO_2 laser. However, they did not confirm this statement in a recent paper on the theory of this laser (Miles and Grow, 1978). A first experimental attempt to use a waveguide with periodic corrugations in an optically pumped submillimeter-wave laser was made by Yamanaka (1976a). Generation of periodic surface corrugations was recently reviewed by Johnson *et al.* (1978).

The DFB can provide low laser thresholds (Wang, 1974b) and high mode selectivity which could be of advantage for transversely excited lasers (CO_2: 10.6 μm; HCN: 337 μm) working at higher pressures as well as for optically pumped lasers (CH_3F: 496 μm; CH_3OH: 570.5 μm). Another impetus for the consideration of DFB structures is the difficulty of finding appropriate mirrors for optically pumped FIR lasers.

For DFB gas lasers using oversized waveguides as resonators, small periodic corrugations will provide small or even insufficient feedback. Unfortunately, deep periodic corrugations and/or strong periodic gain variations represent a difficult theoretical problem. In order to gain insight into the problem of periodic waveguides, we start with the consideration of weak periodic pertubations and gain variations on the basis of Mathieu and Hill equations.

For periodic corrugations of the waveguide and/or gain variations, the amplitude E of the oscillating electric field obeys the equation

$$d^2E/dz^2 + k^2(z)E = d^2E/dz^2 + V(z)E = 0, \qquad (26)$$

with the periodic potential

$$V(z) = V(z + L) = V(z + \pi\beta_0^{-1}) \qquad (27)$$

and the periodic wavenumber

$$k(z) = k(z + \pi\beta_0^{-1}) = n(z)\omega/c + i\alpha(z). \qquad (28)$$

For a DFB laser the index of refraction $n(z)$ and/or the gain $\alpha(z)$ are periodic. The wave propagation is assumed parallel to the z axis.

The above wave equation [Eq. (26)] is a Hill or a Mathieu differential equation. For a waveguide structure without loss or gain $V(z)$ is real. In this case the wave equation corresponds to the Schrödinger equation of an electron in a one-dimensional periodic potential. This equation yields the band structure well known in solid-state physics. The analytic properties of the solutions of this equation for real symmetric potentials $V(z) = V(-z)$ are described in papers by Kohn (1959) and by Meiman (1977).

If the loss or gain due to the laser medium in the waveguide are taken into account, the potential $V(z) = k^2(z)$ is complex. Little is known about the solutions of the wave equation with complex $V(z)$, except for the most recent results obtained by Meiman (1977). Therefore, present theories on DFB lasers are usually restricted to small periodic gain variations and corrugations of the waveguide (Kogelnik and Shank, 1972; Marcuse, 1972; Wang, 1974b; Kneubühl, 1977; Miles and Grow, 1978).

For periodic waveguide corrugations and gain variations, let us consider an index of refraction $n(z)$ and a gain $\alpha(z)$ of the form

$$n(z) = n + n_1 \cos 2\beta_0 z, \qquad \alpha(z) = \alpha + \alpha_1 \cos 2\beta_0 z. \qquad (29)$$

For small corrugations the following conditions are fulfilled:

$$\alpha \ll \beta_0 = 2\pi n/\lambda_0, \qquad n_1 \ll n, \quad \alpha_1 \ll \beta_0. \qquad (30)$$

These assumptions lead to the simple potential

$$V(z) = k^2(z) = \lambda + 2\gamma \cos 2\beta_0 z, \qquad (31)$$

where the parameters λ and γ are defined by

$$\lambda = (\beta + i\alpha)^2, \qquad \gamma = 2\kappa(\beta + i\alpha), \quad |\gamma| \ll |\lambda|, \qquad (32)$$

with $\beta = n\omega/c$ and the coupling constant

$$\kappa = \pi n_1/\lambda_0 + \tfrac{1}{2}i\alpha_1. \qquad (33)$$

Floquet's theorem for the solutions of the Mathieu differential equation is essential for the determination of the dispersion relation of DFB lasers. According to this theorem, all solutions can be represented by

$$E(z) = e^{i\Gamma z} \sum_l a_l e^{+il\beta_0 z} \pm e^{-i\Gamma z} \sum_l a_l e^{-il\beta_0 z}. \qquad (34)$$

The combination of this expression with the wave equation leads to an eigenvalue problem. Since $|\gamma| \ll |\lambda|$ this can be solved with the aid of perturbation theory. Two cases have to be considered:

case I

$$|(l\beta_0 + \Gamma)^2 - \beta_0^2| \gg |\kappa^2|, \qquad l = 0, \pm 2, \pm 4, \ldots \qquad (35a)$$

This condition corresponds to a laser oscillation far from the Bragg frequency. In this case the dispersion relation can be approximated by

$$(\beta + i\alpha)^2 = (l\beta_0 + \Gamma)^2/\{1 - 2\kappa^2/[(l\beta_0 + \Gamma)^2 - \beta_0^2]\} \simeq (l\beta_0 + \Gamma)^2. \qquad (35b)$$

case II

$$|(l\beta_0 + \Gamma)^2 - \beta_0^2| \ll |\kappa^2|. \qquad (36a)$$

In this case the Bragg condition is nearly fulfilled. The following dispersion relation holds:

$$(\beta + i\alpha)^2 = \beta_0^2 + (\Delta\Gamma)^2 + 2\kappa^2 \pm 2\{[\kappa^2 + (\Delta\Gamma)^2][\kappa^2 + \beta_0^2]\}^{1/2}, \qquad (36b)$$

where $\Delta\Gamma = (l\beta_0 + \Gamma) - \beta_0$.

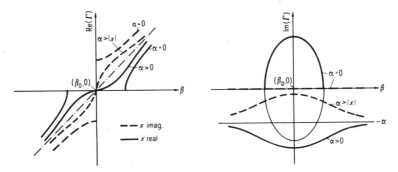

FIG. 10. Dispersion relations $\Gamma(\beta)$ for pure index (κ real) and pure gain (κ imaginary) modulations (Affolter and Kneubühl, 1976a).

Illustrations of dispersion relations are presented in the paper by Affolter and Kneubühl (1976a). Figure 10 shows the dispersion relations $\Gamma = \Gamma(\beta)$ for pure index and pure gain modulation.

Recently Affolter and Kneubühl (1976b, 1977) have considered a FIR waveguide DFB laser using a periodic metal waveguide with strong corrugations as shown in Fig. 11.

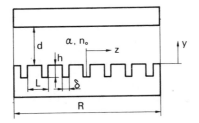

FIG. 11. Metal waveguide structure with deep periodic corrugations.

On the basis of the dispersion relation of the passive ($\alpha = 0$) waveguide (Watkins, 1958) and the theory of lossy periodic waveguides by Butcher (1956), they derived the following dispersion relation:

$$\frac{1}{(\beta + i\alpha)h \tan[(\beta + i\alpha)h]} = \frac{\delta}{L} \sum_{m=-\infty}^{\infty} \frac{\coth(\tau_m d)}{\tau_m h} \frac{\sin(\Gamma_m \delta/2)}{\Gamma_m \delta/2} \qquad (37)$$

with

$$\Gamma_m = \Gamma + (2\pi m/L), \qquad \beta = n\omega/c, \qquad \tau_m^2 = \Gamma_m^2 - (\beta + i\alpha)^2.$$

Here δ is assumed to be small compared to L. The threshold and resonance condition is given in the paper by Affolter and Kneubühl (1976b). In Figs. 12

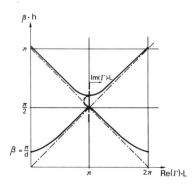

FIG. 12.   Dispersion relation of the passive structure ($\alpha = 0$) shown in Fig. 11 for $h/L = \frac{1}{2}$ (Affolter and Kneubühl, 1977).

and 13 we show the results of a computer evaluation of the dispersion relation. Only the first transverse mode is shown. The curves for the higher modes are similar to the one plotted in Figs. 12 and 13 but shifted towards higher frequencies. As in all periodic structures, the free-space wave number at the band gap center is slightly greater than π/L, where L is the period of modulation. This corresponds to an increase of the phase velocity in periodic media. The cutoff frequency is the same as in planar metallic waveguides with transverse dimension d.

In contrast to DFB lasers with small corrugations or dielectric DFB lasers, a positive feedback does not occur for all values of the slit depth h. If h is equal to a multiple of half the wavelength, the electric field on the boundary between the slots and the waveguide is zero, and the structure acts like a homogeneous waveguide. Therefore we have to choose h according to

$$h = (2n + 1)\lambda/4, \qquad n = 0, 1, 2, 3, \ldots \tag{38}$$

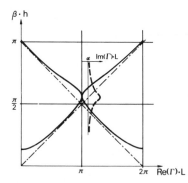

FIG. 13.   Dispersion relation of the active structure ($\alpha > 0$) shown in Fig. 11 for $h/L = \frac{1}{2}$ (Affolter and Kneubühl, 1977).

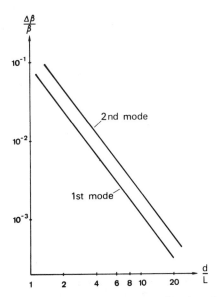

FIG. 14. Relative bandwidth $\Delta\beta/\beta$ of the band gap as a function of the waveguide diameter d for the case: $\alpha = 0$, $\delta/L = 0.2$.

Since all active media for submillimeter-wave lasers have very small emission bands, it is of special interest to know the bandwidth $\Delta\beta$ of the Bragg region. In Fig. 14 we have plotted the relative bandwidth $\Delta\beta/\beta$ of the band gap as a function of the ratio d/L for the case $\alpha = 0$. The bandwidth increases with increasing mode number and decreasing waveguide diameter. From Fig. 14 we see that for a ratio d/L of about 20, corresponding to a waveguide diameter of 5 mm for $L = 248$ μm, the mean accuracy required for the manufacture of the periodic structure has to be on the order of 4×10^{-4}. This is the precision we achieved for our waveguide cut in brass. Furthermore we are able to alter the period L by changing the temperature of the waveguide. A temperature variation of about $50°$ allows a relative length adjustment of 10^{-3}. Therefore this DFB laser should also work for values of d/L greater than 20.

In addition we calculated the maximum value of the imaginary part of the propagation constant Γ as a function of d/L. This is a measure of the strength of the feedback in the Bragg region. Analogous to the bandwidth the coupling increases with increasing mode number and decreasing waveguide diameter. These results are similar to those of Jaggard and Elachi (1976), who calculated the dispersion relation for an unbounded periodic dielectric.

Another scheme for obtaining frequency selectivity for optically pumped submillimeter-wave lasers was also suggested by Affolter and Kneubühl

(1976c). They propose the use of an oversized helical waveguide structure instead of a waveguide with periodic corrugations. Helical waveguides are widely used in microwave traveling-wave tubes to bring down the phase velocity of the electromagnetic wave. In a first approximation this velocity v_{phase} in a helical waveguide is equal to cp/u, where c is the velocity of light, p the pitch, and u the circumference of the helix. In microwave traveling-wave tubes the ratio of p/u is of the order of magnitude $\frac{1}{10}$. Usually the lowest passbands of the helix are used. For optically pumped FIR lasers, oversized helical waveguides with a ratio of pitch to circumference p/u near $\frac{1}{100}$ have to be considered since the free-space wavelength λ is 0.1–1 mm, and the guide radius $a = u/2\pi$ should not be less than 5 mm to guarantee a sufficient gas volume. This type of oversized helical waveguide is shown in Fig. 15.

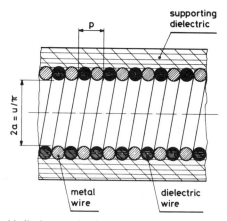

FIG. 15.   Oversized helical waveguide for optically pumped submillimeter-wave lasers.

The waveguide laser differs completely from the traveling-wave tube with respect to the passbands in the dispersion relation of the helix. Whereas the traveling-wave tube makes use of the lower passbands, the waveguide laser has to take advantage of the highest passbands. This is required by the small ratio p/u and by the narrowness of the high passbands. For the helical waveguide of Fig. 15 the dispersion relation is unknown.

However, it should be similar to the dispersion relation of the tape helix which was calculated with the coupled-mode approximation by Pierce and Tien (1954). Figure 16 shows the high-frequency part of the dispersion relation of the tape helix (Pierce and Tien, 1954; Sensiper, 1955). The center frequency $v_n = c/\lambda_n = (2n + 1)c/2u$ of the highest band of the tape helix at $p\Gamma/2\pi = 0.5$ is determined by the condition $(1/2p) - (1/\lambda_n) < (1/u)$. The

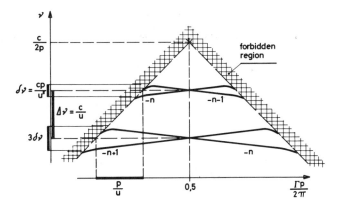

FIG. 16. Dispersion relation $v(\Gamma)$ of the tape helix.

separation between the passbands is about $\Delta v = c/u$, and the approximate bandwidths are $\delta v = cp/u^2$, $3\delta v$, and $5\delta v$, starting with the highest band.

As an example, a helical waveguide with diameter $2a = 16.025$ mm, pitch $p = 0.240$ mm, and corresponding parameters $n = 104$, $\Delta v = 5.96 \times 10^9$ Hz, and $\delta v = 2.84 \times 10^7$ Hz is considered. The fourth band from the top has a center wavelength $\lambda_{n-3} = 496$ μm and an overall width of $7\delta v = 1.99 \times 10^8$ Hz. This band allows frequency selection for the well-known strong CH_3F emission at wavelength 496 μm.

In practice the manufacture of long hollow waveguides with periodic corrugations and helical waveguides with the required precision is difficult. In any case a mechanical or thermal adjustment of the period or pitch to the wavelength of the stimulated emission should be incorporated.

V. Waveguide–Laser Feedback

As shown in Fig. 17, waveguide laser feedback can be realized in several ways. The common approach is to insert the waveguide resonator into a Fabry–Perot resonator using internal (A) or external (B) mirrors. More sophisticated laser engineers apply distributed feedback (C), distributed Bragg reflection (D), or a ring configuration (E). In a distributed feedback (DFB) laser the backward Bragg scattering is confined to the active medium, whereas in a distributed Bragg reflector (DBR) laser backward Bragg scattering is limited beyond the active medium (Wang, 1973, 1974b). In both lasers the backward Bragg scattering is implemented by periodic variations in the waveguide structure or in the active medium (Section IV). The periodic structures play a role in defining the operating frequency of the laser and may be used to couple radiation from the waveguide.

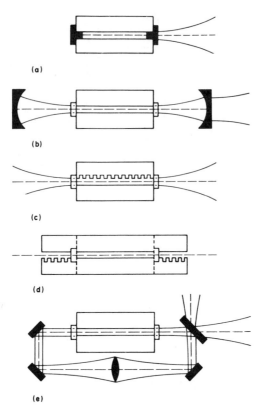

FIG. 17.   Various systems of waveguide laser feedback (Degnan, 1976): (A) Fabry–Perot (internal mirrors); (B) Fabry–Perot (external mirrors); (C) distributed feedback; (D) distributed Bragg reflection; and (E) ring configuration.

A.   WAVEGUIDE FABRY–PEROT RESONATORS WITH INTERNAL MIRRORS

As demonstrated in Section III, wavefronts within hollow oversized waveguides become nonplanar as terms of the order λ/d become significant; d indicates a characteristic inner diameter of the guide. In sufficiently large guides, however, the wavefronts may be considered planar. Thus, efficient feedback may be achieved by placing a well aligned flat mirror at each end of the guide or on tips of plungers which can be moved within the guide for tuning.

This was demonstrated theoretically and experimentally for a waveguide HCN laser by Steffen and Kneubühl (1968b) with the aid of laser resonator interferometry. Since the surfaces of the flat mirrors coincide with the phase-fronts of the waveguide modes, the latter are not coupled by the reflectors.

Hence, the spatial distributions of the guide and resonator modes are identical. The resonant modes of a terminated hollow circular guide have been considered in detail by Steffen and Kneubühl (1968b), using the notation $T^{\pm}_{mnq}$ for resonator modes based on the hybrid EH_{nm} modes of Marcatili and Schmeltzer (1964).

For ideal mirrors, the resonator frequencies are determined by the relations

$$v_{nmq} = (n_0 c/2L)(\lambda_{nm}/\lambda_0)q, \qquad q = 1, 2, 3, 4, \ldots,$$
$$\lambda_{nm} = 2\pi/\beta_{nm} \tag{39}$$

where L indicates the resonator length, n_0 the refractive index of the core, λ_0 the free-space wavelength, λ_{nm} the guide wavelength, and β_{nm} the real part of the propagation constant for the waveguide mode.

In submillimeter-wave lasers using gases at low pressures, the separation of the fundamental and the higher modes is larger than the spectral linewidth of the molecular emission. This allows us to study the mode structure of the waveguide laser resonator with interferometry (Steffen and Kneubühl, 1968b). Here, the resonator length L is varied continuously by moving one resonator mirror with a high-precision screw. Assuming a fixed frequency v_0 for the molecular submillimeter-wave emission, we observe laser emission at the resonator length given by the resonance condition

$$L_{nmq} = \tfrac{1}{2}n_0 \lambda_{nm}q, \qquad q = 1, 2, 3, 4, \ldots. \tag{40}$$

For a long ($q \gg 1$) dielectric hollow circular guide this equation can be approximated (Schwaller et al., 1967; Steffen and Kneubühl, 1968b; Belland et al., 1975b) by

$$L_{nmq} = (n_0 \lambda_0/2)[q + (1/N)(u_{nm}/2\pi)^2] \tag{41}$$

where $N = r_0^2/\lambda_0 L$ is the Fresnel number and u_{nm} the mth root of the Bessel function $J_{n-1}(z)$. Figure 18 shows a laser resonator interferogram obtained from an HCN laser with a hollow circular dielectric waveguide which shows various modes of the 337 and 311 μm emissions (Belland et al., 1975b).

The power loss δ_{nmq} per pass of a waveguide mode is determined by the imaginary component α_{nm} of the propagation constant γ_{nm}

$$\delta_{nmq} = [1 - \exp(-2\alpha_{nm}L)] \simeq 2\alpha_{nm}L. \tag{42}$$

Hence, the resonance width Δv of the waveguide resonator equipped with internal mirrors is given by

$$\Delta v \simeq (c/2\pi L)[\delta_{nmq} + (1/2)\delta_{\text{mirror}}] = (c/\pi)[\alpha_{nm} + (1/4L)\delta_{\text{mirror}}] \tag{43}$$

where δ_{mirror} is the power loss after reflection at the two resonator mirrors.

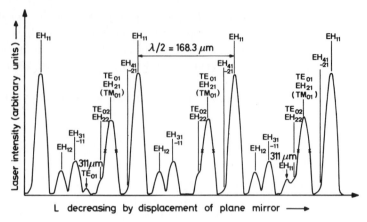

FIG. 18. Resonator interferogram of a waveguide HCN laser obtained by measuring the output power as function of the resonator length L (Belland *et al.*, 1975b, copyright The Institute of Physics, 1975).

Sawatary and Kapany (1970) have photographed the far-field patterns of a number of low-order waveguide modes, e.g., the EH_{11} mode. Later, the near- and far-field radial intensity profiles of low-order, linearly polarized modes launched from a dielectric hollow circular guide were computed by Degnan (1973, 1976). The far-field intensity profiles of the three most powerful modes (EH_{11}, TE_{01}, and EH_{12}) of a waveguide 337-μm HCN laser with a hollow dielectric tube as resonator were measured with a Golay cell by Belland *et al.* (1975b).

B. WAVEGUIDE FABRY–PEROT RESONATORS WITH METALLIC MESHES

The internal Fabry–Perot mirrors have been replaced by metallic meshes in a 337-μm HCN laser by Belland *et al.* (1976) and in optically pumped submillimeter-wave lasers by Tanaka *et al.* (1974) and Wood, R. A. *et al.* (1975, 1976). The use of metallic meshes instead of pinhole mirrors provides uniform output coupling over the entire area of the waveguide. This results in the generation of a well collimated submillimeter-wave output beam and enables the direct observation of laser modes sustained by the laser. This permits output optimization through modification of the grid parameters (Ulrich *et al.*, 1970), and the intensity distributions of waveguide modes can be observed by sweeping the detector across the output beam (Belland *et al.*, 1976; Wood *et al.*, 1976).

Danielewicz *et al.* (1975) have constructed an optically pumped 496-μm CH_3F waveguide laser with a hybrid output mirror consisting of a metallic mesh FIR reflector deposited onto a wedged Si substrate and covered by a multilayer dielectric coating. The latter provides a reflectivity greater than 98% throughout the CO_2 laser bands. The FIR reflectivity is varied by changing the grid parameters or substrate index (Ulrich *et al.*, 1963, 1970). Replacing a hole-coupled reflector by a hybrid mirror at the output end of his laser, Danielewicz *et al.* (1975) achieved a 350-fold increase of the output power.

C. WAVEGUIDE FABRY–PEROT RESONATORS WITH EXTERNAL MIRRORS

In the first optical waveguide laser Smith (1971) used external curved mirrors to image the light radiating from the optical waveguide back into the guide and provide feedback for the laser. Today, this represents one of the basic designs for optical and infrared waveguide lasers. Waveguide lasers of this type show some inherent loss in coupling radiation from the guide into free space and back into the guide.

Abrams (1972) calculated the turnaround coupling loss for the EH_{11} lowest-loss linearly polarized guide mode as a function of mirror radius R and mirror distance z from the guide. He approximated the free-space wave by a Gaussian beam with waist w_0. He found that for $w_0 = 0.6435r_0$ the Gaussian beam best approximates the field distribution of the EH_{11} mode of a waveguide with radius r_0. Furthermore, he demonstrated that the optimum turnaround coupling occurs at a mirror position z, when the radius R of the mirror is equal to the wavefront radius $R' = z + b^2/z$, where b indicates the characteristic distance $\pi w_0^2/\lambda_0$, and λ_0 the free-space wavelength. By numerical calculation he determined the lowest coupling loss, 1.48%, which occurs at $z = b$ for $R = R'$. Finally, he studied the coupling loss when the mirror radius does not match the curvature of the wavefront, i.e., $R' \neq R$, and found a number of optimum mirror positions and radii.

Chester and Abrams (1972) extended the calculations of Abrams (1972) to a symmetric resonator of guide radius r_0 and guide length L with two identical mirrors of radii R positioned symmetrically at identical distances z from the guide ends. They found that the mirror positions z at which low coupling loss occurs are not predictable for any simple wavefront-matching equation. These positions z are sensitive to the guide length L, ratio λ_0/r_0, and mirror radius R. However, they observed that in general, a mirror positioned with z slightly less than R can result in low coupling loss. Physically, this results since a small focused spot is formed at the waveguide entrance, and there is little aperturing of the beam.

Later, Degnan and Hall (1973) presented a general theory of waveguide-laser resonators with finite aperture mirrors which describes the external reflectors by matrices which couple linearly polarized waveguide modes having the same azimutal symmetry. They determined resonator frequencies, resonator efficiency, as well as laser near- and far-field patterns. They found three low loss resonator configurations: (1) large radius of curvature mirrors close to the guide, (2) large radius of curvature mirrors with the center of curvature at the entrance of the guide, and (3) smaller curvature mirrors separated by half their curvature from the guide entrance.

The results of a similar matrix theory for resonators with infinite aperture mirrors were published by Abrams and Chester (1974). They discuss guide losses, coupling losses, and mode shapes for specific cases, with the Fresnel number N of the waveguide ranging from 0.1 to 1 and with various mirror curvatures R and positions z. In addition, they give particularly advantageous resonator parameters for achieving single-mode operation.

The feedback in resonators formed by square or rectangular hollow dielectric waveguides with external spherical mirrors has been calculated recently. Henderson (1976) determined the Gaussian beam waist $w_0 = 0.7032a$ which best approximates the EH_{11} mode of a square hollow dielectric waveguide of half-width a (Fig. 4). Avrillier and Verdonck (1977) discussed the turnaround losses for a hollow rectangular dielectric waveguide resonator equipped with external spherical mirrors. They calculated the coupling loss for the EH_{11} lowest-order mode of a square guide as a function of mirror curvature and position, and demonstrated that some mirror positions and radii provide low coupling losses. Their results are similar to those obtained by Degnan and Hall (1973) for a hollow circular dielectric waveguide resonator.

D. Distributed Feedback and Distributed Bragg Reflection

DFB and DBR lasers do not require external mirrors for the laser feedback. The feedback is provided by Bragg scattering of a forward traveling wave into a backward traveling wave caused by periodic variations of the effective refractive index and/or the gain of the laser medium. Periodic variations of the effective index can be implemented by periodic corrugations of the waveguide.

In Section IV we have discussed the DFB in periodic waveguide structures with and without gain. For weak periodic corrugations or gain variations we calculated dispersion relations by applying perturbation theory to the complex Mathieu equation. We also got entangled in a case of a waveguide with deep periodic corrugations. Now, we wish to focus our attention on the coupled-wave theory which provided the first insight into the properties of DFB and DBR lasers.

The basis of coupled-wave theory is described in the chapter "Matrices and Waves" in the book of Brillouin (1946). The first coupled-wave theory of DFB lasers near laser threshold was published by Kogelnik and Shank (1972). Subsequently, this theory has been refined and extended by several authors: Marcuse (1972), Chinn (1973), De Wames and Hall (1973a,b), Kogelnik et al. (1973), Wang (1973, 1975), Zory (1973), Shubert (1974), Streifer et al. (1975).

The simple coupled-wave model is based on the scalar wave equation [Eq. (26)] for the electric field with the approximative complex periodic potential

$$V(z) = k^2(z) \simeq \beta^2 + 2i\alpha\beta + 4\beta\kappa \cos(2\beta_0 z) \qquad (44)$$

with the notations of Section IV. In general such a periodic pertubation generates an infinite set of diffraction orders. Yet in the vicinity of the Bragg frequency β_0, only two orders are in phase synchronism and of singificant amplitude. All the other orders are neglected in this coupled-mode model. The two significant orders in the DFB structure are two counterrunning waves. These waves grow because of the gain and they feed energy into each other due to Bragg reflection. These waves can be described by complex amplitudes $R(z)$ and $S(z)$. Hence, we write the electric field $E(z)$ as the sum

$$E(z) = R(z) \exp(-i\beta_0 z) + S(z) \exp(+i\beta_0 z). \qquad (45)$$

According to the assumptions manifested in Eq. (30) these amplitudes vary slowly so that their second derivatives can be neglected. If we insert Eq. (45) into the wave equation [Eq. (26)] and compare terms with equal exponentials, we obtain the following pair of coupled wave equations

$$\begin{aligned} -dR/dz + (\alpha - i\delta)R &= i\kappa S \\ +dS/dz + (\alpha - i\delta)S &= i\kappa R, \end{aligned} \qquad (46)$$

where δ indicates a normalized frequency parameter defined by

$$\delta = (\beta^2 - \beta_0^2)/2\beta \simeq \beta - \beta_0. \qquad (47)$$

The general solution of these coupled wave equations is of the form

$$\begin{aligned} R(z) &= R_1 \cos \Gamma z + R_2 \sin \Gamma z \\ S(z) &= S_1 \cos \Gamma z + S_2 \sin \Gamma z \end{aligned} \qquad (48)$$

with the propagation constant $\Gamma(\beta)$ obeying the dispersion relation

$$\Gamma^2 + \kappa^2 + (\alpha - i\delta)^2 = 0. \qquad (49)$$

For vanishing gain this dispersion relation exhibits a gap characteristic for real periodic potentials according to the dispersion relation

$$\Gamma(\beta) \simeq [(\beta - \beta_0)^2 - (\pi n_1/\lambda_0)^2]^{1/2}, \qquad \alpha = \alpha_1 = 0. \qquad (50)$$

The present coupled-wave model assumes a self-oscillating periodic structure of finite length L extending from $z = -L/2$ to $z = +L/2$. Hence, there are no incoming waves, and the internal waves start with zero amplitude at the boundaries of the structure. This implies the following boundary conditions for the wave amplitudes:

$$R(-L/2) = S(+L/2) = 0. \qquad (51)$$

Equations (48) together with the boundary conditions [Eqs. (51)] specify the electromagnetic field in our model of the DFB laser. The corresponding solution yields the self-consistent steady-state field configurations or modes of the periodic structure. Since this procedure represents a linear threshold calculation, the absolute field amplitudes of the modes remain unspecified.

They could be obtained from a nonlinear calculation taking into account gain saturation. For the structure of defined length L and coupling κ, nontrivial solutions are possible only for a discrete set of gain constants α and associated frequencies β. These represent the resonator frequencies of the DFB laser and the corresponding gain values to exceed threshold. These gain values α vary from mode to mode, increasing with frequency spacing $\beta - \beta_0$ from the Bragg frequency β_0.

From the above boundary conditions [Eqs. (51)], we find for the field distributions of the modes

$$R(z) = \pm \sin[\Gamma(z + L/2)]$$
$$S(z) = \pm \sin[\Gamma(z - L/2)]. \tag{52}$$

If we introduce these expressions into the coupled-wave equations [Eqs. (46)] we find the two following equations which define the eigenvalues of the propagation constant Γ:

$$\Gamma = \pm i\kappa \sin(\Gamma L). \tag{53}$$

After determining the eigenvalues of Γ from this equation, we can calculate the corresponding resonant frequencies β and threshold gain constants α by the relation:

$$\alpha - i\beta \simeq -i\beta_0 \pm i\kappa \cos(\Gamma L) = -i\beta_0 \pm \Gamma \cot(\Gamma L). \tag{54}$$

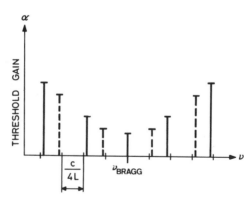

FIG. 19.   Mode spectra and required threshold gains for pure gain (——) and pure index (---) modulation, according to the coupled-wave theory of Kogelnik and Shank (1972).

Consequently, each mode has its characteristic field pattern R and S, characteristic threshold gain α, and resonance frequency β for given length L and coupling κ. Approximations to these equations and numerical calculations are described in detail by Kogelnik and Shank (1972).

Figure 19 illustrates the mode spectra and required threshold gains for pure index and pure gain periodicity. In both spectra the resonances are spaced approximately $c/2nL$ apart. This corresponds to the mode separation in the usual two-mirror laser cavity of length L. For pure index coupling with real κ there is no resonance at the Bragg frequency, whereas for pure gain coupling with imaginary κ there is a resonance exactly at the Bragg frequency. As mentioned above, the threshold gain α increases with frequency spacing from the Bragg frequency.

VI.   Infrared and Submillimeter-Wave Lasers with Hollow Waveguides

A.   HELIUM–XENON AND HELIUM–NEON LASERS

In 1973 Smith and Maloney built an infrared waveguide He–Xe laser operating at 3.5 μm using a circular pyrex tube. It achieved gains of the order of 1000 dB/m at a total pressure of 5.9 Torr. Because of an extremely small saturation parameter the power output remained on the order of microwatts. From a waveguide laser consisting of a hollow slab-guide inside a Fabry–Perot type resonator Anderson et al. (1975) obtained output powers of about 50 μW for the 3.39 μm emission of He–Ne and the 3.5-μm emission of He–Xe.

B.   CARBON DIOXIDE LASERS

Continuous stimulated emission of CO_2 in the infrared at wavelengths around 9.6 and 10.6 μm from waveguide lasers consisting of hollow glass tubes with diameters from 1 to 6 mm and with Fabry–Perot mirrors has been achieved by Bridges et al. (1972), Jensen and Tobin (1972), Degnan et al. (1973), Provorov and Chebotaev (1973, 1975), and Watanabe et al. (1975). In these lasers with flowing gas systems the small signal gain depends strongly on coolant temperature, gas flow rate and pressure, discharge current, and tube diameter. Bridges et al (1972) and Degnan et al. (1973) observed saturation intensities on the order of several kilowatts per squre centimeter as compared to about 0.1 kW/cm^2 for conventional CO_2 lasers. Provorov and Chebotaev (1973) obtained stable continuous laser emissions for gas pressures up to 1 atm for a 2.5 mm-diam glass tube.

High gas temperature degrades the laser performance by producing unwanted collisional deexcitations (Antroprov et al., 1968; Fowler, 1971). In waveguide lasers efficient cooling of the gas is essential, since the optimum current densities are almost two orders of magnitude larger than in conventional laser tubes, whereas the power inputs per unit length are similar in both cases. Therefore, various authors have studied the improvement of gas cooling by replacing the glass tubes by capillaries of BeO (Burkhardt et al., 1972) or Al_2O_3 (Abrams and Bridges, 1973), or by metal guides coated internally with highly reflective insulators (Koepf, 1976). A transverse flow configuration in a planar-waveguide CO_2 laser was designed by McMullen et al. (1974) in order to obtain faster gas exchange and more rapid cooling in the waveguide region. In another waveguide-laser configuration, devised by Papayonaon and Fujisawa (1975), the gas molecules enter a high pressure chamber, seep into a central bore through the pores of a low density BeO tube, and are extracted from the bore by a vacuum pump. From this arrangement they expected reduced pressure gradients in the flow system and consequently, a more uniform gain. Although the porous wall BeO tubes appeared to give about 15% more gain compared to the dense wall capillaries, they also showed considerably larger waveguide losses and reduced thermal conductivity.

Sealed-off waveguide CO_2 lasers show much lower peak gains than the corresponding flow systems. Abrams and Bridges (1973) also found that the saturation intensity varies as the square of the pressure, which indicates that the level lifetimes are dominated by the molecule–molecule collisions. However, for many applications of cw lasers, such as spectroscopy and heterodyne communications, tunability is of greater importance than output power. It was demonstrated (Abrams, 1974a; Beterov et al., 1975; Degnan, 1974) that within a given vibration–rotation transition there is an optimum working pressure with maximum tunability for a particular bore diameter. A sealed-off CO_2 waveguide laser with a BeO square-bore 1-mm^2 capillary achieved a total tuning range of 1200 MHz (Abrams, 1974b), as compared with a doppler-broadened 50-MHz linewidth of a conventional CO_2 laser.

Pulsed-waveguide carbon-dioxide lasers exhibit characteristics which suggest specific applications. The large pressure-broadened linewidth allows mode locking in order to obtain short laser pulses. Mode-locked pulsewidths of 2–3 ns have been reported for flowing gas (Smith et al., 1973) and sealed-off (Abrams, 1974b) waveguide CO_2 lasers at pressures of 150 Torr.

By comparison, conventional mode-locked high-pressure transversely excited atmospheric-pressure (TEA) lasers (Beaulieu, 1970) achieved pulse widths of 0.7 ns (Wood et al., 1970; Abrams and Wood, 1971). The negative resistance of the discharge in a waveguide laser can be combined with an external impedance for the purpose of forming a relaxation oscillator. Thus, Rudko (1974) as well as Zimmermann and Gaddy (1974) obtained laser pulses with repetition frequencies between 4 and 37 kHz. The repetition rate could be changed by varying the power supply voltage or the external impedance.

A waveguide CO_2 TEA laser was first devised by Smith *et al.* (1973). They achieved laser action at pressures up to one atmosphere with a flowing gas mixture of CO_2, N_2, and He. Waveguide TEA lasers are less sensitive to arcs than standard TEA lasers, where they deflect the laser beam causing high optical losses. The preionization from previous discharges results in a more homogeneous and diffuse discharge. In addition, the narrow distance between the electrodes in the small-bore waveguides reduces the required excitation voltage by almost two orders of magnitude compared to conventional TEA lasers. With the same transverse arrangement Wood *et al* (1975a) achieved quasi-continuous gains at 1 atm by preionization of the gas through a train of 20-ns duration 3–8-kV pulses. Smith *et al.* (1976) achieved a quasi-continuous laser output in a TEA laser at pressures up to 0.33 atm, pulsed with a thyratron at a rate of 40 Hz.

The first chemically excited pulsed CO_2 waveguide laser was reported by Poehler *et al* (1975). Pulses with a duration of 20 μs and peak power of 2.5 kW were obtained from gas mixtures of D_2–F_2–CO_2–He at pressures up to 1 atm contained in a 3 mm-bore quartz reactor tube located at a focus of an elliptical cavity. Exothermic chain reactions between D_2 and F_2 were initiated by a Xe flashlamp located at the opposite focus. The reactions form excited DF molecules which transfer their excitation energy to the CO_2 molecules. Since the waveguide laser makes efficient use of the reactor volume, it is expected that their optimization will result in compact chemical lasers yielding energies per unit volume × pressure which rival most conventional configurations.

C. CARBON MONOXIDE LASERS

The first cw operation of a waveguide CO laser was reported by Yusek and Lockhart (1973). They used 1- and 2-mm-bore sealed-off glass capillaries cooled to a temperature of 205 K and obtained 0.3 W at an optimum pressure of 65 Torr. Another sealed-off waveguide CO laser with 2-mm-bore, 140-mm long BeO capillary was devised by Asawa (1974). This laser produced an output power up to 1.1 W at a capillary temperature of 210 K and an optimum pressure of 80 Torr. The output spectrum consists of ten lines in the wavelength region 5.2 to 5.65 μm.

D. ELECTRICALLY EXCITED SUBMILLIMETER-WAVE LASERS

The initial studies on the waveguide characteristics of longitudinally excited submillimeter-wave HCN and ICN lasers by Schwaller *et al.* (1967) and Steffen and Kneubühl (1968b) have already been mentioned. These included theories and experiments on resonances and losses of modes in dielectric waveguide resonators equipped with internal plane Fabry–Perot mirrors as well as investigations of the strong dependence of the laser pulse shapes on small changes of the resonator length. The latter is due to temporal variations of the refractive index in the molecular laser gas caused by the electric discharge (Steffen and Kneubühl, 1968b; Whitbourn et al., 1972; Tait et al., 1973). Schötzau and Kneubühl (1974) demonstrated chemical excitation and suppression of higher dielectric-waveguide modes in a 337-μm HCN laser. This could be achieved by variation of the composition of the laser-active gas mixtures $CH_4:N_2$, $C_2H_6:N_2$, $C_2H_4:N_2$, $C_2H_2:N_2$, and H_2 without any change of other laser parameters. Further investigations on the modes in longitudinally excited HCN lasers were performed by Belland et al. (1975a,b). They measured laser-resonator interferograms and mode-intensity distributions of a longitudinally excited waveguide HCN laser with a plane metal and a plane metal-mesh reflector. They also found that gain and saturation intensity vary inversely with the diameter of their Pyrex tube maintained at a constant temperature of 150°C.

The parameters and chemistry of HCN laser plasmas were studied in detail by Schötzau and Kneubühl (1975a,b) using field probes, chemical probes, and a mass spectrometer. The results show the important role of hydrogen in the HCN laser. Their experiments indicate that H_2 or D_2 is required for the stimulated emission of HCN. Schötzau and Veprek (1975) discussed the influence of chemical processes at the waveguide wall on the radial gain profile of the cw HCN laser. They showed that the HCN molecules formed by chemical reactions on the wall can contribute significantly to the laser gain and cause the experimentally observed constant gain over the cross section.

Characteristic designs of pulsed and cw longitudinally excited waveguide 337-μm HCN lasers and their performance have been reported by Turner and Poehler (1971), Sharp and Wetherell (1972), Belland and Veron (1973), Belland et al. (1976), Vanderkooy and Kang (1976), Bicanic and Dymanus (1976), and Veron et al. (1978). According to our knowledge based on experiments, visits, literature, and private communications, the maximum performance of this type of laser is about 0.3-W cw power, and 0.1-kW pulsed power.

An intense submillimeter-wave source emitting reproducible pulses is desired for plasma heating (Jassby, 1973; Lax and Cohn, 1974), the measurement of electron densities in plasmas, and spectroscopy. For this reason, various authors (Sharp and Wetherell, 1972; Jassby et al., 1973) suggested the application of the previously successful technique of exciting gas-laser emissions by transverse electric discharges to the 337-μm HCN laser. The transverse electric excitation of gas lasers has been known since the discovery of the TEA CO_2 laser by Beaulieu (1970). Since then this technique has proven an effective means of producing laser action in all kinds of high-pressure gas mixtures. A review has been given by Wood (1974).

By replacing the usual electrode structures of transversely excited lasers by a waveguide discharge structure similar to the rectangular metal–dielectric guide described in Section III.F, Adam *et al.* (1973) succeeded in producing the first pure transverse excitation of the 337-μm HCN laser emission without auxiliary discharge. With standard transverse excitation and a closed laser–gas system, the HCN laser emitted pulses of 10–50-μs duration and pulse energies of about 1 mJ for gas pressures up to 5 Torr (Adam and Kneubühl, 1975). Sturzenegger *et al.* (1977) produced homogeneous transverse discharges in an HCN-laser gas mixture at a pressure of 1 atm with the aid of uv preionization, but no 337-μm laser emission was observed.

Recently, Sturzenegger *et al.* (1978) obtained 337-μm laser pulses with energies a factor of 15 higher than those reported by Adam and Kneubühl (1975) from a transversely excited waveguide HCN laser, using a flow system with a gas mixture containing pure HCN at pressures up to 15 Torr.

E. OPTICALLY PUMPED SUBMILLIMETER-WAVE LASERS

Chang and Bridges (1970) discovered the first optically pumped submillimeter-wave laser emission by exciting the CH_3F molecule with 9.55-μm CO_2 laser radiation. In the meantime, several hundred new optically pumped FIR laser emissions have been observed, which are listed in reviews by Chang (1974, 1977a), Radford (1975), Yamanaka (1976b), Rosenbluh *et al.* (1976), Röser and Schultz (1977), and Gallagher *et al.* (1977). Röser and Schultz (1977) also included a special list of molecules in interstellar space, whose stimulated submillimeter-wave emissions have been optically pumped in the laboratory. Gallagher *et al.* (1977) discuss the line positions of the stimulated emissions with respect to the atmospheric transmission spectrum.

Three years after the discovery of optically pumped laser emissions by Chang and Bridges (1970), Hodges and Hartwick (1973) introduced a waveguide resonator in an optically pumped FIR methyl-alcohol laser and obtained milliwatt-level cw laser oscillations on a number of rotational transitions in the 40–200-μm wavelength region.

In their laser they focused the pump beam from a conventional cw CO_2 laser through a coupling hole into a 70-cm waveguide resonator. The FIR laser emission left the resonator through a second coupling hole at the other end of the resonator. The output varied linearly in the pump power range 0.5–4 W. They found that, compared to the performance of a brass tube of comparable diameter and length, a quartz tube of 9 mm inner diameter exhibited smaller output powers and allowed considerably fewer high-order transverse modes to oscillate. This result is consistent with smaller surface impedance propagation losses in metal guides at longer wavelengths and in agreement with the investigations by Yamanaka (1977). According to Degnan (1976), one-way losses in a dielectric tube with a refractive index of about $n = 1.5$ become substantial ($>2\%$) when $\lambda^2 L/r_0^3 = (\lambda/r_0)N > 0.05$, where L is the length and r_0 the radius of the tube. This relation imposes an upper limit on the FIR wavelength which can oscillate in a dielectric tube resonator of given diameter.

Extensive studies on losses and polarization effects in waveguides suited for optically pumped submillimeter-wave lasers have been performed by Yamanaka and co-workers (Yamanaka and Yoshinaga, 1974; Yamanaka et al., 1975; Yamanaka, 1977). They calculated attenuation constants of small-sized low-loss hollow waveguides on the basis of existing theories, e.g., cylindrical and rectangular copper guides, and cylindrical guides made of fused quartz. Their data show that a hollow fused-quartz guide with 10 mm inner diameter is adequate for wavelengths shorter than 100 μm but too lossy for longer wavelengths. On the other hand, a cylindrical copper guide with the same diameter has negligible losses for the TE_{01} mode in the wavelength region between 10 and 1000 μm.

Although the generation of a rich spectrum of stimulated FIR emissions has not proven difficult, the production of intense radiation by optical pumping of polar molecules presents a difficult task (Hodges, 1978). This difficulty originates in the excitation process and the small energy difference ($\Delta E < kT$) between laser states. Energy storage is not possible, since the laser state lifetimes are essentially equal. In order to approach the quantum limit corresponding to one FIR photon per pumping photon, a detailed knowledge of the fundamental laser processes is required. Hence, laser emission characteristics have been extensively analyzed using rate equation models (e.g., Temkin, 1977). Good agreement is reported in certain cases. However, quantum mechanical two-photon effects have recently been shown to play an important role in characterizing the laser performance (Seligson et al., 1977; Petuchowski et al., 1977; Wiggins et al., 1977).

When the laser is off resonance, FIR radiation may be emitted in a two-photon process corresponding to stimulated Raman emission (Petuchowski et al., 1977; Wiggins et al., 1977). Furthermore, the dynamic ac-Stark effect can significantly modify the emission characteristics of cw and pulsed lasers (Chang, 1977b).

The development of cw optically pumped FIR lasers has concentrated on waveguide resonators and improved output coupling techniques (Yamanaka, 1977; Koepf, 1977; Koepf and McAvoy, 1977). The main problem in all types of optically pumped FIR lasers is the confinement of the infrared pump radiation while coupling out the far infrared. At 119- and 447-μm wavelength, cw output powers of 0.1–0.5 W have been observed. Quasi-cw operation with a 200-μs pulsewidth has produced 2 W at 119 μm and 0.5 to 1 W on a variety of other wavelengths (Hodges, 1978).

The realization of MW pulse power and a simultaneous narrow emission line width of about 25 MHz requires a master-oscillator power-amplifier (MOPA) combination and a Fox–Smith resonator geometry for the oscillator (De Temple, 1977). Several groups are reporting now measuring laser pulses at 358-μm (D_2O) and 496-μm (CH_3F) wavelength with output powers of 0.2–1 MW, corresponding to pulse energies of 10–50 mJ (Cohn et al., 1976; Evans et al., 1977; Drozdowicz et al., 1977; Brown et al., 1977; Woskoboinikow et al., 1978).

VII. Future Aspects

In the near future the development of infrared and submillimeter-wave gas lasers with long low-loss waveguide resonators will allow the laser scientists and engineers to take full advantage of the large pressure-broadened linewidths with respect to highly tunable cw lasers or subnanosecond mode-locked pulsed lasers. Compact waveguide gas lasers may serve as compact efficient radiation sources in a large variety of applications, e.g., as local oscillators in heterodyne systems for astrophysical observations from stratospheric gondolas or space laboratories.

Apart from theoretical considerations, the problem of the construction of DFB and DBR gas lasers in the infrared and submillimeter-wave ranges seems to be still unsolved. There are, however, strong hopes that with a deeper insight into the characteristics of waveguides with strong periodic corrugations and with further efforts in optical pumping this aim may be achieved in the near future.

Finally, it should be mentioned that the development of new low-loss waveguides with cross sections adapted to specific requirements of various pumping schemes of infrared and submillimeter-wave lasers represents another rewarding area of future research.

ACKNOWLEDGMENTS

The authors wish to thank Dr. B. Adam, Baden; Dr. K. J. Button, Cambridge, Mass.; Prof. Dr. P. D. Coleman, Urbana; Dr. E. J. Danielewicz, Stuttgart; Dr. D. T. Hodges, Los Angeles; Prof. Dr. H. Weber, Bern; Prof. Dr. M. Yamanaka, Osaka; Prof. Dr. R. Jost, Prof. Dr. E. Lukosz, and Dipl. Phys. Ch. Sturzenegger, Zurich, for information, discussions, and suggestions. The research on submillimeter-wave waveguide lasers by the Infrared Physics Group of the Solid State Physics Laboratory at the ETH Zurich is supported by the Swiss National Science Foundation and the ETH Zurich.

274   F. K. KNEUBÜHL AND E. AFFOLTER

REFERENCES

Abrams, R. L. (1972). *IEEE J. Quantum Electron.* **QE-8**, 838–843.
Abrams, R. L. (1974a). *Appl. Phys. Lett.* **25**, 304–306.
Abrams, R. L. (1974b). "The Laser Spectroscopy" (R. G. Brewer and A. Mooradian, eds.). Plenum Press, New York.
Abrams, R. L., and Bridges, W. B. (1973). *IEEE J. Quantum Electron.* **QE-9**, 940–946.
Abrams, R. L., and Chester, A. N. (1974). *Appl. Opt.* **13**, 2117–2125.
Abrams, R. L., and Wood, O. R. (1971). *Appl. Phys. Lett.* **19**, 518–520.
Adam, B., and Kneubühl, F. K. (1975). *Appl. Phys.* **8**, 281–291.
Adam, B., Schötzau, H. J., and Kneubühl, F. K. (1973). *Phys. Lett.* **45A**, 365–366.
Affolter, E., and Kneubühl, F. K. (1976a). *J. Appl. Math. Phys. (Z. Angew. Math. Phys.)* **27**, 512–516.
Affolter, E., and Kneubühl, F. K. (1976b). *Phys. Lett.* **58A**, 91–92.
Affolter, E., and Kneubühl, F. K. (1976c). *Phys. Lett.* **58A**, 183–184.
Affolter, E., and Kneubühl, F. K. (1977). *J. Appl. Math. Phys. (Z. Angew. Math. Phys.)* **28**, 1171–1175.
Anderson, D. B., *et al.* (1975). Rockwell Int. Corp. Anaheim, California, Rep. AFAL-TR 75–53.
Antropov, E. T., Silinbekchurin, I. A., Sobolev, N. N., and Sokovikov, V. V. (1968). *IEEE J. Quantum Electron.* **QE-4**, 790–796.
Argence, E., and Kahan, Th. (1964). "Théorie des guides et cavités electromagnetiques." Dunod, Paris.
Asawa, C. K. (1974). *Appl. Phys. Lett.* **24**, 121–123.
Avrillier, S., and Verdonck, J. (1977). *J. Appl. Phys.* **48**, 4937–4941.
Batt, R. J., Bradley, H. L., Doswell, A., and Harris, D. J. (1974). *IEEE Trans. Microwave Theory Tech.* **MTT-22**, 1089–1094.
Beattie, J. R. (1955). *Phil. Mag.* **46**, 235–245.
Beaulieu, A. J. (1970). *Appl. Phys. Lett.* **16**, 504–505.
Belland, P., and Véron, D. (1973). *Opt. Commun.* **9**, 146–148.
Belland, P., Véron, D., and Whitbourn, L. B. (1975a). *Proc. Conf. Infrared Phys.*, Zurich p. C193.
Belland, P., Véron, D., and Whitbourn, L. B. (1975b). *J. Phys.* **D8**, 2113–2122.
Belland, P., Pigot, C., and Véron, D. (1976). *Phys. Lett.* **56A**, 21–22.
Bergstein, L., and Schachter, H. (1964). *J. Opt. Soc. Am.* **54**, 887–903.
Beterov, I. M., Provorov, A. S., and Chebotaev, V. P. (1975). *Sov. J. Quantum Electron.* **5**, 257–258.
Bicanic, D. D., and Dymanus, A. (1976). *Infrared Phys.* **16**, 601–604.
Borgnis, F. E., and Papas, C. H. (1958). *Encycl. Phys.* **16**, 285–422.
Bridges, T. J., Burkhardt, E. G., and Smith, P. W. (1972). *Appl. Phys. Lett.* **20**, 403–405.
Brillouin, L. (1946). "Wave Propagation in Periodic Structures." McGraw-Hill, New York.
Brown, F., Hislop, P. D., and Tarpinian, J. O. (1977). *IEEE J. Quantum Electron.* **QE-13**, 445–446.
Burke, J. J. (1970). *Appl. Opt.* **9**, 2444–2452.
Burkhardt, E. G., Bridges, T. J., and Smith, P. W. (1972). *Opt. Commun.* **6**, 193–195.
Burlamacchi, P., and Pratesi, R. (1973a). *Appl. Phys. Lett.* **22**, 334–335.
Burlamacchi, P., and Pratesi, R. (1973b). *Appl. Phys. Lett.* **23**, 475–476.
Burlamacchi, P., Pratesi, R., and Salimeni, R. (1973). *J. Appl. Phys.* **44**, 4248–4250.
Burlamacchi, P., Pratesi, R., and Salimbeni, R. (1974a). *Opt. Commun.* **11**, 109–111.
Burlamacchi, P., Pratesi, R., and Ronchi, L. (1974b). *Opto- Electronics* **6**, 465–472.
Butcher, P. N. (1956). *Proc. IEE* **103**, 301–306.

Chang, T. Y. (1974). *IEEE Trans. Microwave Theory Tech.* **MTT-22**, 983–988.
Chang, T. Y. (1977a). *Topics Appl. Phys.* **16**, 215–274.
Chang, T. Y. (1977b). *IEEE J. Quantum Electron.* **QE-13**, 937–942.
Chang, T. Y., and Bridges, T. J. (1970). *Opt. Commun.* **1**, 423–426.
Chang, W. S., Muller, M. W., and Rosenbaum, F. J. (1974). "Laser Applications" (M. Ross, ed.), Vol. 2. Academic Press, New York.
Chester, A. N., and Abrams, R. L. (1972). *Appl. Phys. Lett.* **21**, 576–578.
Chinn, S. R. (1973). *IEEE J. Quantum Electron.* **QE-9**, 574–580.
Chu, L. J. (1938). *J. Appl. Phys.* **9**, 538–591.
Cohn, D. R., Button, K. J., Temkin, R. J., and Drozdowicz, Z. (1976). *Infrared Phys.* **16**, 429–434.
Collin, R. E. (1960). "Field Theory of Guided Waves." McGraw-Hill, New York.
Danielewicz, E. J., Plant, T. K., and DeTemple, T. A. (1975). *Opt. Commun.* **13**, 366–369.
Degnan, J. J. (1973). *Appl. Opt.* **12**, 1026–1030.
Degnan, J. J. (1974). *J. Appl. Phys.* **45**, 257–262, 3223.
Degnan, J. J. (1976). *Appl. Phys.* **11**, 1–33.
Degnan, J. J., and Hall, D. R. (1973). *IEEE J. Quantum Electron.* **QE-9**, 901–910.
Degnan, J. J., Walker, H. E., McElroy, J. H., and McAvoy, N. (1973). *IEEE J. Quantum Electron.* **QE-9**, 489–491.
DeTemple, T. A. (1977). *Proc. SPIE* **105**, 11–16.
De Wames, R. E., and Hall, W. F. (1973a). *Appl. Phys. Lett.* **23**, 28–30.
De Wames, R. E., and Hall, W. F. (1973b). *J. Appl. Phys.* **44**, 3638–3640.
Drozdowicz, Z. *et al.* (1977). *IEEE J. Quantum Electron.* **QE-13**, 413–417.
Evans, D. E., Sharp, L. E., Peebles, W. A., and Taylor, G. (1977). *IEEE J. Quantum Electron.* **QE-13**, 54–58.
Fowler, M. C. (1971). *Appl. Phys. Lett.* **18**, 175–178.
Gallagher, J. J., Blue, M. D., Bean, B., and Perkowitz, D. (1977). *Infrared Phys.* **17**, 43–55.
Garmire, E. (1976). *Appl. Opt.* **15**, 3037–3039.
Garmire, E., McMahon, T., and Bass, M. (1976a). *Appl. Opt.* **15**, 145–150.
Garmire, E., McMahon, T., and Bass, M. (1976b). *Appl. Phys. Lett.* **29**, 254–256.
Garmire, E., McMahon, T., and Bass, M. (1977). *Appl. Phys. Lett.* **31**, 92–94.
Heilblum, M., and Harris, J. H. (1975). *IEEE J. Quantum Electron.* **QE-11**, 75–83.
Henderson, D. M. (1976). *Appl. Opt.* **15**, 1066–1070.
Hodges, D. T. (1978). *Digest Int. Conf. Submillimetre Waves Their Appl., 3rd, Guildford* pp. 18–20.
Hodges, D. T., and Hartwick, T. S. (1973). *Appl. Phys. Lett.* **23**, 252–253.
Jaggard, D. L., and Elachi, C. (1976). *J. Opt. Soc. Am.* **66**, 674–682.
Jassby, D. L. (1973). *J. Appl. Phys.* **44**, 919–920.
Jassby, D. L., Marhic, M. E., and Regan, P. R. (1973). *Appl. Opt.* **12**, 1403–1404.
Jensen, R. E., and Tobin, M. S. (1972). *Appl. Phys. Lett.* **20**, 508–510.
Johnson, L. F., Kammlott, G. W., and Ingersoll, K. A. (1978). *Appl. Opt.* **17**, 1165–1181.
Karbowiak, A. E. (1955). *Proc. IEE* **102B**, 698–708.
Karbowiak, A. E. (1964). *Proc. IEE* **111**, 1781–1788.
Kneubühl, F. K. (1977). *J. Opt. Soc. Am.* **67**, 959–963.
Koepf, G. (1976). *IEEE J. Quantum Electron.* **QE-12**, 9–15.
Koepf, G. A. (1977). *IEEE J. Quantum Electron.* **QE-13**, 732–734.
Koepf, G. A., and McAvoy, N. (1977). *IEEE J. Quantum Electron.* **QE-13**, 418–421.
Kogelnik, H. (1975). *IEEE Trans. Microwave Theory Tech.* **MTT-23**, 2–16.
Kogelnik, H., and Li, T. (1966). *Appl. Opt.* **5**, 1550–1567.
Kogelnik, H., and Shank, C. V. (1972). *J. Appl. Phys.* **43**, 2327–2335.

Kogelnik, H., Shank, C. V., and Bjorkholm, J. E. (1973). *Appl. Phys. Lett.* **22**, 135–137.
Kohn, W. (1959). *Phys. Rev.* **115**, 809–821.
Krammer, H. (1976). *IEEE J. Quantum Electron.* **QE-12**, 505–507.
Krammer, H. (1977). *Appl. Opt.* **16**, 2163–2165.
Kurokawa, K. (1969). "An Introduction to the Theory of Microwave Circuits." Academic Press, New York.
Laakmann, K. D., and Steier, W. H. (1976). *Appl. Opt.* **15**, 1334–1340.
Lax, B., and Cohn, D. R. (1974). *IEEE Trans. Microwave Theory Tech.* **MTT-22**, 1049–1052.
Lenham, A. P., and Treherne, D. M. (1966). *J. Opt. Soc. Am.* **56**, 683–685.
Marcatili, E. A. J. (1969). *Bell Syst. Tech. J.* **48**, 2071–2102.
Marcatili, E. A., and Schmeltzer, R. A. (1964). *Bell Syst. Tech. J.* **43**, 1783–1809.
Marcuse, D. (1972). *IEEE J. Quantum Electron.* **QE-8**, 661–669.
Marcuse, D. (1974). "Theory of Dielectric Optical Waveguides." Academic Press, New York.
Marcuwitz, N. (1949). "Waveguide Handbook." McGraw-Hill, New York.
Marcuwitz, N. (1956). *IRE Trans. Antennas Prop.* **AP-4**, 192–194.
McMullen, J. D., Anderson, D. B., and Davis, R. L. (1974). *J. Appl. Phys.* **45**, 5084–5087.
Meiman, N. N. (1977). *J. Math. Phys.* **18**, 834–848.
Miles, R. O., and Grow, R. W. (1976). *J. Opt. Soc. Am.* **66**, 292.
Miles, R. O., and Grow, R. W. (1978). *IEEE J. Quantum Electron.* **QE-14**, 275–283.
Miller, S. E. (1954). *Bell Syst. Tech. J.* **33**, 1209–1265.
Mott, N. F., and Jones, H. (1936). "The Theory of Properties of Metals and Alloys." Oxford Univ. Press (Clarendon), London and New York.
Nishihara, H., Inoue, T., and Koyama, J. (1974). *Appl. Phys. Lett.* **25**, 391–393.
Ohlmann, R. C., Richards, P. L., and Tinkham, M. (1958). *J. Opt. Soc. Am.* **48**, 531–533.
Panish, M. B. (1975). *IEEE Trans. Microwave Theory Tech.* **MTT-23**, 20–30.
Papayoanou, A., and Fujisawa, A. (1975). *Appl. Phys. Lett.* **26**, 158–160.
Petuchowski, S. J., Rosenberger, A. T., and De Temple, T. A. (1977). *IEEE J. Quantum Electron.* **QE-13**, 476–481.
Pierce, J. R., and Tien, P. K. (1954). *Proc. IRE* **42**, 1389–1396.
Poehler, T. O., Walker, R. E., and Leight, J. W. (1975). *Appl. Phys. Lett.* **26**, 560–561.
Provorov, A. S., and Chebotaev, V. P. (1973). *Sov. Phys.-Dokl.* **18**, 56–57.
Provorov, A. S., and Chebotaev, V. P. (1975). *Sov. J. Quantum Electron.* **5**, 413–419.
Radford, H. E. (1975). *IEEE J. Quantum Electron.* **QE-11**, 213–214.
Ramo, S., Whinnery, J. R., and van Duzer, T. (1965). "Fields and Waves in Communication Electronics." Wiley, New York.
Rosenbluh, M., Temkin, R. J., and Button, K. J. (1976). *Appl. Opt.* **15**, 2635–2644.
Röser, H. P., and Schultz, G. V. (1977). *Infrared Phys.* **17**, 531–536.
Rudko, R. I. (1974). *IEEE Quantum Electron.* **QE-10**, 497–498.
Sawatari, T., and Kapany, N. S. (1970). *J. Opt. Soc. Am.* **60**, 132–133.
Schötzau, H. J., and Kneubühl, F. K. (1974). *Phys. Lett.* **46A**, 415–416.
Schötzau, H. J., and Kneubühl, F. K. (1975a). *IEEE J. Quantum Electron.* **QE-11**, 817–822.
Schötzau, H. J., and Kneubühl, F. K. (1975b). *Appl. Phys.* **6**, 25–30.
Schötzau, H. J., and Veprek, S. (1975). *Appl. Phys.* **7**, 271–277.
Schwaller, P., Steffen, H., Moser, J. F., and Kneubühl, F. K. (1967). *Appl. Opt.* **6**, 827–829.
Seligson, D., Ducloy, M., Leite, J. R. R., Sanchez, A., and Feld, M. S. (1977). *IEEE J. Quantum Electron.* **QE-13**, 468–472.
Sensiper, S. (1955). *Proc. IRE* **53**, 149–161.
Sharp, L. E., and Wetherell, A. T. (1972). *Appl. Opt.* **11**, 1737–1741.
Shevchenko, V. V. (1971). "Continuous Transitions in Open Waveguides." Golem Press, Boulder, Colorado.

Shubert, R. (1974). *J. Appl. Phys.* **45**, 209–215.
Smith, P. W. (1971). *Appl. Phys. Lett.* **19**, 132–134.
Smith, P. W., and Maloney, P. J. (1973). *Appl. Phys. Lett.* **22**, 667–669.
Smith, P. W., Adams, C. R., Maloney, P. J., and Wood, O. R. II, (1976). *Opt. Commun.* **16**, 50–53.
Smith, P. W., Maloney, P. J., and Wood, O. R. II, (1973). *Appl. Phys. Lett.* **23**, 524–526.
Snitzer, E. (1961). *J. Opt. Soc. Am.* **51**, 491–498.
Steffen, H., and Kneubühl, F. K. (1967). Private communication.
Steffen, H., and Kneubühl, F. K. (1968a). *Phys. Lett.* **27A**, 612–613.
Steffen, H., and Kneubühl, F. K. (1968b). *IEEE J. Quantum Electron.* **QE-4**, 992–1008.
Streifer, W., Burnham, R. D., and Scifres, D. R. (1975). *J. Appl. Phys.* **46**, 946–948.
Sturzenegger, Ch., Adam, B., and Kneubühl, F. K. (1977). *IEEE J. Quantum Electron.* **QE-13**, 473–475.
Sturzenegger, Ch., Adam, B., Vetsch, H., and Kneubühl, F. K. (1978). *Digest Conf. Submillimet. Waves Their Appl. 3rd, Guildford*, pp. 27–28.
Tait, G. D., Whitbourn, L. B., and Robinson, L. C. (1973). *Phys. Lett.* **46A**, 239–240.
Tamir, T. (ed.) (1975). *Top. Appl. Phys.* **7**, "Integrated Optics." Springer-Verlag, New York.
Tanaka, A., Tanimoto, A., Murata, N., Yamanaka, M., and Yoshinaga, H. (1974). *Jpn. J. Appl. Phys.* **13**, 1491–1492.
Taylor, H. F., and Yariv, A. (1974). *Proc. IEEE* **62**, 1044–1060.
Temkin, R. J. (1977). *IEEE J. Quantum Electron.* **QE-13**, 450–454.
Tobin, M. S., and Jensen, R. E. (1976). *Appl. Opt.* **15**, 2023–2024.
Tobin, M. S., and Jensen, R. E. (1977). *IEEE J. Quantum Electron.* **QE-13**, 481–484.
Turner, R., and Poehler, T. O. (1971). *J. Appl. Phys.* **42**, 3819–3826.
Ulrich, R., Renk, K. F., and Genzel, L. (1963). *IEEE Trans. Microwave Theory Tech.* **MTT-11**, 363–371.
Ulrich, R., Bridges, T. J., and Pollack, M. A. (1970). *Appl. Opt.* **9**, 2511–2516.
Unger, H. G. (1954). *Arch. Elek. Uebertr.* **8**, 241–252.
Vanderkooy, J., and Kang, C. S. (1976). *Infrared Phys.* **16**, 627–637.
Véron, D., Belland, P., and Beccaria, M. J. (1978). *Digest Conf. Submillimetre Waves Their Appl. 3rd, Guildford* pp. 182–183.
Wang, S. (1973). *J. Appl. Phys.* **44**, 767–780.
Wang, S. (1974a). *Opt. Commun.* **10**, 149–153.
Wang, S. (1974b). *IEEE J. Quantum Electron.* **QE-10**, 413–427.
Wang, S. (1975). *Appl. Phys. Lett.* **26**, 89–91.
Watanabe, S., Ito, H., and Inaba, H. (1975). *Proc. Int. Conf. Infrared Phys. (CIRP), Zurich* p. C208.
Watkins, D. A. (1958). "Topics in Electromagnetic Theory." Wiley, New York.
Weinstein, L. A. (1964). *Zh. Tekhn. Fiz.* **34**, 154–161.
Weinstein, L. A. (1969). "Open Resonators and Open Waveguides." Golem Press, Boulder, Colorado.
Whitbourn, L. B., Robinson, L. C., and Tait, G. D. (1972). *Phys. Lett.* **38A**, 315–317.
Wiggins, J. D., Drozdowicz, Z., and Temkin, R. J. (1977). *IEEE J. Quantum Electron.* **QE-14**, 23–30.
Wood, O. R. II, (1974). *Proc. IEEE* **62**, 355–397.
Wood, O. R. II, Abrams, R. L., and Bridges, T. J. (1970). *Appl. Phys. Lett.* **17**, 376–378.
Wood, O. R. II, Smith, P. W., Adams, C. R., and Maloney, P. J. (1975). *Appl. Phys. Lett.* **27**, 539–541.
Wood, R. A., Pidgeon, C. R., Brignall, N., Al Berkdar, F., and Dunnett, W. (1975). *Opt. Commun.* **14**, 301–303.

Wood, R. A., Brignall, N., Pidgeon, C. R., and Al Berkdar, F. (1976). *Infrared Phys.* **16**, 201–205.

Woskoboinikow, P., Mulligan, W., Praddaude, H. C., Cohn, D. R., and Lax, B. (1978). *Digest Conf. Submillitre Waves Their Appl. 3rd, Guildford* pp. 31–32.

Yamanaka, M. (1976a). Private communication.

Yamanaka, M. (1976b). *Rev. Laser Eng. (Jpn.)* **3**, 253–294.

Yamanaka, M. (1977). *J. Opt. Soc. Am.* **67**, 952–958.

Yamanaka, M., and Yoshinaga, H. (1974). *Digest Int. Conf. Submillimetre Waves Their Appl. Atlanta* p. 26(IEEE Catalog No. 74CHO 856-5MTT).

Yamanaka, M., Tsuda, H., and Mitani, S. (1975). *Opt. Commun.* **15**, 426–428.

Yusek, R., and Lockhart, G. (1973). CLEA Washington, Paper 16.1

Zeidler, G. (1971). *J. Appl. Phys.* **42**, 884–885.

Zimmermann, J., and Gaddy, O. L. (1974). *IEEE J. Quantum Electron.* **QE-10**, 92–93.

Zory, P. (1973). *Appl. Phys. Lett.* **22**, 125–128.

CHAPTER 7

Free Electron Lasers and Stimulated Scattering from Relativistic Electron Beams

P. Sprangle,   Robert A. Smith,   and   V. L. Granatstein

1. Introduction

The possibility of developing lasers in which the active medium is a stream of free electrons has recently evoked much interest. The potential advantages are numerous and include both continuous frequency tunability (e.g., through the variation of electron energy) and very high power operation, since no damage can occur to the lasing medium as happens with liquid or solid-state lasers. Furthermore, it is hoped that the high efficiencies which characterize free electron generators of microwave radiation can eventually be realized with free electron lasers.

In the past few years, the first operation of high peak power free electron lasers (FEL's) has been reported. They have generated seven kilowatts (Deacon *et al.*, 1977) and megawatt power levels in the infrared (Granatstein *et al.*, 1977). Furthermore, recent theoretical results indicate that much larger powers should be achievable. The operative mechanism in these experiments was stimulated backscattering of a low frequency pump wave from a relativistic electron beam, leading to amplification of coherent radiation at a high frequency, corresponding to a double doppler shift of the pump frequency.

The low frequency pump wave may take the form of a static spatially periodic magnetic field or a propagating electromagnetic wave. In the case that the pump wave is a static periodic magnetic field of period l, the frequency of the backscattered wave will be given by $(1 + v_z/c)\gamma_z^2 v_z(2\pi/l)$, which for a highly relativistic beam becomes $\sim 4\pi\gamma_z^2 c/l$, where v_z is the axial velocity of the electrons and $\gamma_z = (1 - v_z^2/c^2)^{-1/2}$. If the pump field is an electromagnetic pump at frequency ω_0 and wave number $k_0 = \omega_0/c$, propagating antiparallel to the electrons, the upshifted backscattered frequency will be $(1 + v_z/c)^2\gamma_z^2\omega_0 \simeq 4\gamma_z^2\omega_0$. For either type of incident pump wave, the scattered wave frequency can be varied simply by changing the electron beam energy or the incident pump wavelength.

Stimulated scattering of photons by an electron ensemble was first predicted by Kapitza and Dirac (1933). The first analysis of high frequency radiation from a relativistic electron beam passing through a periodic magnetic field (undulator) was made by Motz (1951). The proposal for generating short wavelength radiation by stimulated scattering of an electromagnetic pump wave from a relativistic electron beam is due to Pantell et al. (1968).

Experiments demonstrating stimulated scattering with an output in the infrared have been carried out at Stanford University. These experiments utilize the relatively low current electron beam in a linear accelerator ($I \simeq 1$ A, 24 MV $< V <$ 43 MV). They are in the single-particle scattering regime and hence do not involve collective electron oscillations. The single pass gain is low, and the process is strongly dependent on the finite length of the interaction region. The first Stanford experiment (Elias et al., 1976) demonstrated wave amplification at a wavelength of 10.6 μm using a static periodic 2.4 kG helical magnetic pump with a 3.2-cm period; a 7% increase in power was achieved in a 5.2 m interaction length. The second Stanford experiment (Deacon et al., 1977) produced laser oscillations in an optical cavity with peak power of 7 kW at a wavelength of 3.4 μm; 0.01% of the electron beam energy was converted into radiation.

The analysis by Motz (1951) is applicable to the single-particle scattering process in the Stanford experiments. Subsequent analyses by Madey (1971) and by Sukhatme and Wolff (1973) included the importance effect of the finite length of the interaction region. The treatment of Sukhatme and Wolff (1973) was quantum mechanical, as were analyses by Madey (1971) and Madey et al., (1973). However, recent analyses (Hopf et al., 1976a; Kroll and McMullin, 1978) have shown that a classical treatment of the stimulated scattering process is appropriate. Additional, detailed analyses of the Stanford experiments have also appeared in the literature (Colson, 1976, 1977; Bernstein and Hirshfield, 1978; Sprangle and Granatstein, 1978), as have considerations of extending operation to shorter wavelengths up to x-ray regime (Hasegawa *et al.*, 1976; Elias *et al.*, 1976).

With intense relativistic electron beams, collective effects play an important role in the scattering process. The theoretical analyses (Sprangle and Granatstein, 1974; Sprangle *et al.*, 1975) show that for intense magnetized relativistic electron beams, the pump wave may decay into either a space charge beam mode or a cyclotron beam mode depending on, among other things, the pump frequency. The axial magnetic field can be used to increase the effective strength of the pump field by adjusting the electron cyclotron frequency to equal roughly the pump frequency in the beam frame. The resonance produced in this way can substantially enhance the gain of the scattered wave.

Experimental and theoretical research demonstrating stimulated scattering in the submillimeter regime have been carried out at the Naval Research Laboratory (NRL). In experiments using intense relativistic electron beams ($I \sim 30$ kA, 1 MV $< V < 3$ MV), strong submillimeter radiation was first reported by Granatstein *et al.* (1974). Analyses of this process by Sprangle and Granatstein (1974) and Sprangle *et al.* (1975) show that the submillimeter radiation was generated by stimulated Raman-type scattering. That is, one of the scattered waves was a beam space charge wave enhanced by a resonance between the pump frequency and the electron cyclotron frequency. A follow-up experiment designed to enhance the stimulated Raman scattering process generated a megawatt of radiation in a superradiant oscillator operating at a wavelength of 400 μm (Granatstein *et al.*, 1977).

Experiments at Columbia University using a static, rippled, magnetic-field pump with outputs in the centimeter (Efthimion and Schlesinger, 1977) and millimeter wave regime (Marshall *et al.*, 1977; Gilgenbach *et al.*, 1978) clearly distinguished between the cases where the idler was a space charge mode and a beam cyclotron mode. Efficiencies as large as 0.2% were measured in these experiments. Recently a joint Columbia–NRL experiment (Mc-Dermott *et al.*, 1978) has reported operation of a stimulated Raman scattering laser which employs a quasi-optical cavity; laser output was 1 MW at 400 μm and line narrowing to $\Delta\omega/\omega = 2\%$ was observed (compared with $\Delta\omega/\omega \gtrsim 10\%$ in the superradiant oscillator case).

Theoretical treatments of stimulated Raman scattering in plasma were first applied to laser-pellet fusion studies as well as to ionospheric experiments (Silin, 1965, 1967; DuBois and Goldman, 1965, 1967; Nishikawa, 1968; Kaw and Dawson, 1971; Perkins and Kaw, 1971; Forslund *et al.*, 1973; Manheimer and Ott, 1974; Drake *et al.*, 1974). Relativistic analyses with application to the generation of short wavelength radiation using electron beams are now in the literature, treating both the linear regime (Sprangle and Granatstein, 1974; Miroshnichenko, 1975; Sprangle *et al.*, 1975; Kroll and McMullin, 1978) and the nonlinear regime (Kwan *et al.*, 1977; Sprangle and Drobot, 1978). Efficiencies greater than 10% have been calculated, but no efficiency approaching this level has been measured to date; experiments have not been driven into saturation. However, e-folding lengths on the order of 10 cm have been calculated and correspond closely to experimental measurements (Granatstein *et al.*, 1977).

The present paper confines itself to free electron lasers based on stimulated scattering from relativistic electron beams. Other mechanisms involving relativistic electron beams which may lead to free electron lasers have also been studied. While none of these has to date produced the high power in the submillimeter and infrared regimes that characterize the stimulated scattering work, the research is in much too early a stage to ignore the other possible mechanisms. For reference we cite the following mechanisms which have been considered as the basis for free electron sources of coherent high frequency radiation: (1) interaction of a relativistic electron beam with periodic slow wave structures or with dielectric media (Coleman, 1961; Walsh *et al.*, 1977; Schneider and Spitzer, 1974; Gover and Yariv, 1978); and (2) scattering of electromagnetic waves from the front of a relativistic electron beam or from a moving ionization front (Granatstein *et al.*, 1976; Pasour *et al.*, 1977; Buzzi *et al.*, 1977; Lampe *et al.*, 1978).

The linear regime has been discussed in a survey by Granatstein and Sprangle (1977) and in comprehensive treatments by Kroll and McMullin (1977) and Hasegawa (1977). The present paper presents an original reformulation of stimulated scattering theory that attempts to encompass all previous treatments, and compares the theory with available experimental data. In addition, we present a semi-qualitative discussion of nonlinear saturation. Also, the linear treatment differs from previous analyses by performing a perturbation expansion about an exact electron beam equilibrium in a general way.

The details of the physical mechanism associated with FEL's depend, to some extent, on the type of scattering process under consideration, which in turn is a function of the system parameters, such as beam energy, temperature, and density, the length of the interaction region, strength of pump field, etc. There are characteristics of the physical mechanism, however, which are common to all the various scattering processes. These common characteristics can best be described classically in the beam frame of reference.

In the following discussion, quantities in the beam frame will be written with primes. In the beam frame we stipulate that the existing electron equilibrium is perturbed by a low frequency density wave; the existence of this wave will be justified later. Only waves propagating along the z axis, i.e., the direction of the beam velocity in the laboratory frame, will be considered. This longitudinal perturbation has a frequency and wavenumber given by (ω', k'). The introduction of a large amplitude, high frequency incident electromagnetic pump E'_0 at (ω'_0, k'_0) forces the electrons to oscillate at a frequency $\omega'_0 \gg \omega'$ in the direction along E'_0 with a maximum velocity given by $v'_{0\perp} = |e|E'_0/(m_0\omega'_0)$. This transverse oscillation velocity $v'_{0\perp}$, perpendicular to k'_0, couples to the density wave, thus inducing a transverse current at frequency $\omega'_+ = \omega' + \omega'_0$ and wave number $k'_+ = k' + k'_0$. This current now generates an electromagnetic wave at (ω'_+, k'_+). The generated or scattered electromagnetic field consists of backscattered waves propagating antiparallel to the incident pump wave. Forward scattered waves are also induced, but will not be considered because they are downshifted in frequency when transformed back to the laboratory frame. The pump and backscattered wave couple through the $v' \times B'$ term in the Lorentz force equation, resulting in a longitudinal force at (ω', k'). This induced longitudinal force, also called the ponderomotive or radiation pressure force, will reinforce the original longitudinal wave. The backscattered electromagnetic wave is, therefore, unstable, resulting in stimulated emission of radiation.

The ponderomotive potential associated with the (ω', k') wave may be comparable to or may dominate the self space charge potential wave set up by the beam. If the frequency of the pump wave is far above any of the characteristic frequencies of the beam we find that $\omega'_0 = -ck'_0$ and $\omega'_+ = ck'_+$. The backscattered wave frequency in the laboratory is $\omega_+ = \gamma_z(\omega'_0 + v_z k'_0) = (1 + v_z/c)\gamma_z\omega'_0 = (1 + v_z/c)^2\gamma_z^2\omega_0$. For a highly relativistic beam the backscattered frequency is $\omega_+ \approx 4\gamma_z^2\omega_0$.

In Section II we discuss in detail the physical model to be analyzed, which consists of a relativistic electron beam of arbitrary intensity entering and propagating through a static, periodic, circularly polarized magnetic pump field. The general analytic formalism is presented. The calculation is relativistically covariant and consists of a Vlasov–Maxwell perturbation analysis. The perturbation is performed about an exact electron beam equilibrium distribution, formed from constants of the motion in the magnetic pump field. The general analysis culminates in expressions for the driving current densities which, through the wave equations, are sources of the backscattered fields. The time dependences of the scattered fields are harmonic, i.e., temporal steady state conditions exist. Spatial transient effects due to the boundary of the interaction region at $z = 0$ are retained by choosing the scattered fields to have an arbitrary spatial dependence. In Section III we obtain spatial growth rates and estimates for the saturation efficiencies and field amplitudes of various scattering processes: Case (1) is a transient, low gain, single-particle regime and applies to the recent FEL experiments at Stanford University; Case (2) is a high gain, single-particle, thermal beam regime, which is equivalent to the well known stimulated Compton scattering process; Case (3a) is a high gain, cold beam, weak pump scattering regime where collective space charge effects play a dominant role; and Case (3b) is the strong pump limit of Case (3a). In Case (3b) collective space charge effects are dominated by the ponderomotive forces. The experiments performed at the NRL and Columbia University correspond to the regime covered by Case (3a). These processes have been listed in order of increasing interaction strength. Section IV contains a discussion and illustrates an example which applies to generation of coherent visible radiation. Section V gives a description of the pertinent experimental work; and conclusions regarding future work, both experimental and theoretical, are presented in Section VI.

II.   Physical Model and Analysis

We consider a completely nonneutralized electron beam, the motion of which is treated fully relativistically. Only spatial variations along the z axis will be considered for the electron beam, pump field, and scattered fields. The scattered fields consist of electromagnetic as well as electrostatic waves. The amplitude of the static magnetic pump is taken to increase adiabatically in space, reaching a constant value for $z > 0$, as shown in Fig. 1. The reason for including this feature in our model will become clear when the form for the equilibrium electron distribution function is chosen. We choose as the pump wave a magnetic field of the form

$$\mathbf{B}_0 = B_0(\hat{\mathbf{e}}_x \cos k_0 z + \hat{\mathbf{e}}_y \sin k_0 z) \tag{1}$$

where $k_0 = 2\pi/l$, and l is the period of the static field. The pump amplitude is considered to be a function of z for $z < 0$ and constant for $z > 0$. Although the field in Eq. (1) does not satisfy $\nabla \times \mathbf{B}_0 = 0$, we may consider it to be a good approximation near the axis ($r = 0$) of the exact helically symmetric field that is curl-free and varies as $I_0(k_0 r)$. If the beam radius is taken to be

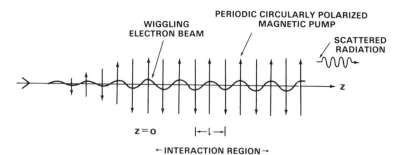

FIG. 1.   Schematic of the FEL model. The unmodulated electron beam enters the interaction region from the left. The pump field builds up adiabatically and reaches a constant amplitude for $z > 0$.

much smaller then $l/2\pi$, the actual field can be well represented by Eq. (1). In our model the self electric and magnetic fields of the beam are neglected.

A. GENERAL FORMALISM

The pump field in Eq. (1) is derived from the vector potential

$$\mathbf{A}_0 = -\mathbf{B}_0/k_0 \tag{2}$$

where $\mathbf{B}_0 = \mathbf{V} \times \mathbf{A}_0$. The steady state equations of motion for the relativistic beam particles in the field of Eq. (2) are

$$\partial p_x/\partial z = (|e|/c)\,\partial A_{0x}/\partial z,$$
$$\partial p_y/\partial z = (|e|/c)\,\partial A_{0y}/\partial z, \tag{3}$$
$$p_z\,\partial p_z/\partial z = -(|e|/c)(p_x\,\partial A_{0x}/\partial z + p_y\,\partial A_{0y}/\partial z).$$

Equations (3) yield the following three constants of the motion for the individual beam particles:

$$\alpha(p_x, z) = p_x - (|e|/c)A_{0x}(z) = p_x + (|e|B_0/ck_0)\cos k_0 z,$$
$$\beta(p_y, z) = p_y - (|e|/c)A_{0y}(z) = p_y + (|e|B_0/ck_0)\sin k_0 z, \tag{4}$$
$$u(\mathbf{p}) = (p_x^2 + p_y^2 + p_z^2)^{1/2},$$

where α and β are the canonical momenta in the x and y directions, and u is the magnitude of the total momentum. These constants of the motion will be used later to construct the equilibrium distribution function. The electric and magnetic fields of the scattered waves are denoted by $\mathbf{E}(z, t)$ and $\mathbf{B}(z, t)$, and given in terms of the potentials $\phi(z, t)$ and $\mathbf{A}(z, t) = (A_x, A_y, 0)$ by

$$\mathbf{E}(z, t) = -(\partial\phi/\partial z)\hat{\mathbf{e}}_z - (1/c)(\partial\mathbf{A}/\partial t), \qquad \mathbf{B}(z, t) = \mathbf{V} \times \mathbf{A}. \tag{5}$$

We shall perform a perturbation Vlasov analysis by expanding the electron distribution function to first order in the potentials ϕ and $\mathbf{A}$ about the exact equilibrium. The relativistic form for the Vlasov equation is

$$\partial f/\partial t + v_z\,\partial f/\partial z - |e|[\mathbf{E} + \mathbf{v} \times (\mathbf{B}_0 + \mathbf{B})/c] \cdot \partial f/\partial \mathbf{p} = 0.$$

The electron distribution function is written as

$$f(\mathbf{p}, z, t) = f^{(0)}(\mathbf{p}, z) + f^{(1)}(\mathbf{p}, z, t), \tag{6}$$

where $f^{(0)}$ and $f^{(1)}$ are solutions of the equilibrium and first order perturbed Vlasov equations, respectively:

$$v_z\,\partial f^{(0)}/\partial z - (|e|/c)\mathbf{v} \times \mathbf{B}_0 \cdot \partial f^{(0)}/\partial \mathbf{p} = 0, \tag{7a}$$

$$\partial f^{(1)}/\partial t + v_z\,\partial f^{(1)}/\partial z - (|e|/c)\mathbf{v} \times \mathbf{B}_0 \cdot \partial f^{(1)}/\partial \mathbf{p}$$
$$= |e|(\mathbf{E} + \mathbf{v} \times \mathbf{B}/c) \cdot \partial f^{(0)}/\partial \mathbf{p}, \tag{7b}$$

where $\mathbf{p} = \gamma m_0 \mathbf{v}$ and $\gamma = (1 + p^2/m_0^2 c^2)^{1/2}$.

It proves convenient at this point to make a transformation from the independent variables (p_x, p_y, p_z, z, t) to new independent variables (α, β, u, z, t). Thus we write the equilibrium and first order components of the electron distribution function as

$$
\begin{aligned}
g^{(0)}(\alpha, \beta, u) &= f^{(0)}(\mathbf{p}, z), \\
g^{(1)}(\alpha, \beta, u, z, t) &= f^{(1)}(\mathbf{p}, z, t).
\end{aligned}
\tag{8}
$$

In terms of these new independent variables, Eq. (7b) takes the form

$$
\partial g^{(1)}/\partial t + (p_z/\gamma m_0)\, \partial g^{(1)}/\partial z = \mathbf{F} \cdot \mathbf{L} g^{(0)},
\tag{9}
$$

where

$$
\mathbf{F}(\alpha, \beta, u, z, t) = |e|(\mathbf{E} + \mathbf{p} \times \mathbf{B}/\gamma m_0 c),
$$

$$
\mathbf{L}(\alpha, \beta, u, z) = \hat{\mathbf{e}}_x\, \partial/\partial\alpha + \hat{\mathbf{e}}_y\, \partial/\partial\beta + (\mathbf{p}/u)\, \partial/\partial u,
$$

and the dependent variables are

$$
p_x(\alpha, z) = \alpha + (|e|/c)A_{0x}(z), \qquad p_y(\beta, z) = \beta + (|e|/c)A_{0y}(z),
$$

$$
p_z(\alpha, \beta, u, z) = [u^2 - p_x^2(\alpha, z) - p_y^2(\beta, z)]^{1/2}, \qquad \gamma(u) = (1 + u^2/m_0^2 c^2)^{1/2}.
$$

Note that the positive branch of the dependent variable $p_z(\alpha, \beta, u, z)$ will be chosen in order to represent a beam propagating in the positive z direction. The advantage of the transformation in Eq. (4) is clearly seen in Eq. (9), in which the Lorentz force term, corresponding to $\mathbf{v} \times \mathbf{B}_0 \cdot \partial f^{(1)}/\partial\mathbf{p}$ in Eq. (7b), has been transformed away. We also exploit the symmetry of the model by transforming to the vector basis $(\hat{\mathbf{e}}_+, \hat{\mathbf{e}}_-, \hat{\mathbf{e}}_z)$, which is given in terms of $(\hat{\mathbf{e}}_x, \hat{\mathbf{e}}_y, \hat{\mathbf{e}}_z)$ by

$$
\hat{\mathbf{e}}_\pm = \tfrac{1}{2}(\hat{\mathbf{e}}_x \pm i\hat{\mathbf{e}}_y).
\tag{10}
$$

Noting Eq. (5) we may now write the scattered fields $\mathbf{E}$ and $\mathbf{B}$ in the interaction region as

$$
\begin{aligned}
\mathbf{E}(z, t) = {}& [-\tfrac{1}{2}(\partial\tilde{\phi}(z)/\partial z)\hat{\mathbf{e}}_z + i(\omega/c)\tilde{A}_+(z)\hat{\mathbf{e}}_+ \\
& + i(\omega/c)\tilde{A}_-(z)\hat{\mathbf{e}}_-]e^{-i\omega t} + \text{c.c.}
\end{aligned}
\tag{11}
$$

$$
\mathbf{B}(z, t) = -i[(\partial\tilde{A}_+(z)/\partial z)\hat{\mathbf{e}}_+ - (\partial\tilde{A}_-(z)/\partial z)\hat{\mathbf{e}}_-]e^{-i\omega t} + \text{c.c.},
$$

where $\phi(z, t) = \tfrac{1}{2}\tilde{\phi}(z)e^{-i\omega t} + \text{c.c.}$, $\mathbf{A}_\pm(z, t) = \tilde{A}_\pm(z)\hat{\mathbf{e}}_\pm\, e^{-i\omega t} + \text{c.c.}$, and c.c. denotes the complex conjugate. The chosen time dependence of the fields in Eqs. (11) is appropriate after all the temporal transients have decayed away and the spatial dependence along the z axis has been left arbitrary. In this new basis representation the force vector $\mathbf{F}$ and the vector operator $\mathbf{L}$ introduced in Eq. (9) take the forms

$$
\begin{aligned}
\mathbf{F}(\alpha, \beta, u, z, t) = {}& \tilde{F}_+(\alpha, \beta, u, z)e^{-i\omega t}\hat{\mathbf{e}}_+ + \tilde{F}_-(\alpha, \beta, u, z)e^{-i\omega t}\hat{\mathbf{e}}_- \\
& + \tilde{F}_z(\alpha, \beta, u, z)e^{-i\omega t}\hat{\mathbf{e}}_z + \text{c.c.}
\end{aligned}
\tag{12a}
$$

and

$$L(\alpha, \beta, u, z) = L_+(\alpha, \beta, u, z)\hat{e}_+ + L_-(\alpha, \beta, u, z)\hat{e}_- + L_z(\alpha, \beta, u, z)\hat{e}_z, \quad (12b)$$

where

$$\tilde{F}_\pm = (|e|/c)[i\omega - v_z(\alpha, \beta, u, z)\,\partial/\partial z]\tilde{A}_\pm(z),$$

$$\tilde{F}_z = |e|/2\{-\partial\tilde{\phi}(z)/\partial z + [\gamma(u)m_0 c]^{-1}$$
$$\times [p_-(\alpha, \beta, z)\,\partial\tilde{A}_+(z)/\partial z + p_+(\alpha, \beta, z)\,\partial\tilde{A}_-(z)/\partial z]\},$$

$$L_\pm = \partial/\partial\alpha \mp i\,\partial/\partial\beta + [p_\pm(\alpha, \beta, z)/u]\,\partial/\partial u,$$

$$L_z = [p_z(\alpha, \beta, u, z)/u]\,\partial/\partial u,$$

$$v_z(\alpha, \beta, u, z) = p_z(\alpha, \beta, u, z)/\gamma(u)m_0,$$

$$p_\pm(\alpha, \beta, z) = p_x \mp ip_y = \alpha \mp i\beta - (|e|B_0/ck_0)e^{\mp ik_0 z}.$$

The perturbed part of the distribution function may also be written in the form

$$g^{(1)}(\alpha, \beta, u, z, t) = \tilde{g}^{(1)}(\alpha, \beta, u, z)e^{-i\omega t} + \text{c.c.} \quad (13)$$

The Vlasov equation for $g^{(1)}$ now becomes

$$[-i\omega + v_z(\alpha, \beta, u, z)\,\partial/\partial z]\tilde{g}^{(1)}e^{-i\omega t} + \text{c.c.}$$
$$= \mathbf{F}\cdot\mathbf{L}g^{(0)} = \tilde{H}(\alpha, \beta, u, z)e^{-i\omega t} + \text{c.c.}, \quad (14)$$

where

$$\tilde{H}(\alpha, \beta, u, z) = \frac{|e|}{2c}\left\{\left[\left(i\omega - v_z\frac{\partial}{\partial z}\right)\tilde{A}_+\right]\left(\frac{\partial}{\partial\alpha} + i\frac{\partial}{\partial\beta}\right)g^{(0)}\right.$$
$$+ \left[\left(i\omega - v_z\frac{\partial}{\partial z}\right)\tilde{A}_-\right]\left(\frac{\partial}{\partial\alpha} - i\frac{\partial}{\partial\beta}\right)g^{(0)}$$
$$\left. + \left[i\omega(p_-\tilde{A}_+ + p_+\tilde{A}_-) - cp_z\frac{\partial\tilde{\phi}}{\partial z}\right]\frac{1}{u}\frac{\partial g^{(0)}}{\partial u}\right\}.$$

The general solution of Eq. (14) is

$$\tilde{g}^{(1)}(\alpha, \beta, u, z) = \int_0^z dz'\, M(\alpha, \beta, u, z, z')\tilde{H}(\alpha, \beta, u, z'), \quad (15)$$

where we have assumed that the beam entered the interaction region unperturbed, i.e., $\tilde{g}^{(1)}(\alpha, \beta, u, 0) = 0$, and where

$$M(\alpha, \beta, u, z, z') = [1/v_z(\alpha, \beta, u, z')]\exp[i\omega\tau(\alpha, \beta, u, z, z')],$$

$$\tau(\alpha, \beta, u, z, z') = \int_{z'}^z \frac{dz''}{v_z(\alpha, \beta, u, z'')}$$

Rearranging Eq. (15), we may write the Fourier coefficient of the perturbed distribution function as

$$\tilde{g}^{(1)}(\alpha, \beta, u, z) = \left[\tilde{G}_+(\alpha, \beta, u, z)\left(\frac{\partial}{\partial \alpha} + i\frac{\partial}{\partial \beta}\right) + \tilde{G}_-(\alpha, \beta, u, z)\left(\frac{\partial}{\partial \alpha} - i\frac{\partial}{\partial \beta}\right) \right.$$

$$\left. + \tilde{G}_z(\alpha, \beta, u, z)\frac{\partial}{\partial u} \right] g^{(0)}(\alpha, \beta, u), \tag{16}$$

where

$$\tilde{G}_\pm = (|e|/2c) \int_0^z dz'\, M(\alpha, \beta, u, z, z')[i\omega - v_z(\alpha, \beta, u, z')\,\partial/\partial z']\tilde{A}_\pm(z'), \tag{17a}$$

$$\tilde{G}_z = (|e|/2cu) \int_0^z dz'\, M(\alpha, \beta, u, z, z')[i\omega p_-(\alpha, \beta, z')\tilde{A}_+(z')$$

$$+ i\omega p_+(\alpha, \beta, z')\tilde{A}_-(z') - cp_z(\alpha, \beta, u, z')\,\partial\tilde{\phi}(z')/\partial z']. \tag{17b}$$

Careful inspection of the definition of $\tilde{G}_\pm$ shows that the integrand in Eq. (17a) is a perfect differential; thus, we can perform the integration over z' and obtain

$$\tilde{G}_\pm(\alpha, \beta, u, z) = -(|e|/2c)\{\tilde{A}_\pm(z) - \tilde{A}_\pm(0)\exp[i\omega\tau(\alpha, \beta, u, z, 0)]\}. \tag{18}$$

The expression in Eq. (15) for $\tilde{g}^{(1)}$ determines the first order perturbation of the electron distribution by the scattered fields with arbitrary axial space dependence, correct to all orders in the pump amplitude.

The perturbing driving current density $\mathbf{J}(z, t)$ which excites the scattered fields can be found from $g^{(1)}$ and is written in general as

$$\mathbf{J}(z, t) = [\tilde{J}_+(z)\hat{\mathbf{e}}_+ + \tilde{J}_-(z)\hat{\mathbf{e}}_- + \tilde{J}_z(z)\hat{\mathbf{e}}_z]e^{-i\omega t} + \text{c.c.}, \tag{19}$$

where

$$\begin{bmatrix} \tilde{J}_\pm(z) \\ \tilde{J}_z(z) \end{bmatrix} = \frac{-|e|}{m_0} \int_{-\infty}^{\infty} \int_{-\infty}^{\infty} \int_0^{\infty} \frac{1}{\gamma(u)} \begin{pmatrix} p_\pm(\alpha, \beta, z) \\ p_z(\alpha, \beta, u, z) \end{pmatrix}$$

$$\times\, \tilde{g}^{(1)}(\alpha, \beta, u, z)\frac{u}{p_z(\alpha, \beta, u, z)}\, d\alpha\, d\beta\, du, \tag{20}$$

and we have used the fact that

$$d^3p = [u/p_z(\alpha, \beta, u, z)]\, d\alpha\, d\beta\, du.$$

The general framework of the formalism is completed by relating self-consistently the scattered fields in Eqs. (11) to the driving current density given by Eq. (19). This is done through the wave equations for $\phi(z, t)$ and $A_\pm(z, t)$ which are given by

$$\partial^2 \phi/\partial t\, \partial z = 4\pi J_z, \tag{21a}$$

$$(\partial^2/\partial z^2 - (1/c^2)\, \partial^2/\partial t^2)A_\pm = -(4\pi/c)J_\pm. \tag{21b}$$

It should be mentioned at this point that the terms $\tilde{G}_+$ which contribute to $g^{(1)}$ and thus to J do not appear in any previous analyses on FELs to our knowledge. This is due to the fact the previous analyses do not use the exact equilibrium distribution function in their treatment of the problem.

As already mentioned, the ponderomotive potential wave which results from the coupling of the pump and scattered wave is responsible for the longitudinal bunching of the beam electrons. The modulated beam together with the pump field produces a current density which drives the scattered wave. The most straightforward way to appreciate the bunching process associated with the ponderomotive potential is to consider the form of the Hamiltonian for a particle in the combined fields of the pump and scattered waves. The Hamiltonian of such a particle with zero transverse canonical momentum is given by

$$H = \{c^2 p_z^2 + |e|^2[A_0(z) + A(z, t)]^2 + m_0^2 c^4\}^{1/2} - |e|\phi(z, t)$$

where $A_0(z)$, $A(z, t)$, and $\phi(z, t)$ are found from Eqs. (2) and (5). Since $|A(z, t)| \ll |A_0(z)|$ we can expand the Hamiltonian to first order in $A(z, t)$ and write it in the form

$$H = \{c^2 p_z^2 + |e|^2 A_0^2(z) + m_0^2 c^4\}^{1/2} - |e|[\phi(z, t) + \phi_{pond}(z, t)],$$

where $\phi_{pond}(z, t)$ is defined as the ponderomotive potential, given by

$$\phi_{pond}(z, t) = -|e|A_0(z) \cdot A(z, t)/\gamma m_0 c^2,$$

and

$$\gamma = (1 + p_z^2/m_0^2 c^2 + |e|^2 A_0^2/m_0^2 c^4)^{1/2}.$$

In this expanded form we see that the transverse component of the scattered field given by $A(z, t)$, when coupled to the pump field, plays a role similar to that of the space charge potential $\phi(z, t)$. The ponderomotive potential $\phi_{pond}(z, t)$, has the effect of axially bunching the beam. This bunching process in turn gives rise to the space charge scalar potential $\phi(z, t)$. Hence, the ponderomotive potential drives the space charge potential. For future reference, we note that the ponderomotive potential can be put into the form

$$\phi_{pond}(z, t) = \tfrac{1}{2}\tilde{\phi}_{pond}(z)e^{-i\omega t} + \text{c.c.}, \tag{22a}$$

where

$$\tilde{\phi}_{pond}(z) = (2\Omega_0/\gamma c k_0)(e^{ik_0 z}\tilde{A}_+(z) + e^{-ik_0 z}\tilde{A}_-(z)) \qquad (22b)$$

and $\Omega_0 = |e|B_0/m_0 c$.

B. EVALUATION OF THE CURRENT DENSITY

In order to obtain an explicit expression for the driving current density given in Eq. (19) we need to specify a form for the equilibrium distribution function. As was mentioned earlier, the pump magnetic field amplitude $B_0(z)$ builds up adiabatically and reaches a constant value inside the interaction region. We assume that the electron beam is generated far to the left of the interaction region, where the pump field vanishes. In general the beam will be produced with a thermal momentum spread in the transverse as well as the longitudinal direction. However, the transverse momentum spread, $P_{\perp th}$, may be neglected if it is much less than the ordered transverse momentum $|e|B_0/ck_0$ induced by the pump field. Unless the pump amplitude is unusually small, the condition, $P_{\perp th} \ll |e|B_0/ck_0$ can be easily satisfied. Then the transverse momentum is proportional only to the pump amplitude and the solutions of the equations of motion are accurately given by $\alpha = \beta = 0$ in Eqs. (4). Thus, for $P_{\perp th} \ll |e|B_0/ck_0$, we may choose the equilibrium distribution function to be of the form

$$g^{(0)}(\alpha, \beta, u) = n_0 \, \delta(\alpha) \, \delta(\beta)g_0(u), \qquad (23)$$

where n_0 is the ambient beam density far to the left of the interaction region. The normalization of $g_0(u)$ is such that

$$\int_0^\infty du \, [u/u_z(u)]g_0(u) = 1,$$

where $u_z(u) = p_z(\alpha = 0, \beta = 0, u, z) = (u^2 - |e|^2 B_0^2/c^2 k_0^2)^{1/2}$. We may remark that for $\alpha = \beta = 0$, each particle has constant axial momentum. This is rather important property of the pump field in Eq. (1). Since the equilibrium axial momentum and velocity do not oscillate as a function of z, the pump field does not introduce an effective longitudinal spread in velocity which would limit the effectiveness of the scattering process. The appropriate functional form of $g_0(u)$ will depend on the scattering process being considered; we shall leave it general for the moment. Using the distribution in Eq. (23), it is straightforward to evaluate the Fourier amplitudes of the driving current density in Eq. (20). The details of this calculation are given in Appendix 1. Defining

$$\Psi_\pm(z) \equiv [(\omega/v_z) \pm k_0]z,$$

we find

$$
\tilde{J}_{\pm}(z) = \frac{-\omega_b^2}{4\pi} \int_0^\infty \left[\frac{u}{\gamma u_z c} \left\{ \left(1 + \frac{\beta_\perp^2}{2} \right) [\tilde{A}_{\pm}(z) - \tilde{A}_{\pm}(0) e^{i\omega z/v_z}] \right. \right.
$$

$$
+ \frac{\beta_\perp^2}{2} \frac{\omega}{k_0 v_z} e^{i\Psi_\mp(z)} [\tilde{A}_+(0)(1 - e^{ik_0 z}) - \tilde{A}_-(0)(1 - e^{-ik_0 z})]
$$

$$
+ \frac{\beta_\perp^2}{2} e^{i\Psi_\mp(z)} [\tilde{A}_\mp(z) e^{-i\Psi_\mp(a)} - \tilde{A}_\mp(0)^{\mp ik_0 z}] \bigg\}
$$

$$
+ \frac{m_0 \beta_\perp}{2} e^{i\Psi_\mp(z)} \int_0^z \left\{ \frac{\partial \tilde{\phi}(z')}{\partial z'} e^{-i\omega z'/v_z} \right.
$$

$$
\left. \left. + i \frac{\omega \beta_\perp}{c} [\tilde{A}_+(z') e^{-i\Psi_-(z')} + \tilde{A}_-(z') e^{-i\Psi_+(z')}] \right\} dz' \frac{\partial}{\partial u} \right] g_0 \, du \qquad \text{(24a,b)}
$$

$$
\tilde{J}_z(z) = -\frac{\omega_b^2}{8\pi} e^{i\omega z/v_z} \int_0^\infty \left[\frac{u}{\gamma u_z c} \frac{\omega}{k_0 v_z} \right.
$$

$$
\times \beta_\perp [\tilde{A}_+(0)(e^{ik_0 z} - 1) - \tilde{A}_-(0)(e^{-ik_0 z} - 1)]
$$

$$
- m_0 \int_0^z \left\{ \frac{\partial \tilde{\phi}(z')}{\partial z'} + \frac{i\omega \beta_\perp}{c} [\tilde{A}_+(z') e^{ik_0 z'} + \tilde{A}_-(z') e^{-ik_0 z'}] \right\}
$$

$$
\times e^{-i\omega z'/v_z} \, dz' \frac{\partial}{\partial u} \bigg] g_0(u) \, du, \qquad \text{(24c)}
$$

where $\omega_b = (4\pi |e|^2 n_0/m_0)^{1/2}$ is the beam plasma frequency, $\beta_\perp = \Omega_0/\gamma k_0 v_z$ the transverse beam velocity normalized to v_z, $\Omega_0 = |e| B_0/m_0 c$, and $v_z = u_z/\gamma m_0$.

It is informative to consider at this point the current density $\mathbf{J}(z, t)$ in the absence of the pump field in order to find the pump-free eigenmodes of the system. Setting $B_0 = 0$ we find from Eqs. (24) that

$$
\tilde{J}_{\pm}(z)|_{B_0=0} = -\frac{\omega_b^2}{4\pi c} \int_0^\infty \frac{1}{\gamma} [\tilde{A}_{\pm}(z) - \tilde{A}_{\pm}(0) e^{i(\omega/v_z)z}] g_0 \, du,
$$

$$
\text{(25)}
$$

$$
\tilde{J}_z(z)|_{B_0=0} = \frac{\omega_b^2}{8\pi} m_0 \int_0^\infty \int_0^z \frac{\partial \tilde{\phi}(z')}{\partial z'} e^{-i(\omega/v_z)(z'-z)} \, dz' \frac{\partial}{\partial u} g_0 \, du,
$$

where $v_z|_{B_0=0} = u/\gamma m_0$. For the moment we will consider the eigenmodes for $z \gg 0$. In this case we can take $\tilde{A}_{\pm}(z)$ and $\tilde{\phi}(z)$ to be plane waves of the form $\tilde{A}_{\pm}(z) = \tilde{A}_{\pm}(0) \exp(ik_{\pm} z)$ and $\tilde{\phi}(z) = \tilde{\phi}(0) \exp(ikz)$, where $k_{\pm}$ and k are real constants. Neglecting the free streaming terms in Eqs. (24), i.e., $\exp[i(\omega/v_z)z]$, because they eventually phase mix away, we obtain for $z \gg 0$

$$\tilde{J}_{\pm}(z) = -\frac{\omega_b^2}{4\pi c} \int_0^\infty \frac{g_0 \, du}{\gamma} \tilde{A}_{\pm}(z),$$

$$\tilde{J}_z(z) = \frac{-\omega_b^2}{8\pi} m_0 \omega \int_0^\infty \frac{\partial g_0/\partial u \, du}{\omega - v_z k} \tilde{\phi}(z). \qquad (26)$$

Combining the current densities given by Eqs. (19) and (26) with the wave equations [Eqs. (21)], we arrive at the conventional dispersion relations for the transverse electromagnetic field and the electrostatic field:

$$D(\omega, k_{\pm})\tilde{A}_{\pm}(z) = 0, \qquad [1 + \chi(\omega, k)]\tilde{\phi}(z) = 0, \qquad (27)$$

where χ is the electron susceptibility

$$\chi(\omega, k) = \frac{\omega_b^2}{k} \int \frac{m_0 \, \partial g_0/\partial u \, du}{\omega - v_z k} \qquad (28a)$$

and

$$D(\omega, k_{\pm}) = \omega^2 - c^2 k_{\pm}^2 - \omega_b^2 \int_0^\infty g_0/\gamma \, du. \qquad (28b)$$

It is now possible to show that for the doppler-upshifted backscattered wave we can consider the A_+ or A_- wave separately. In the presence of the pump field it will be shown that the products $D(\omega, k_{\pm})\tilde{A}_{\pm}(z)$ are no longer zero, but are equal to a small coupling term proportional to B_0^2. In order for $\tilde{A}_+(z)$ to be excited it is necessary for $D(\omega, k_+) \approx 0$; likewise, for $\tilde{A}_-(z)$ to be excited, $D(\omega, k_-)$ must be very small. It can be shown by carefully scrutinizing Eqs. (24) that k_+ and k_- are connected by the relationship $k_+ = k_- - 2k_0$. Because of this relationship, the functions $D(\omega, k_+)$ and $D(\omega, k_-)$ cannot be simultaneously zero for the same backscattered upshifted frequency. Therefore, A_+ and A_- are independently excited. Furthermore, by simply changing the sign of ω and either k_+ or k_- we see that the behavior of the two waves is identical. Hence, in the remainder of this paper we shall arbitrarily take the wave at (ω, k_+) to be resonant, i.e., $D(\omega, k_+) \approx 0$, and the (ω, k_-) wave to be nonresonant, i.e., $D(\omega, k_-) \neq 0$. This amounts to keeping only the $\tilde{A}_+$ wave and neglecting the $\tilde{A}_-$ wave. We will now evaluate the growth rates and saturation level for various scattering regimes.

III.   Gain and Saturation Level in Various FEL Regimes

A.   CASE 1. LOW GAIN, TENUOUS COLD BEAM LIMIT

As our first example we will consider the low gain, (i.e., "short cavity") scattering process in which collective effects are unimportant. This limit corresponds to the parameter regime of the recent experiments carried out at

Stanford University with highly relativistic (≤ 48 MeV), low current (≤ 2.4A) electron beams. Collective effects are manifested by the space charge potential ϕ, which arises from the charge bunching under the action of the pondermotive potential ϕ_{pond}. In the very tenuous beam limit the scalar potential is much less than the pondermotive potential. It can be shown that collective effects can be neglected when the magnitude of the electron susceptibility χ is small compared to unity. In this limit we can set $\phi = 0$ and represent the scattered electromagnetic wave in the form [see Eq. (11)]

$$\tilde{A}_+(z) = \tilde{A}_+(0) \exp\left[i \int_0^z k_+(z') \, dz'\right], \tag{29}$$

where $k_+(z) \equiv k_{0+} + \delta k(z)$, $k_{0+} \gg |\delta k|$ is real and constant, and δk is a complex, slowly varying function of z. Using this form for the field we find, from Eq. (24a),

$$\begin{aligned}
\tilde{J}_+(z) = -\frac{\omega_b^2}{4\pi c} \int_0^\infty du \Bigg\{ & \frac{u}{\gamma u_z}\left[\left(1 + \frac{\beta_\perp^2}{2}\right)(1 - e^{-i\psi(u,z)})\right. \\
& - \frac{\beta_\perp^2}{2}\frac{\omega}{k_0 v_z} e^{-i\psi(u,z)}(1 - e^{-ik_0 z})\Bigg] \\
& + i\frac{\beta_\perp^2}{2}\omega m_0 e^{-i[\psi(u,z)+k_0 z]} \int_0^z dz' \, e^{i[\psi(u,z')+k_0 z']} \frac{\partial}{\partial u}\Bigg\} \\
& \times g_0 \tilde{A}_+(0) \exp\left[i \int_0^z k_+(z') \, dz'\right],
\end{aligned} \tag{30}$$

where $\psi(u,z) = \int_0^z k_+(z') \, dz' - \omega z/v_z$. In the context of Eqs. (29) and (30) the wave equation [Eq. (21b)] can be used to derive the following dispersion relation:

$$k_{0+}^2 - \omega^2/c^2 + 2k_{0+}\delta k(z) = -(\omega_b^2/c^2)\mu(\omega, k_{0+}, z), \tag{31}$$

where

$$\begin{aligned}
\mu(\omega, k_{0+}, z) = \int_0^\infty \Bigg\{ & \frac{u}{\gamma u_z}\left[\left(1 + \frac{\beta_\perp^2}{2}\right)(1 - e^{-i(k_{0+}-\omega/v_z)z})\right. \\
& - \frac{\beta_\perp^2}{2}\frac{\omega}{v_z k_0} e^{-i(k_{0+}-\omega/v_z)z}(1 - e^{-ik_0 z})\Bigg] \\
& - \frac{1}{2}\beta_\perp^2 m_0 v_z \omega \frac{(1 - e^{-i(k-\omega/v_z)z})}{\omega - v_z k}\frac{\partial}{\partial u}\Bigg\} g_0 \, du
\end{aligned} \tag{32}$$

and $k = k_{0+} + k_0$.

Note that in obtaining Eq. (31) we used the facts that in the present scatter-ing limit, $|\delta k(z)| \ll k_{0+}$, $\int_0^z \delta k(z')\, dz' \ll 1$ and $\delta k(z)$ is a very "weak" function of z, so that $\psi(u, z)$ has been approximated in the right hand side (RHS) of Eq. (32) by $(k_{0+} - \omega/v_z)z$. Solving for the imaginary part of Eq. (31), we find

$$
\begin{aligned}
\mathrm{Im}\,(\delta k) = \frac{-\omega_b^2/c^2}{2k_{0+}} \int_0^\infty & \left[\frac{\beta_\perp^2 m_0 \omega}{2} \frac{\sin[(\omega/v_z - k)z]}{(\omega/v_z - k)} \frac{\partial}{\partial u} \right. \\
& - \frac{u}{\gamma u_z} \left\{ \left(1 + \frac{\beta_\perp^2}{2}\right) \sin[(\omega/v_z - k_{0+})z] \right. \\
& \left. + \frac{\beta_\perp^2}{2} \frac{\omega}{v_z k_0} (\sin[(\omega/v_z - k_{0+})z] - \sin[(\omega/v_z - k)z]) \right\} \bigg] g_0\, du. \quad (33)
\end{aligned}
$$

The total gain in amplitude of the scattered wave over the interaction length L is given by

$$
\begin{aligned}
G_L = & -\int_0^L \mathrm{Im}[\delta k(z')]\, dz' \\
= & \frac{\omega_b^2/c^2}{2k_{0+}} \int_0^\infty \left[\frac{\beta_\perp^2}{4} m_0 \omega L^2 \left\{ \frac{\sin[(k - \omega/v_z)L/2]}{(k - \omega/v_z)L/2} \right\}^2 \frac{\partial}{\partial u} \right. \\
& - \frac{u}{\gamma u_z} \left\{ \frac{\beta_\perp^2}{2} \frac{\omega}{v_z k_0} L \frac{\sin^2[(k - \omega/v_z)L/2]}{(k - \omega/v_z)L/2} \right. \\
& \left. - \left(1 + \frac{\beta_\perp^2}{2} \frac{\omega}{v_z k_0}\right) L \frac{\sin^2[(k_{0+} - \omega/v_z)L/2]}{(k_{0+} - \omega/v_z)L/2} \right\} \bigg] g_0\, du. \quad (34)
\end{aligned}
$$

Equation (34) is a complicated expression for G_L which depends on $g_0(u)$. If, however, g_0 is such that the phase of the sine functions in Eq. (34) does not vary appreciably over the range of v_z for which g_0 is nonzero, then to evaluate the integral we may take

$$
g_0(u) = (u_z/u)\, \delta(u - u_0). \quad (35)
$$

The condition for the applicability of Eq. (35) is that the axial thermal velocity spread V_{th} of the beam satisfy

$$
V_{th} \ll \pi c/\gamma_{z0}^2 k_0 L = \lambda_s c L. \quad (36)
$$

The thermal energy spread E_{th} of the beam in general is given by

$$
E_{th} = \gamma_0 \gamma_{z0}^2 m_0 v_{z0} V_{th}, \quad (37)
$$

and for a highly relativistic beam, the inequality [Eq. (36)] may be rewritten as

$$
E_{th}/E_0 \ll \gamma_{z0}^2 \lambda_s/L, \quad (38)
$$

where $E_0 = (\gamma_0 - 1)m_0 c^2 \simeq \gamma_0 m_0 c^2$ is the kinetic energy of the particles. In the experiments to date this condition is satisfied by roughly one order of magnitude.

For a highly relativistic beam, we may approximate $v_z \simeq c$, $u_z \simeq u$, $k_{0+} \simeq 2\gamma_z^2 k_0$. Then using Eq. (35), the total gain becomes

$$G_L = \left(\frac{\xi}{2\gamma_{z0}}\right)^2 k_0 L \left\{ \frac{\beta_{0\perp}^2 \gamma_{z0}^2}{2} \left[(k_0 L)^2 \frac{\partial}{\partial \theta_0} \left(\frac{\sin \theta_0}{\theta_0}\right)^2 + 2 \frac{\sin^2 \theta_0}{\theta_0} \right] \right.$$
$$\left. - (1 + \beta_{0\perp}^2 \gamma_{z0}^2) \frac{\sin^2(\theta_0 + k_0 L/2)}{(\theta_0 + k_0 L/2)} \right\}$$
$$\equiv \left(\frac{\xi}{2\gamma_{z0}}\right)^2 k_0 L F(\theta_0, k_0 L, \beta_{0\perp} \gamma_{z0}), \tag{39}$$

where $\xi = \omega_b/\gamma_0^{1/2} c k_0$ is the beam strength parameter, $\gamma_0 = \gamma(u_0)$, $\gamma_{z0} = \gamma_z(u_0)$, $\beta_{0\perp} = \beta_\perp(u_0)$, and $\theta_0 = (\omega/v_{z0} - k_{0+} - k_0)L/2$. In deriving Eq. (39) we have used the fact that

$$\partial/\partial u = -(\omega/v_z^2)(L/2)(u/m_0 \gamma u_z \gamma_z^2) \, \partial/\partial \theta_0.$$

In Figs. (2a,b) we exhibit the function $F(\theta_0, k_0 L, \beta_{0\perp} \gamma_{z0})$ for the cases where $L = 160l$ and $L = l$. If the interaction region consists of a large number of pump periods ($k_0 L \gg 1$) the first term of the gain expression in Eq. (39) is dominant. For this case

$$G_L \approx (\xi^2/8)\beta_{0\perp}^2(k_0 L)^3(\partial/\partial \theta_0)(\sin \theta_0/\theta_0)^2. \tag{40}$$

The function $\partial/\partial \theta_0(\sin \theta_0/\theta_0)^2$ has a maximum value of 0.54 when $\theta_0 = -1.3$; therefore, the maximum value of gain is

$$G_{L_{max}} \approx (\xi/4)^2 \beta_{0\perp}^2(k_0 L)^3. \tag{41}$$

These expressions for the gain are similar to results found previously by Motz (1951), Sukhatme and Wolff (1973), Colson (1976), Hopf et al. (1976a), and Kroll and McMullin (1978).

A detailed calculation of the efficiency in the FEL low gain regime was presented by Hopf et al., who found that the efficiency was approximately given by

$$\eta = (\lambda_s/L)\gamma^2 \tag{42}$$

where $\lambda_s = 2\pi/k_+$. It is possible to obtain this result from general arguments based on electron trapping in the longitudinal wave associated with the electron beam. (The total longitudinal wave in general consists of both the space charge potential and ponderomotive potential waves.) The following discussion, which will be used to estimate saturation efficiencies as well as the

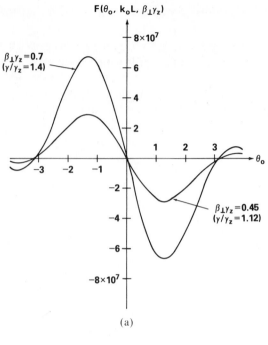

(a)

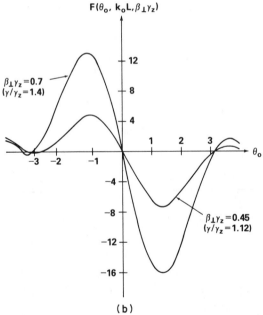

(b)

FIG. 2. The gain function $F(\theta_0, k_0 L, \beta_\perp \gamma_{z0})$ corresponding to the low gain, tenuous cold beam scattering limit. The interaction length for (a) is 160 pump periods ($k_0 L = 320\pi$) and for (b) 1 pump period ($k_0 L = 2\pi$).

296

saturated electromagnetic field amplitude, may be applied to all cases involving scattering from a cold beam in which trapping saturates the process.

For maximum spatial gain, the phase velocity $v_{ph} = \omega/\mathrm{Re}(k)$ of the longitudinal wave and the directed axial beam velocity v_{z0} are related, in the linear development of the instability, by an expression of the form

$$v_{ph} - v_{z0} = -\Delta v \tag{43}$$

where $\Delta v > 0$ is a small term which depends on the particular scattering regime being considered. In the low gain, tenuous beam limit we have just discussed, for example, the gain G_L maximizes at $\theta_0 = -1.3$; this condition may be written in the form of Eq. (43). In the high gain regimes to be discussed later, in which the scattering process may be described by a dispersion relation homogeneous in space, a condition of the form of Eq. (43) will be seen to describe the condition of maximum coupling between the beam and the backscattered wave. Since wave growth will occur for $\Delta v > 0$, we can interpret this condition as stating that the phase velocity of the longitudinal wave must be less than the axial beam velocity in order for beam kinetic energy to be converted into wave energy. The wave amplitude increases, at the expense of the beam kinetic energy, until the beam electrons become trapped at the bottoms of the potential wells of the longitudinal wave. At this point the wave reaches its maximum amplitude and the average beam kinetic energy reaches a minimum. The average axial beam speed at saturation, $v_{z,s}$, is given approximately by the trapping condition

$$v_{ph} - v_{z,s} = +\Delta v \tag{44}$$

Therefore, at saturation the kinetic energy lost by the beam corresponds to a decrease in axial beam speed of $v_{z0} - v_{z,s} = 2\,\Delta v$. It should be noted that when the beam electrons are trapped at the bottom of the potential wells of the wave they are modulated spatially but are fairly monoenergetic. The decrease in beam energy can be equated to the increase in wave energy. In this way the efficiency and saturated field amplitude of the scattering process can be estimated.

In the case of a relativistic electron beam the change in kinetic energy at saturation, when the beam particles are deeply trapped, is

$$\Delta E_{K.E.} = [\gamma(v_{z0}) - \gamma(v_{z,s})]m_0 c^2$$

$$\simeq 2[\partial\gamma(v_z = v_{z0})/\partial v_z]\,\Delta v\, m_0 c^2 = 2\gamma_0\gamma_{z0}^2 m_0 v_{z0}\,\Delta v. \tag{45}$$

The efficiency is then given by

$$\eta = \Delta E_{K.E.}/(\gamma_0 - 1)m_0 c^2 \simeq 2\gamma_{z0}^2\,\Delta v/c, \tag{46}$$

where the last expression in Eq. (46) is for a highly relativistic beam, $v_{z0} \approx c$. The amplitude of the scattered electromagnetic fields at saturation can now be obtained by the following argument.

In the steady state, conservation of total energy flux states that

$$(\partial/\partial z)[nv_z(\gamma - 1)m_0 c^2 \hat{e}_z + (c/4\pi)\mathbf{E} \times \mathbf{B}] = 0.$$

Integrating from the input of the interaction region ($z = 0$) to the point where the scattered field is maximum ($z = z_{sat}$) gives

$$|\tilde{A}_+(z = z_{sat})|^2 \approx |\tilde{A}_+(z = 0)|^2 + (4\pi n_0 c^2/\omega^2)\,\Delta E_{K.E.}$$

$$\approx |\tilde{A}_+(z = 0)|^2 + [4\pi n_0/(\omega/c)^2]\eta\gamma_0 m_0 c^2. \quad (47)$$

In obtaining $|\tilde{A}_+(z = z_{sat})|$ we used the fact that the fields were given by Eqs. (11), $\tilde{A}_- = 0$, $\mathbf{E} \times \mathbf{B} \approx (\omega^2/c^2)|\tilde{A}_+(z)|^2$, and $v_{z0} = v_{z,s} \approx c$. The magnitude of the transverse electric field at saturation becomes

$$|\mathbf{E}|_{sat} = (\omega/c)|\tilde{A}_+(z = z_{sat})|. \quad (48)$$

The results for the efficiency and saturated transverse electric field in Eqs. (46) and (48) apply to those scattering processes in which electron trapping is responsible for saturation. To apply these results to the low gain scattering process in Case 1, we note that the condition for maximum gain over a distance $z_{sat} = L$ is given by $\theta_0 = -1.3$, which when put in the form of Eq. (43) becomes

$$v_{ph} - v_{z0} = (-2.6/kL)v_{z0} = -\Delta v. \quad (49)$$

Substituting Eq. (49) into Eq. (46) the estimated value of efficiency is

$$\eta \approx 5.2\gamma_{z0}^2/kL \approx \gamma_{z0}^2 \lambda_s/L, \quad (50)$$

in agreement with Eq. (42), where $\lambda_s = 2\pi/k_{0+} \approx 2\pi/k$ is the wavelength of the scattered wave. The assumption implicit in the above argument is that the length L and the energy lost by the electrons are matched so that the trapping occurs at the end of the system. The energy lost by the electrons, however, goes into the scattered wave and depends upon the amplitude of the input signal. To find the field amplitude necessary to effectuate this energy change in the length L we note that in the low gain limit $G_L \ll 1$, we have $|\tilde{A}_+(z = z_{sat} = L)| \approx |\tilde{A}_+(z = 0)|(1 + G_L)$, and thus Eq. (48) gives

$$|\tilde{A}_+(z = 0)|^2 = [2\pi n_0/(\omega/c)^2](\eta/G_L)\gamma_0 m_0 c^2.$$

At maximum gain conditions we have, using Eq. (41),

$$|\tilde{A}_+(z = 0)| = \sqrt{5.2}(\gamma_0/\gamma_{z0}^2)(m_0 c^2/|e|)/\beta_{0\perp}(k_0 L)^2. \quad (51)$$

Since $|\tilde{A}_+(z = z_{\mathrm{sat}})| \approx |\tilde{A}_+(z = 0)|$, the saturated transverse electric field amplitude can be estimated from Eqs. (48) and (51) to be

$$|\mathbf{E}|_{\mathrm{sat}} \approx 2\sqrt{5.2}(m_0 c^2/|e|)k_0\gamma_0/\beta_{0\perp}(k_0 L)^2. \tag{52}$$

Equation (52) gives the required input signal necessary to achieve the efficiency in Eq. (46) at maximum gain conditions. This input signal level is required for optimum efficiency. If the input signal is weaker than this value, saturation of the wave will not occur within the length L. If it is stronger, saturation will occur before the beam reaches the end of the interaction region, resulting in smaller total gain.

B. CASE 2. HIGH GAIN, TENUOUS THERMAL BEAM LIMIT

We now consider the high gain limit with a tenuous thermal beam; this limit is the relativistic analog of stimulated Compton scattering. We again neglect the space charge potential but now assume that the wave e-folds many times, hence $|\tilde{A}(z)| \gg |\tilde{A}(0)|$ except near $z = 0$. Since for this case

$$-\mathrm{Im} \int_0^z k_+(z')\,dz' \gg 1,$$

the driving current is found from Eq. (30) to take the form

$$\tilde{J}_+(z) = -\frac{\omega_b^2}{4\pi c}\int_0^\infty \left[\frac{u}{\gamma u_z}\left(1 + \frac{\beta_\perp^2}{2}\right) - \frac{\beta_\perp^2}{2}\omega v_z m_0 \frac{\partial/\partial u}{\omega - v_z k}\right]g_0\,du\,\tilde{A}_+(0)e^{ik_+ z} \tag{53}$$

where k_+ is complex and independent of z. Substituting Eq. (53) into Eq. (21b) gives the dispersion relation

$$k_+^2 - \frac{\omega^2}{c^2} = -\frac{\omega_b^2}{c^2}\int_0^\infty \left[\frac{u}{\gamma u_z}\left(1 + \frac{\beta_\perp^2}{2}\right)g_0(u) - \frac{\beta_\perp^2}{2}\omega v_z m_0 \frac{\partial g_0/\partial u}{\omega - v_z k}\right]du, \tag{54}$$

where $k = k_+ + k_0 \equiv k_r + ik_i$. In the Compton scattering regime the phase velocity of the ponderomotive wave is resonant with a small fraction of the thermal beam electrons (see Fig. 3). We now consider the case where the thermal width of the particle distribution function is large compared to the width of the wave distribution $(\omega - v_z k)^{-1}$. If the thermal velocity spread of the relativistic beam is such that

$$V_{\mathrm{th}} \gg |ck_i/k_r|, \tag{55}$$

then the pump wave scatters from single particles. When the inequality [Eq. (55)] is satisfied we can take the imaginary parts of both sides of Eq. (54) and use the fact that

$$\mathrm{Im}[1/(\omega - v_z k)] = -(\pi/|k_r|)\delta(v_z - \omega/k_r)$$

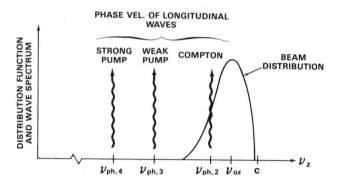

FIG. 3.   Pictorial representation of phase velocities of the longitudinal waves and electron beam distribution function. When the phase velocities lie far outside the beam distribution, the beam may be considered cold.

as k_i approaches zero from negative values. Because

$$v_z = c(u^2 - u_{0\perp}^2)^{1/2}/(u^2 + m_0^2 c^2)^{1/2},$$

we can write

$$\delta(v_z - \omega/k_r) = (\gamma\gamma_z^2 m_0 u_z/u)\,\delta(u - u_{\rm ph}) \tag{56}$$

where $\partial v_z/\partial u = u/(\gamma\gamma_z^2 m_0 u_z)$, $u > 0$, $u_{\rm ph} \equiv \gamma_{\rm ph}(\beta_{\rm ph}^2 m_0^2 c^2 + u_{0\perp}^2)^{1/2}$, $u_z = (u^2 - u_{0\perp}^2)^{1/2}$, $\beta_{\rm ph} \equiv \omega/k_r c$, $\gamma_{\rm ph} \equiv (1 - \beta_{\rm ph}^2)^{-1/2}$, and $u_{0\perp}^2 = (|e| B_0/ck_0)^2$. The imaginary part of k, from Eq. (54) using Eq. (56), is given by

$$k_i = (-\omega_b^2/4k_r c^2)(\pi/|k_r|)\beta_\perp^2 v_z m_0\,\omega(\gamma\gamma_z^2 m_0 u_z/u)\,\partial g_0/\partial u|_{u=u_{\rm ph}}. \tag{57}$$

For a drifting electron beam with average total momentum u_0, we can take $g_0(u)$ to be Maxwellian-like of the form

$$g_0(u) = (u_z/u\sqrt{2\pi U_{\rm th}})e^{-(u-u_0)^2/(2U_{\rm th}^2)}, \tag{58}$$

such that $\int_0^\infty du\, g_0 u/u_z = 1$. Substituting Eq. (58) into Eq. (57), we get for a highly relativistic beam the result

$$k_i = (\omega_b^2/8k_0)\sqrt{\pi}\beta_\perp^2(\gamma m_0^2/U_{\rm th}^2)xe^{-x^2}|_{u=u_{\rm ph}} \tag{59}$$

where $x = (u - u_0)/2^{1/2}U_{\rm th}$. The function xe^{-x^2} has a minimum value of -0.43 when $x = -1/2^{1/2}$, i.e., $u_{\rm ph} = u_0 - U_{\rm th}$. Since for a highly relativistic beam the thermal energy spread $E_{\rm th}$ is very nearly equal to $cU_{\rm th}$, the maximum

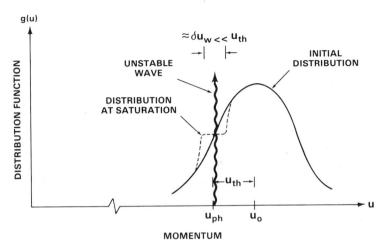

FIG. 4. Schematic of quasi-linear plateau formation in saturation of relativistic stimulated Compton scattering limit.

spatial growth rate becomes

$$|k_i|_{\text{max}} \approx (\sqrt{\pi}/20)\xi^2 k_0 \beta_{0\perp}^2 (E_0/E_{\text{th}})^2, \tag{60}$$

where, as before, $E_0 \approx \gamma_0 mc^2$, and $\xi = \omega_b/\gamma_0^{1/2} ck_0$.

These results for the growth rate agree with those of Lin and Dawson (1975), Hasegawa *et al.* (1976), Kroll and McMullin (1978), and Sprangle and Drobot (1978).

In order to estimate the efficiency in this stimulated Compton regime, we first note that only a small fraction of the beam particles are resonant with the unstable wave. This fraction is determined by the width of the spectrum of the spatially growing wave. The width of the wave spectrum in velocity space is given to a good approximation by

$$\delta v_w \approx |2ck_i/k| \ll V_{\text{th}}. \tag{61}$$

In momentum space (see Fig. 4) this width becomes

$$\delta u_w \approx \partial u/\partial v_z|_{v_z=v_{z0}} \delta v_w \approx |2\gamma_0 \gamma_{z0}^2 m_0 ck_i/k| \approx |\gamma_0 m_0 ck_i/k_0| \ll U_{\text{th}}. \tag{62}$$

As the backscattered electromagnetic wave grows, a small fraction of the beam particles, those with momenta between $u_{\text{ph}} - \delta u_w/2$ and $u_{\text{ph}} + \delta u_w/2$, become trapped in the potential associated with the ponderomotive wave. As this happens, the distribution function begins to flatten out near $u = u_{\text{ph}}$ and $\partial g/\partial u$ decreases until it vanishes. A plateau is then formed, and the distribution becomes stable with respect to the initially unstable wave because

$\partial g/\partial u|_{u=u_{\mathrm{ph}}} = 0$ (see Fig. 4). The change in kinetic energy between the initial and final distribution functions is given by

$$\delta E_{\mathrm{K.E.}} = \frac{n_0 m_0 c^2}{\sqrt{2\pi}\, U_{\mathrm{th}}} \int_{u_{\mathrm{ph}} - \delta u_{\mathrm{w}}/2}^{u_{\mathrm{ph}} + \delta u_{\mathrm{w}}/2} (\gamma - 1)[e^{-(u-u_0)^2/2U_{\mathrm{th}}^2} - e^{-1/2}]\, du, \quad (63)$$

where we have used the fact that $u_{\mathrm{ph}} = u_0 - U_{\mathrm{th}}$ for the maximum growth rate. Since $\delta u_{\mathrm{w}} \ll U_{\mathrm{th}}$, evaluation of Eq. (63) gives

$$\delta E_{\mathrm{K.E.}} \approx \frac{n_0}{\sqrt{2\pi}} e^{-1/2} \left(\frac{\delta u_{\mathrm{w}}}{2U_{\mathrm{th}}}\right)^2 \left(\frac{\partial \gamma}{\partial u}\right)_{u=u_{\mathrm{ph}}} \delta u_{\mathrm{w}}\, m_0 c^2 \approx \frac{n_0 c}{4\sqrt{2\pi}} \left(\frac{\delta u_{\mathrm{w}}}{U_{\mathrm{th}}}\right)^3 U_{\mathrm{th}}, \quad (64)$$

where $\delta u_{\mathrm{w}}/U_{\mathrm{th}}$ is evaluated from Eq. (62) with k_i given by Eq. (60) and is given by

$$\delta u_{\mathrm{w}}/U_{\mathrm{th}} = (\sqrt{\pi}/20)\xi^2 \beta_{0\perp}^2 (E_0/E_{\mathrm{th}})^3 \ll 1. \quad (65)$$

The efficiency, using Eq. (64), is given by

$$\eta = \delta E_{\mathrm{K.E.}}/n_0(\gamma_0 - 1)m_0 c^2 = (1/4\sqrt{2\pi})(\delta u_{\mathrm{w}}/U_{\mathrm{th}})^3 E_{\mathrm{th}}/E_0. \quad (66)$$

Using the expression for the efficiency in Eq. (66) and Eq. (47) together with (48), we find that the electric field amplitude at saturation is

$$|\mathbf{E}|_{\mathrm{sat}} = (m_0 c^2/|e|)\gamma_0 \xi k_0[(\delta u_{\mathrm{w}}/U_{\mathrm{th}})^3 E_{\mathrm{th}}/E_0\, 4\sqrt{2\pi}]^{1/2} \quad (67)$$

The requirement on the beam thermal velocity is stated in Eq. (55). In terms of the ratio E_{th}/E_0, this condition becomes

$$E_{\mathrm{th}}/E_0 \gg [(\sqrt{\pi}/40)\xi^2 \beta_{0\perp}^2]^{1/3}, \quad (68)$$

where Eqs. (55), (60), and (37) have been used.

C. CASE 3. HIGH GAIN, DENSE COLD BEAM LIMIT

In this limit, we shall retain the space charge scalar potential ϕ which arises from charge bunching by the ponderomotive potential. Because the scattered fields are assumed to e-fold many times within the interaction region we may neglect boundary terms at $z = 0$ and choose $\tilde{\phi}$ and $\tilde{A}_+$ to have the form [see Eq. (11)]

$$\tilde{\phi}(z) = \tilde{\phi}(0)e^{ikz}, \qquad \tilde{A}_+(z) = \tilde{A}_+(0)e^{ik_+z}, \quad (69)$$

where k_+ and k are complex constants.

Substituting the fields in Eqs. (69) into Eqs. (24) and taking the high gain limit, i.e., $|\exp(ikz)| \gg 1$ and $|\exp(ik_+z)| \gg 1$, the driving current densities become

$$
\tilde{J}_+(z) = -\frac{\omega_b^2}{8\pi} \int_0^\infty \left\{ \left[\frac{u}{\gamma u_z c}(2 + \beta_\perp^2) - \frac{\beta_\perp^2 m_0 v_z \omega/c}{\omega - v_z(k_+ + k_0)}\frac{\partial}{\partial u} \right] \tilde{A}_+(0)e^{ik_+z} \right.
$$

$$
\left. - \frac{\beta_\perp m_0 v_z k\, \partial/\partial u}{\omega - v_z k} \tilde{\phi}(0)e^{i(k-k_0)z} \right\} g_0\, du, \tag{70a}
$$

$$
\tilde{J}_z(z) = -\frac{\omega_b^2}{8\pi} \int_0^\infty m_0 \left[\frac{v_z k}{\omega - v_z k} \tilde{\phi}(0)e^{ikz} \right.
$$

$$
\left. + \frac{\beta_\perp v_z \omega/c}{\omega - v_z(k_+ + k_0)} \tilde{A}_+(0)e^{i(k_+ + k_0)z} \right] \frac{\partial g_0}{\partial u}\, du. \tag{70b}
$$

We see immediately from Eqs. (70) that the selection rule between the wave numbers of the three waves is

$$
k = k_+ + k_0. \tag{71}
$$

The coefficients of the driving currents $\tilde{J}_+$ and $\tilde{J}_z$, using Eq. (70), can be written in the form

$$
\tilde{J}_+(z) = -\frac{\omega_b^2}{8\pi c} \left\{ \left[2\left\langle \frac{u}{\gamma u_z} \right\rangle + \left(\frac{\Omega_0}{ck_0}\right)^2 S(\omega, k) \right] \tilde{A}_+(0) \right.
$$

$$
\left. - \frac{\Omega_0}{ck_0}\frac{c^2 k^2}{\omega_b^2} \chi_\gamma(\omega, k)\tilde{\phi}(0) \right\}e^{ik_+z},
$$

$$
\tilde{J}_z(z) = -\frac{\omega k}{8\pi} \left[\chi(\omega, k)\tilde{\phi}(0) + \frac{\Omega_0}{ck_0}\chi_\gamma(\omega, k)\tilde{A}_+(0) \right]e^{ikz}, \tag{72}
$$

where

$$
\langle(\cdots)\rangle = \int_0^\infty du\, g_0(u)(\cdots),
$$

$$
S(\omega, k) = \langle uc^2/\gamma^3 u_z v_z^2 \rangle - \omega c \int_0^\infty [m_0 c(\partial g_0/\partial u)/\gamma^2 v_z(\omega - v_z k)]\, du,
$$

$$
\chi(\omega, k) = (\omega_b^2/k) \int_0^\infty [m_0(\partial g_0/\partial u)/(\omega - v_z k)]\, du,
$$

$$
\chi_\gamma(\omega, k) = (\omega_b^2/k) \int_0^\infty [m_0(\partial g_0/\partial u)/\gamma(\omega - v_z k)]\, du.
$$

Substituting Eq. (69) into the wave equations (21) gives

$$\tilde{\phi}(0)e^{ikz} = \frac{-\Omega_0}{ck_0} \frac{\chi_\gamma(\omega, k)}{1 + \chi(\omega, k)} \tilde{A}_+(0)e^{ikz},$$

$$\left[D(\omega, k_+) - \frac{\omega_b^2}{2} \left(\frac{\Omega_0}{ck_0} \right)^2 S(\omega, k) \right] \tilde{A}_+(0)e^{ik_+z} = -\frac{c^2k^2}{2} \frac{\Omega_0}{ck_0} \chi_\gamma(\omega, k)\tilde{\phi}(0)e^{ik_+z}$$

(73)

where the first of Eqs. (73) can be put into an illuminating form by noting that to a good approximation γ can be taken out of the integral in the definition of $\chi_\gamma(\omega, k)$, since it is a slowly varying function of u. With this approximation together with the definition of the ponderomotive potential in Eq. (22) we find that

$$\tilde{\phi}(z) = \frac{\chi(\omega, k)/2}{1 + \chi(\omega, k)} \tilde{\phi}_{pond}(z),$$

(74)

where $\tilde{\phi}(z) = \tilde{\phi}(0) \exp(ikz)$. In Eq. (74) we see explicitly the fact that the ponderomotive potential drives the space charge potential.

Combining the two Eqs. (73) we arrive at the fully relativistic high gain dispersion relation

$$D(\omega, k_+) = (\omega_b^2/2)(\Omega_0/ck_0)^2 Q(\omega, k),$$

(75)

where

$$Q(\omega, k) = S(\omega, k) + (c^2k^2/\omega_b^2)\chi_\gamma^2(\omega, k)/[1 + \chi(\omega, k)].$$

Again, we take a monoenergetic electron beam represented by the distribution function

$$g_0 = (u_z/u) \delta(u - u_0)$$

(76)

where u_0 is the total beam momentum. For such a beam we find that

$$\chi(\omega, k) = -\omega_b^2/(\gamma_{z0}^2 \gamma_0)/(\omega - v_{z0}k)^2,$$

$$\chi_\gamma(\omega, k) = -(\omega_b^2/\gamma_0^2)(k - v_{z0}\omega/c^2)/k(\omega - v_{z0}k)^2,$$

$$S(\omega, k) = -(\omega^2 - c^2k^2)/\gamma_0^3(\omega - v_{z0}k)^2,$$

$$Q(\omega, k) = -D(\omega, k)/\gamma_0^3(\omega - v_{z0}k)^2[1 + \chi(\omega, k)],$$

$$D(\omega, k) = \omega^2 - c^2k^2 - \omega_b^2/\gamma_0;$$

and the dispersion relation becomes

$$D(\omega, k_+) = -(\omega_b^2/2\gamma_0^3)(\Omega_0/ck_0)^2 D(\omega, k)/[(\omega - v_{z0}k)^2 - \omega_b^2/\gamma_{z0}^2\gamma_0], \quad (77)$$

where $\gamma_0 = \gamma(u_0)$, $v_{z0} = v_z(u_0)$, and $\gamma_{z0} = \gamma_z(u_0)$.

Note that in the absence of the pump ($\Omega_0 = 0$) we recover the uncoupled dispersion relations [Eqs. (27)] for both the electromagnetic and electrostatic waves. In the presence of the pump, however, these two modes are coupled. Since the electromagnetic waves in the presence of the pump field approximately satisfy the pump-free dispersion relation $D(\omega, k_+) \approx 0$, we can replace $D(\omega, k)$ on the RHS of Eq. (77) by $-2kk_0c^2$. The dispersion relation in Eq. (77) may now be recast into the form

$$[k - (k_0 + K)][k - (k_0 - K)][k - (\omega/v_{z0} + \kappa)][k - (\omega/v_{z0} - \kappa)]$$
$$= -\alpha^2 k_0 k, \tag{78}$$

where

$$K \equiv (\omega^2/c^2 - \omega_b^2/\gamma_0 c^2)^{1/2} > 0,$$

$$\kappa \equiv \omega_b/(v_{z0}\gamma_{z0}\gamma_0^{1/2}),$$

$$\alpha^2 \equiv (\omega_b^2/\gamma_0^3 v_{z0}^2)(\Omega_0/ck_0)^2 = (\xi\beta_{0\perp}k_0)^2,$$

$$\beta_{0\perp} = \Omega_0/v_{z0}k_0, \quad \text{and} \quad \xi = \omega_b/\sqrt{\gamma_0}ck_0.$$

We may now discern two limits for the solution of Eq. (78), depending upon the pump strength.

1.   Weak Pump Regime

When the pump amplitude is smaller than a critical value, which we shall determine below, we may take the interacting longitudinal and transverse waves to be approximately normal modes of the system. The positive-energy, fast space charge mode of the beam is described by $k \simeq \omega/v_{z0} - \kappa$, while the negative-energy slow space charge wave has $k \simeq \omega/v_{z0} + \kappa$. It is the interaction of the slow wave with the pump that leads to instability of the backscattered electromagnetic waves. The coupling of the negative-energy space charge wave and the transverse electromagnetic wave is depicted in Fig. 5. Thus, in the case of a weak pump we can set

$$k = k_0 + K + \delta k = \omega/v_{z0} + \kappa + \delta k, \tag{79}$$

where $|\delta k| \ll |k_0 + K|$, $k_0 \ll K$, $\omega/v_{z0} \gg \kappa$, and $K \approx \omega/v_{z0}$. Substituting Eq. (79) into Eq. (78) we find that the spatial growth rate for a highly relativistic beam is

$$\delta k = -i(\alpha/2)(k_0/\kappa)^{1/2} = -(i/2)\beta_{0\perp}(\gamma_{z0}\xi)^{1/2}k_0. \tag{80}$$

This result [Eq. (80)] has been found previously by Sprangle et al. (1975), Kwan et al. (1977), Kroll and McMullin (1978), and Sprangle and Drobot (1978), under the assumption $|\delta k| \ll 2\kappa$, which leads to the constraint upon the pump amplitude

$$\beta_{0\perp} \ll \beta_{crit} \equiv 4(\xi/\gamma_{z0}^3)^{1/2}. \tag{81}$$

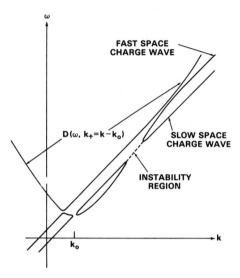

FIG. 5.　Coupled dispersion diagrams of the beam waves and transverse electromagnetic waves. Instability occurs near the intersection between the slow beam mode and transverse electromagnetic mode.

The difference between the phase velocity of the longitudinal wave and the axial beam velocity is

$$v_{ph} - v_{z0} = -\Delta v = -(\xi/2\gamma_{z0}^3)c.$$

Using this result in the efficiency expression of Eq. (46), we find (Kwan *et al.*, 1977; Sprangle and Drobot, 1978)

$$\eta = \xi/\gamma_{z0}, \tag{82}$$

with the corresponding saturated field amplitude, from Eqs. (47) and (48),

$$|\mathbf{E}|_{sat} = (m_0 c^2/|e|)k_0(\gamma_0/\gamma_{z0}^{1/2})\xi^{3/2}. \tag{83}$$

The monoenergetic beam assumption is valid if the axial thermal velocity of the beam V_{th} is much less than $\Delta v = v_{z0} - v_{ph}$, so that the wave is not resonant with any beam particles (see Fig. 3). Since $E_{th} = \gamma_0\gamma_{z0}^2 m_0 c\, V_{th}$ for relativistic beams, the thermal energy spread must satisfy the requirement that

$$E_{th}/E_0 \ll \xi/2\gamma_{z0} = \eta/2. \tag{84}$$

2. Strong Pump Regime

In the weak pump regime, where the longitudinal wave satisfies $1 + \chi(\omega, k) \simeq 0$, we see from Eq. (74) that the magnitude of the scalar potential $\tilde{\phi}(z)$, owing to charge separation, is much greater than the magnitude of the ponderomotive potential $\tilde{\phi}_{pond}(z)$. Thus the space charge collective effects dominate the interaction. When the pump amplitude is sufficiently high, however, the ponderomotive forces may modify the electrostatic wave dispersion so that $|\chi(\omega, k)| \ll 1$. In this case the space charge collective effects are negligible. Thus to consider the strong pump regime we take

$$|\delta k| \gg 2\kappa \qquad (85)$$

in Eq. (79). Substituting Eq. (79) into Eq. (78), the dispersion relation reduces to

$$(\delta k)^3 = -\xi^2 \beta_{0\perp}^2 k_0^3/2 \qquad (86)$$

and the unstable root is (Kroll and McMullin, 1978; Sprangle and Drobot, 1978)

$$\delta k = [(1 - i\sqrt{3})/2^{4/3}](\xi\beta_{0\perp})^{2/3}k_0. \qquad (87)$$

From the result in Eq. (87), we see that the strong pump condition [Eq. (85)] is the inverse of Eq. (81), i.e.,

$$\beta_{0\perp} \gg \beta_{crit} = 4(\xi/\gamma_{z0}^3)^{1/2}. \qquad (88)$$

The phase velocity of the unstable longitudinal wave is

$$v_{ph} = \omega/\text{Re}(k) = v_{z0} - \text{Re}(\delta k)/k = v_{z0} - (\xi\beta_{0\perp})^{2/3}c/2^{1/3}4\gamma_{z0}^2. \qquad (89)$$

From Eqs. (46) and (48) we find that the corresponding efficiency and saturated amplitude are given by

$$\eta = (\xi\beta_{0\perp}/4)^{2/3} \qquad (90)$$

and

$$|\mathbf{E}|_{sat} = (m_0 c^2/|e|)k_0(\xi^4\gamma_0^3\beta_{0\perp}/4)^{1/3}, \qquad (91)$$

respectively (Sprangle and Drobot, 1978). The energy spread requirement to use a cold beam distribution is given by

$$E_{th}/E_0 \ll \tfrac{1}{2}(\xi\beta_{0\perp}/4)^{2/3} = \tfrac{1}{2}\eta. \qquad (92)$$

The general behavior of the growth rate and efficiency as a function of $\beta_{0\perp}$ (pump strength), in both the strong and weak pump regimes, are shown schematically in Fig. 6. In the weak pump regime the spatial growth rate is proportional to the pump strength, while the efficiency is independent of the pump. For the strong pump case both the growth rate and the efficiency vary as $\beta_{0\perp}^{2/3}$. Higher growth rates and efficiencies, therefore, make the strong pump limit a more attractive mechanism for generating high frequency radiation. In Table I we summarize the results of analyses of Cases 1–3 for a highly relativistic beam ($v_z \simeq c$).

TABLE I

SUMMARY OF FEL REGIMES[a]

FEL regimes	Spatial growth rate, $-\text{Im}(k)$	Efficiency η	Scattered E-field, $|\mathbf{E}|_{\text{sat}}$	Thermal spread, E_{th}/E_0
Case 1: low gain, $-\int_0^L \text{Im}(k)\,dz \ll 1$ tenuous cold beam	$\left(\dfrac{\xi\beta_{0\perp}}{4}\right)^2 (k_0 L)^3$ integrated over L	$\dfrac{\pi}{k_0 L}$	$\dfrac{4.5}{\pi^2}\dfrac{m_0 c^2}{|e|} k_0 \gamma_0 \beta_{0\perp}^{-1}\eta^2$	$\ll \eta$
Case 2: high gain, $-\text{Im}(k)z \gg 1$ tenuous thermal beam	$\dfrac{\sqrt{\pi}}{20}\left(\xi\beta_{0\perp}\dfrac{E_0}{E}\right)^2 k_0$	$\dfrac{1}{4\sqrt{2\pi}}\left(\dfrac{\delta u_w}{U_{\text{th}}}\right)^3 \dfrac{E_{\text{th}}}{E_0}$	$\dfrac{m_0 c^2}{|e|} k_0 \gamma_0 \xi \eta^{1/2}$	$\gtrsim \left(\dfrac{\sqrt{\pi}}{40}\xi^2\beta_{0\perp}^2\right)^{1/3}$
Case 3a: high gain, $-\text{Im}(k)z \gg 1$ weak pump, $\beta_{0\perp} \ll \beta_{\text{crit}}$, dense cold beam	$\dfrac{\sqrt{\gamma_{z0}}}{2}\xi^{1/2}\beta_{0\perp} k_0$	$\dfrac{\xi}{\gamma_{z0}}$	$\dfrac{m_0 c^2}{|e|} k_0 \gamma_0 \xi^{1/2}\eta$	$\ll \eta$
Case 3b: high gain, $-\text{Im}(k)z \gg 1$ strong pump, $\beta_{0\perp} \gg \beta_{\text{crit}}$, dense cold beam	$\dfrac{\sqrt{3}}{2^{4/3}}(\xi\beta_{0\perp})^{2/3} k_0$	$2^{-4/3}(\xi\beta_{0\perp})^{2/3}$	$\dfrac{m_0 c^2}{|e|} k_0 \gamma_0 \xi \eta^{1/2}$	$\ll \eta$

[a] The parameters are defined as: $\beta_{z0} = v_{z0}/c \approx 1$, $\xi = \omega_b/\gamma_{z0}^{1/2} c k_0$, $k_0 = 2\pi/l$, $\omega_b = (4\pi|e|^2 n_0/m_0)^{1/2}$, $\Omega_0 = |e|B_0/m_0 c$, $\beta_{0\perp} = \Omega_0/\gamma_0 c k_0$, $\beta_{\text{crit}} = 4(\xi/\gamma_{z0}^3)^{1/2}$, $\gamma_0 = (1 - \beta_{z0}^2 - \beta_{0\perp}^2)^{-1/2}$, $\gamma_{z0} = (1 - \beta_{z0}^2)^{-1/2}$, $E_0 = (\gamma_0 - 1)m_0 c^2$, $E_{\text{th}} = \gamma_{z0}^2 \gamma_0 m_0 c V_{\text{th}}$, and $\delta u_w/U_{\text{th}} = (\sqrt{\pi}/20)\xi^2\beta_{0\perp}^2 (E_0/E_{\text{th}})^3$.

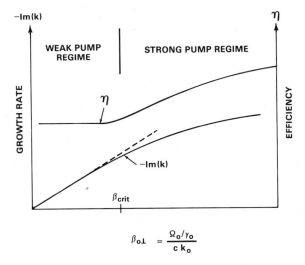

FIG. 6.   Schematic of spatial growth rate and efficiency in the high gain, dense cold beam limit as a function of the pump field strength characterized by $\beta_{0\perp}$.

IV.   Discussion: A Visible Light Free Electron Laser

In this section we apply the theory to a specific example of a high power FEL at optical wavelengths. We first note two considerations important for high current beams: (i) the effect of the beam itself on the pump magnetic field; and (ii) the induced energy spread in the beam due to its self electrostatic potential. These considerations place an upper limit on the beam current.

With regard to (i) we note that the beam equilibrium motion in the external pump field contains a transverse velocity component $\beta_{0\perp}c$. The beam motion, therefore, induces a spatially periodic, diamagnetic field that tends to oppose the applied pump field. It may easily be shown that with an equilibrium distribution

$$g_0(\alpha, \beta, u) = n_0(u_z/u)\,\delta(\alpha)\,\delta(\beta)\,\delta(u - u_0),$$

the diamagnetic field amplitude B_d is of order

$$|B_d| \approx \xi^2 B_0,$$

where ξ is the beam strength parameter defined above: $\xi = \omega_b/(\gamma_0^{1/2}ck_0)$. Thus, in order to neglect the perturbations to the equilibrium orbit owing to the diamagnetic field, we require $\xi \ll 1$.

With regard to point (ii), we note that an unneutralized electron beam has an inherent energy spread owing to the variation of the electrostatic potential within the beam. For a cylindrical beam of radius r_0 the potential difference between the central axis and the outer edge of the beam is

$$\Delta\phi = \pi|e|n_0 r_0^2.$$

Associated with this potential difference is a kinetic energy change across the beam, given by

$$\Delta E = \Delta\gamma m_0 c^2 = |e|\,\Delta\phi.$$

This variation in kinetic energy may be interpreted as indicating a thermal energy spread in the beam, of relative magnitude

$$\Delta E/E_0 = |e|\,\Delta\phi/(\gamma_0 - 1)m_0 c^2 \simeq (\xi k_0 r_0/2)^2,$$

where we have taken $\gamma_0 \gg 1$. For sufficiently high beam densities this inherent energy spread may invalidate the cold beam approximation.

As noted above, the strong pump scattering requirement (Case 3b) is characterized by the highest spatial growth rates and efficiencies, and is thus the most promising case for application. Therefore, we shall choose parameters appropriate to this regime.

We take a periodic magnetic pump field of period $l = 0.75$ cm and amplitude $B = 7.5$ kG. The electron beam current I_0 is taken to be 2 kA, with beam radius 0.1 cm and energy 45 Mev ($\gamma_0 = 90$). The transverse velocity induced by the pump is

$$v_{0\perp} = \beta_{0\perp}c = |e|B_0/\gamma_0 m_0 ck_0 = 6.0 \times 10^{-3}c,$$

where $k_0 = 2\pi/l$. The radian plasma frequency of the beam is $\omega_b = 2.0 \times 10^{11}$ sec^{-1}, and the pump strength parameter $\xi = \omega_b/(\gamma_0^{1/2}ck_0) = 8.4 \times 10^{-2}$. The longitudinal Lorentz factor $\gamma_{z0} = 80$, giving $\beta_{crit} = 4(\xi/\gamma_{z0}^3)^{1/2} = 1.62 \times 10^{-3} \ll \beta_{0\perp}$, verifying that these parameters are indeed characteristic of the strong pump regime. The wavelength of the scattered wave is $\lambda_s = l/2\gamma_{z0}^2 = 0.58$ μm, in the visible region of the spectrum. From the formulas of Case 3b, summarized in Table I, we find a spatial growth rate $|-\mathrm{Im}\,k| = \sqrt{3}(\xi\beta_{0\perp}/4)^{2/3}k_0 = 3.6 \times 10^{-2}$ cm^{-1}, corresponding to an e-folding length of 28 cm. Thus, very high gains may be achieved in a single pass over a total distance of several meters. The single-pass efficiency is given by

$$\eta = (\xi\beta_{0\perp}/4)^{2/3} = 0.25\%.$$

With regard to the constraints on the current imposed by the neglect of the induced diamagnetic field and the beam quality necessary to apply the cold beam approximation, the first of these is well satisfied as $\xi^2 \ll 1$, and the second marginally satisfied as $\Delta E/E_0 = (\xi r_0 k_0/2)^2 = 0.12\%$, to be compared with the efficiency of 0.25%. The radiated power P_s, at 0.58 μm, is given by $P_s = \eta I_0 E_0 = 230$ MW. The parameters of this example are summarized in Table II.

TABLE II

ILLUSTRATION OF VISIBLE RADIATION SOURCE FOR $\lambda_s = 0.58\ \mu m$
IN THE STRONG PUMP SCATTERING REGIME

Magnetic pump field parameters

Pump wavelength	$l = 0.75$ cm
Pump magnetic field amplitude	$B_0 = 7.5$ KG

Electron beam parameters

E-beam current	$I_0 = 2$ kA
E-beam energy ($\gamma_0 = 90$)	$E_0 = 45$ MeV
E-beam power	$P_0 = 90$ GW
E-beam radius	$r_0 = 0.1$ cm
Axial beam gamma	$\gamma_{z0} = 80$
Energy spread due to self beam potential	$\Delta E/E_0 = 0.12\%$

Scattered visible radiation parameters

Radiation wavelength	$\lambda_s = l/(2\gamma_{z0}^2) = 0.58\ \mu m$
e-folding length of radiation	$L_e = 28$ cm
Single pass efficiency	$\eta = 0.25\%$
Radiation power	$P_s = 230$ MW

Finally, we note that it is in principle possible to increase substantially the single-pass efficiency by recovering part of the beam energy with a depressed collector and by continuously decreasing the phase velocity of the pondermotive wave in order to delay trapping, which saturates the process. The decrease in phase velocity can be achieved by adiabatically decreasing the wavelength of the pump field. In this way energy will be extracted from the beam until a greater degree of thermalization is attained, with correspondingly higher efficiency. Extrapolating the results of present technology to the FEL, we might expect the combination of beam recovery and phase-velocity tapering to yield increases of roughly an order of magnitude in the single-pass efficiency. As of this writing (mid-1978), however, detailed analyses of these effects have not been completed; these will be the subjects of continued investigations.

V. Laboratory Experiments

Data on laboratory studies of stimulated scattering from relativistic electron beams are summarized in Table III. The five experiments listed divide naturally into two regimes; i) the low gain regime investigated using

TABLE III

A Comparison of Experimental Studies of Stimulated Scattering from Relativistic Electron Beams

Scattering regime	Experimental configuration	Pump wave	Electron beam	Output radiation	Performing organization
(1) Low gain	Amplifier filling factor $= 2 \times 10^{-2}$	Magnetic wiggler $l = 3.2$ cm $B_0 = 2.4$ kG $L = 5.2$ m	24 MeV 0.07 A, peak $n_0 = 2 \times 10^9$ cm^{-3} $E_{th}/E_0 = 0.1\%$	$\lambda_s = 10.6$ μm $G_L = 0.035$ (7% energy gain)	Stanford University (Elias et al., 1976)
(2) Low gain	Oscillator with optical cavity, mirror trans. $= 1.5\%$	Magnetic wiggler $l = 3.2$ cm $B_0 = 2.4$ kG $L = 5.2$ m	43 MeV 2.6 A peak $E_{th}/E_0 = 0.05\%$ 0.06 mm mrad	$\lambda_s = 3.4$ μm $\Delta\lambda_s/\lambda_s = 0.2\%$ 7 kW peak	Stanford University (Deacon et al., 1977)
(3) High gain, collective	Superradiant oscillator	Electromagnetic wave $l = 2$ cm $\mathscr{E}_0 = 4 \times 10^4$ V/cm $L = 0.36$ cm	2 MeV 30 kA peak $n_0 = 3 \times 10^{12}$ cm^{-3} $\Delta E/E_0 = 4\%$	$\lambda_s = 400$ μm $G_L > 2$ 1 MW peak	Naval Research Laboratory (Granatstein et al., 1977)
(4) High gain, collective	Superradiant oscillator	Magnetic wiggler $l = 0.6$ cm $B_0 = 0.5$ kG $L = 0.36$ m	0.86 MeV 5 kA peak $n_0 = 3 \times 10^{11}$ cm^{-3} $\Delta E/E_0 = 2\%$	$\lambda_s = 1500$ μm 8 MW peak	Columbia University (Marshall et al., 1977; Gilgenbach et al., 1978; Gilgenbach, 1978)
(5) High gain, collective	Oscillator with optical cavity, mirror trans. $= 2\%$	Magnetic wiggler $l = 0.8$ cm $B_0 = 0.4$ kG $L = 0.4$ m	1.2 MeV 25 kA peak $n_0 = 4 \times 10^{12}$ cm^{-3} $\Delta E/E_0 = 3\%$	$\lambda_s = 400$ μm $\Delta\lambda_s/\lambda_s = 2\%$ 1 MW peak	NRL and Columbia (McDermott et al., 1978)

the superconducting linear accelerator at Stanford University [(1) and (2) in Table III]; and ii) the high gain collective regime investigated using intense relativistic electron beam accelerators at NRL and Columbia University [(3), (4), and (5) in Table III].

The low gain regime studies were characterized by relatively high electron energies (24–43 MeV) and correspondingly short output wavelength (3–11 μm). Line broadening of spontaneous emission was dominated by finite length effects rather than by spread in electron energy since $l/L > 2E_{th}/E_0$ (see Eq. 38). The current was relatively small (0.1–2.6 A), corresponding to a low value of amplitude gain; in an amplifier experiment (1), $G_L = 0.035 \ll 1$. In an oscillator experiment (2), a peak output power of 7 kW was realized, which corresponded to 0.006 % of the energy in the electron beam; average output power was 0.36 W.

In contrast, the intense beam studies had lower electron energy (0.86–2 MeV) and correspondingly longer output wavelengths (400–1500 μm). Line broadening was dominated by electron energy spread since $2E_{th}/E_0 > l/L$. (E_{th} was estimated to equal the electrostatic energy shear across the beam.) The current was much larger (5–40 kA), making possible the participation of collective electron beam modes in the stimulated scattering process, and resulting in large gain; in one superradiant oscillator experiment (3), the measured amplitude gain was $G_L > 2$. In a second superradiant oscillator experiment (4), the strength of the pump was increased to ensure saturation of wave growth, and a peak power output of 8 MW was measured; this was three orders of magnitude larger than the saturated peak power in the low gain oscillator experiment, corresponding both to larger peak beam power and to a somewhat larger efficiency (0.2 %). As yet, there is no continuous power capability in the intense beam experiments, the experiments being characterized by a single electron pulse of 10–50-nsec duration.

In the intense beam experiments two types of pump wave were investigated: a magnetic wiggler (4, 5) as was used in the low gain experiments and, alternatively, a powerful electromagnetic wave (3). The intense beam experiments also had an externally imposed uniform axial magnetic field that was large enough so that magnetic resonances in the output power were observed.

Some appreciation for the significance of the experimental results listed in Table III may be obtained by comparing the peak powers achieved with other existing capabilities for generating high power coherent radiation. Two distinct technologies are currently being used. Conversion of electron kinetic energy into microwaves by coupling beam waves to electromagnetic waves on a slow wave structure is a highly developed technology at centimeter and millimeter wavelengths. This technology is characterized by numerous device types which can be scaled or tuned to cover a wide frequency range.

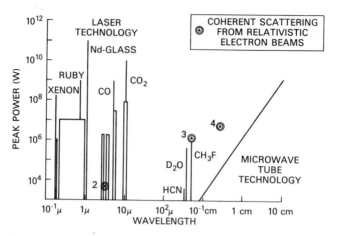

FIG. 7. A comparison of technologies available for the generation of high power coherent radiation. The three data points correspond to the preliminary coherent scattering results and are numbered as in Table III.

As shown by the solid line in Fig. 7, however, the available power decreases rapidly as wavelength is shortened.

At optical and near infrared wavelengths, the technology of lasers based on atomic and molecular transitions has been successful in generating very high power using a variety of media. The number of atomic or molecular transitions which result in laser action are limited, however, and high power is generally available only in narrow bands. Some tunable molecular systems do exist (e.g., dye lasers) but these are restricted in their power output because large average power operation results in decomposition of the active medium (i.e., the dye).

In contrast, the preliminary results obtained with the FEL point to the possibility of developing high power, tunable systems operating over a wavelength range extending from the millimeter to the optical. The data points indexed as 3 and 4 in Fig. 7 correspond to the intense beam scattering results listed in Table III as (3) and (4), respectively. The low gain oscillator result obtained with the Stanford superconducting linear accelerator (2) is shown by the third data point. A more detailed description of the various stimulated scattering experiments follows.

A. Experiments in the Low Gain Regime at Infrared Wavelengths

The experimental arrangement corresponding to (1) in Table III is depicted in Fig. 8. The output from a 50-kW, 10.6-μm laser was amplified upon interacting with a beam of 24-MeV electrons passing through a wiggler magnetic field with period $l = 3.2$ cm. The axial electron energy in the interaction region

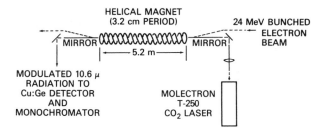

FIG. 8. Experimental arrangement in free electron amplifier study in the low gain regime. [Reproduced from Elias *et al.* (1976).]

was evidently $\gamma_{z0} = 39$ to satisfy $l/2\gamma_{z0}^2 = \lambda_s$; this may be compared with a total energy of $\gamma_0 = 48$.

As shown by Kroll and McMullin (1978), the experimental parameters indicate that the interaction took place in the low gain, tenuous cold beam regime; thus, the expressions for Case 1 in Table I are applicable. Gain was measured to be a linear function of current for $5 < I < 70$ mA. The maximum energy gain measurement was 7%, corresponding to a stimulated emission of 60 keV/electron or an efficiency of 0.25%. The measured gain is somewhat larger than, but of the same order as, the gain calculated from the expression for $-\int_0^L \mathrm{Im}(k)\,dz$ in Case 1 of Table I when this is multiplied by a filling factor of 2×10^{-2}, which is the ratio of the electron beam area to the area of the applied 10.6-μm radiation. The measured efficiency approaches the theoretical value of $\pi/k_0 L = 0.31$%, indicating that the process was close to saturation.

The spontaneous power emitted at 10.6 μm is plotted as a function of electron energy in Fig. 9 together with gain measured when the CO_2 laser radiation was applied. The full width at the $1/e$ points of the spontaneous power curve is 0.4%, which agrees well with a theoretical broadening of 0.3% due to the finite length of the interaction region. The shape of the gain curve, showing inverse dependence on the first derivative of the spontaneous emission curve, also agrees with theory.

The experimental arrangement corresponding to the oscillator in (2) of Table III is shown in Fig. 10. The same helical magnet undulator was used as with the amplifier experiment shown in Fig. 8. However, energy of the electrons was increased to 43 MeV so that stimulated emission occurred at the shorter wavelength of 3.4 μm. The electrons are formed into bunches only 1.3 mm in length with a spacing between bunches of approximately 25 m. The spacing between the resonator mirrors is carefully adjusted so that the round-trip bounce time for the 3.4-μm radiation is equal to the time interval between electron pulses entering the cavity.

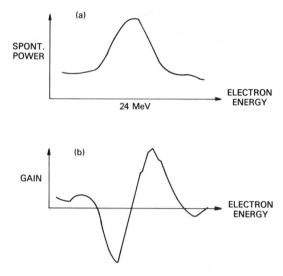

FIG. 9.   (a) Spontaneous power at 10.6 μm as a function of electron energy, and (b) the gain in energy of 10.6 μm radiation imposed from the CO_2 laser as a function of electron energy. [Reproduced from Elias *et al.* (1976).]

The emission spectrum of the laser oscillator above threshold is shown together with the spectrum of the spontaneous radiation emitted by the electron beam in Fig. 11. The 0.9 % linewidth of the spontaneous emission should be compared with theoretical linewidth of 0.6 % due to the finite length of the helical undulator. Lasing in the optical resonator reduces linewidth to 0.2 %.

The peak output power was 7 kW, corresponding to 0.006 % of the beam energy. Assuming this experiment to be in the same parameter regime as the amplifier experiment of Elias *et al.*, one would expect that the applicable theoretical efficiency would again be $\pi/k_0 L = 0.3$ %. Perhaps the lower experimental efficiency was due to a choice of output window transmissivity that was not optimum. (See Note added in proof, p. 325.)

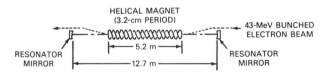

FIG. 10.   Experimental arrangement in free electron laser oscillator study in the low gain regime. [Reproduced from Deacon *et al.* (1977).]

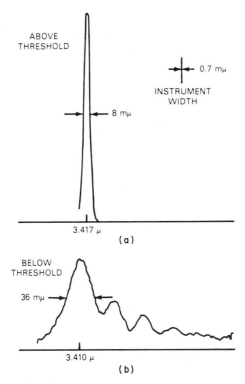

FIG. 11. Emission spectra of laser oscillator (a) above lasing threshold and (b) spontaneous radiation emitted by electron beam. [Reproduced from Deacon *et al.* (1977).]

B. SUBMILLIMETER GENERATION IN THE HIGH GAIN, COLLECTIVE REGIME WITH AN ELECTROMAGNETIC PUMP

The experimental arrangement corresponding to (3) in Table III is shown schematically in Fig. 12. A voltage pulse (-2 MV, 50 ns) was applied to a cold cathode immersed in a uniform axial magnetic field $\langle B \rangle \approx 13$ kG. A resultant annular electron beam was injected into a drift tube along the lines of $\langle B \rangle$. Electrostatic forces reduced electron kinetic energy so that $\gamma_0 = 3.9$.

At one point in the drift tube, the electron beam was passed through a nonadiabatic perturbation in the applied magnetic field, which converted a large fraction of the electron streaming energy into energy transverse to the axis of the drift tube. The electrons with large transverse energy then reacted unstably with an electromagnetic mode of the drift tube [cyclotron maser process as described in Hirshfield and Granatstein (1977)] and produced a 2 cm pump wave having a power of 10–100 MW. This pump wave was reflected off a metallic plate and traveled back up the drift tube toward the

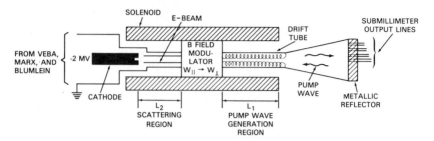

FIG. 12. Experimental arrangement in study of stimulated scattering in the high gain, collective regime with electromagnetic pump wave (Granatstein *et al.*, 1977). The VEBA electron accelerator system is described in Parker and Ury (1975).

cathode. When the pump wave encountered the cold streaming electrons near the cathode, stimulated scattering occurred, resulting in backscattered submillimeter waves.

The submillimeter output radiation was monitored through an array of metallic tubes passing through the reflecting plate. Each tube passed the radiation through wave filters and into a pyroelectric detector. The submillimeter output power was optimized by varying both the axial position of the magnetic modulator and the level of $\langle B \rangle$. The total submillimeter radiation incident on the metallic reflector was shown to be peaked at $\lambda_s = 400 \ \mu\text{m}$ and to have a power of 1 MW.

Dependence of the submillimeter radiation on the position of the magnetic modulator is shown in Fig. 13. As the modulator was moved downstream from the cathode, the scattering region L_2 was lengthened while simultaneously the region L_1 in which the pump wave is generated was shortened. Thus, one expected an initial increase in the submillimeter output as the scattering became stronger, followed by a decreasing submillimeter output as the pump wave was weakened. In Fig. 13, a plot of submillimeter power vs L_2 shows the expected behavior, with the submillimeter output having a maximum at $L_2 = 30$ cm. The rising portion of the curve in Fig. 13 has a slope indicating an amplitude gain for the stimulated scattering of $G_L > 2$ or equivalently a spatial growth rate of $|\text{Im}(k)| > 0.02 \ \text{cm}^{-1}$.

Only a lower bound is determined by the data of Fig. 13, since the pump wave amplitude decreases as L_2 is made larger. This lower bound on spatial growth rate may be compared with the theoretical value $-\text{Im}(k) = 0.04$ cm^{-1} calculated from the expression for Case 3a in Table I, and using the expression for $\beta_{0\perp}$ from Sprangle *et al.* (1975) (viz,

$$\beta_{0\perp} = |e| \mathcal{E}_0 / \gamma_0 m_0 (\omega_0 - \langle \Omega \rangle / \gamma_0),$$

where $\mathcal{E}_0$ is the electric field and $\langle \Omega \rangle = |e| \langle B \rangle / m_0 c$. For the parameters of the experiment in Fig. 10, we calculate $\beta_{0\perp} = .02$ and $\beta_{\text{crit}} = 0.4$, while

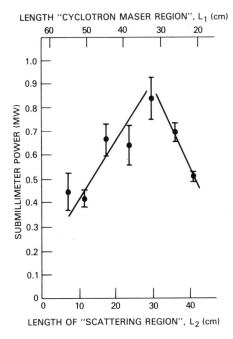

FIG. 13. Radiated power at 400 μm vs. length of scattering region (Granatstein *et al.*, 1977).

$E_{th}/E_0 = 0.04$ and $\xi/\gamma_{z0} = 0.13$; thus, the condition $\beta_{0\perp} \ll \beta_{crit}$ is well satisfied while $E_{th}/E_0 \ll \xi/\gamma_{z0}$ is marginally satisfied. Hence, one is justified in using the theoretical expressions for the high gain collective scattering regime of Case 3a.

Dependence of the submillimeter power on $\langle B \rangle$, the solenoidal magnetic field strength, is shown in Fig. 14. A maximum is seen at $\langle B \rangle = 13$ kG with a half-power width of 3 kG. The pump wave power at $\lambda_s = 2$ cm was relatively constant in the range 10 kG $< \langle B \rangle < 16$ kG, so that the behavior depicted in Fig. 14 corresponds to resonant interactions in the stimulated scattering process.

The experimental configuration in Fig. 12 is limited by the fact that both pump wave generation and stimulated scattering occur in the same electron beam. This does not allow for separate optimization of each process, and complicates interpretation of the experimental results. To overcome these disadvantages, the new experimental configuration in Fig. 15 is being assembled at NRL. Two synchronized pulsed-power systems are being employed. The pulse-forming network from a 600-kV flash x-ray system (peak power = 3.6 GW) will be used to drive an S-band magnetron that will furnish a multigigawatt pump wave at $\lambda_0 = 9.4$ cm.

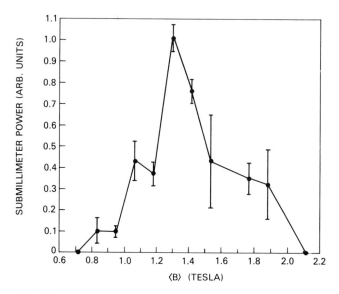

FIG. 14. Submillimeter output radiation at 400 μm vs. strength of uniform axial magnetic field (Granatstein *et al.*, 1977).

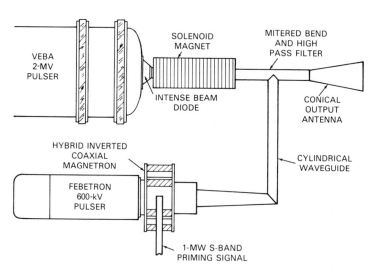

FIG. 15. Planned stimulated scattering experiment with electromagnetic pump wave. Operation in the high gain, strong pump regime is expected (Case 3b in Table I).

A larger 2.5-MV accelerator (peak power 25 GW) will provide the annular electron beam ($A_b = 7$ cm^2) in which stimulated scattering may take place. The backscattered output radiation is expected to be at $\lambda_s = 1700$ μm. Given the magnitude of microwave power available from the magnetron, this stimulated scattering experiment is expected to be the first operating in the strong pump limit ($\beta_{0\perp} > \beta_{crit}$). The expressions for Case 3b in Table I are applicable, and using them, one calculates a spatial growth rate $-\text{Im}(k) = 0.4$ cm^{-1} and an efficiency $\eta = 11\%$.

The investigations of stimulated scattering using electromagnetic pump waves are of practical importance since pump wavelengths can be realized which are much shorter than those which are possible with a magnetic wiggler. Thus for a given output wavelength there is a reduced requirement on accelerator voltage. However, as is apparent from the discussion above, the production of strong electromagnetic pump waves is far more difficult technologically than the production of a magnetic wiggler field.

C. Experiments in the High Gain, Collective Regime with a Magnetic Wiggler

The experimental arrangement corresponding to (4) in Table III is depicted in Fig. 16. As in V.B above, an intense electron beam of annular cross section was propagated axially down a drift tube along the lines of a solenoidal magnetic field $\langle B \rangle$. However, in this case, a ripple component was added to the solenoidal field over a distance typically $L = 40$ cm. The radial component of the ripple field B_0 could be varied from 0 to 500 G, and ripple periods of 6 and 8 mm were used. From the parameters in Table III, one calculates

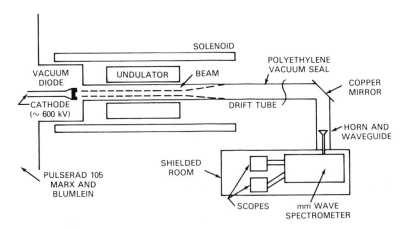

FIG. 16. Experimental arrangement in study of stimulated scattering in the high gain, collective regime with magnetic undulator pump. [Reproduced from Gilgenbach *et al.* (1978).]

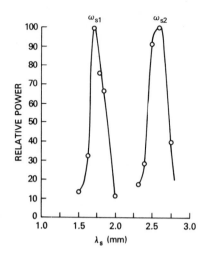

FIG. 17.   Spectrum of output radiation. Each peak is arbitrarily normalized to 100 on vertical scale. $l = 8$ mm, $\gamma_{z0} = 1.8$, $\langle B \rangle = 7.6$ kG for peak at ω_{S2}, and $n_0 = 3 \times 10^{11}$ cm^{-3}. [Reproduced from Gilgenbach *et al.* (1978).]

$\beta_{0\perp} = 0.01$, $\beta_{crit} = 0.35$, and $\xi/\gamma_{z0} = 0.03$ while $E_{th}/E_0 = 0.02$. Thus $\beta_{0\perp} \ll \beta_{crit}$ is well satisfied, but $E_{th}/E_0 \lesssim \xi/\gamma_{z0}$ rather than $\ll \xi/\gamma_{z0}$. One therefore expects operation to have been marginally in the high gain, collective regime.

The emission spectrum was resolved using a grating spectrometer of special design. A typical spectrum is shown in Fig. 17. As shown in the figure, it was found that the emission spectrum consisted of peaks at two distinct wavelengths, which were interpreted as corresponding to the frequencies

$$\omega_{S1} = (1 + \beta_{z0})\gamma_{z0}(\gamma_{z0}k_0 v_{z0} - \omega_b/\gamma_0^{1/2})$$

and

$$\omega_{S2} = (1 + \beta_{z0})\gamma_{z0}^2(v_{z0}k_0 - |e|\langle B \rangle/m_0 c\gamma_0).$$

The shorter wavelength peak at ω_{S1} would then correspond to a stimulated scattering process involving a space charge beam mode as an idler. The longer wavelength peak at ω_{S2} would correspond to a scattering process with a cyclotron mode idler. Indeed it was found that the longer wavelength peak would be tuned in frequency by changing $\langle B \rangle$ while ω_{s1} remained constant.

Marshall *et al.* (1977) measured power in the wave at ω_{S2} to be 8 MWs. This was found experimentally to represent a saturated power level (Gilgenbach, 1978) and corresponds to an efficiency of 0.2%. The theoretical analysis presented in this paper does not apply to the case of a cyclotron mode idler. The power measured at ω_{S1} was 2 MW, but this was not established to be a saturated value.

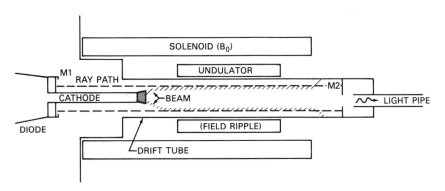

FIG. 18. Experimental arrangement in study of high gain free electron laser with quasi-optical cavity. M_1 and M_2 designate mirrors. [Reproduced from McDermott et al. (1978).]

The experiment of Marshall et al. (1977) shown in Fig. 16 functioned as a superradiant oscillator. Experiments have now been carried out (McDermott et al., 1978) using mirrors to form a quasi-optical cavity. The experimental arrangement corresponding to (5) in Table III is shown in Fig. 18. The separation between the mirrors was 1.5 m while the time duration of the electron beam pulse was approximately 40 ns, so that radiation would execute only four round-trip bounces between the mirrors while the beam was on. Nevertheless, evidence of lasing was observed with coherence of the radiation measured as $\Delta\lambda_s/\lambda_s = 2\%$ compared with $\Delta\lambda_s/\lambda_s \gtrsim 10\%$ for the superradiant oscillator. According to McDermott et al. (1978), the stimulated scattering process involved a space charge wave idler, while scattering from the cyclotron idler was damped.

VI. Conclusions

Free electron lasers based on stimulated scattering from intense electron beams hold the promise of making available very high power, moderately coherent, widely tunable sources over a wavelength range from millimeters down to optical wavelengths and beyond. This paper contains design examples of a 1700-μm laser with peak power of several gigawatts operating at 11% efficiency and a 0.58-μm laser with power of hundreds of megawatts operating with 0.25% efficiency.

Clearly, much investigation will be required before this potential can be fully evaluated. The experiments which have been carried out to date give support to the linear theory, but carefully designed experimental investigations of the nonlinear regime still remain to be executed.

Also, accelerator technology studies are required to evaluate the average power potential. In connection with FEL's based on linear accelerators, the use of storage rings is under study. For FEL's based on intense electron beams, there is great interest in the studies which have been initiated on repetitively pulsing intense beam accelerators and on recovering energy in the spent electron beam through depressed-collector techniques.

The theoretical studies, while presently more advanced than the experiments, are also far from complete. It should be of great value to relax some of the model idealizations, and include consideration of oblique angle scattering, energy shear in the electron beam, the presence of a uniform magnetization, and finite transverse beam geometry. Most important of all, the study of axially nonuniform systems (e.g., tapering the periodicity of the magnetic wiggler) may indicate efficiencies greatly enhanced over those calculated above for uniform systems.

Appendix

We indicate briefly here the details of the derivation of Eqs. (24) from Eq. (20), within the context of the distribution function [Eq. (23)]. Substituting the expression for $\tilde{g}^{(1)}$ given in Eq. (16) with

$$g^{(0)}(\alpha, \beta, u) = n_0 \, \delta(\alpha) \, \delta(\beta) \, g_0(u)$$

into Eq. (20) and integrating by parts gives

$$
\begin{pmatrix} \tilde{J}_{\pm}(z) \\ \tilde{J}_z(z) \end{pmatrix} = \frac{|e| n_0}{m_0} \int_0^{\infty} \frac{u}{\gamma(u)} \left\{ \left(\frac{\partial}{\partial \alpha} + i \frac{\partial}{\partial \beta} \right) \left[\begin{pmatrix} p_{\pm}(\alpha, \beta, z) \\ p_z(\alpha, \beta, u, z) \end{pmatrix} \frac{\tilde{G}_{+}(\alpha, \beta, u, z)}{p_z(\alpha, \beta, u, z)} \right] \right.
$$
$$
+ \left(\frac{\partial}{\partial \alpha} - i \frac{\partial}{\partial \beta} \right) \left[\begin{pmatrix} p_{\pm}(\alpha, \beta, z) \\ p_z(\alpha, \beta, u, z) \end{pmatrix} \frac{\tilde{G}_{-}(\alpha, \beta, u, z)}{p_z(\alpha, \beta, u, z)} \right]
$$
$$
\left. - \begin{pmatrix} p_{\pm}(\alpha, \beta, z) \\ p_z(\alpha, \beta, u, z) \end{pmatrix} \frac{\tilde{G}_z(\alpha, \beta, u, z)}{p_z(\alpha, \beta, u, z)} \frac{\partial}{\partial u} \right\} \Bigg|_{\alpha = \beta = 0} g_0(u) \, du. \tag{A1}
$$

In order to evaluate the current densities in Eq. (A1) the following relations are needed:

$$p_{\pm}(\alpha, \beta, z) = p_x \mp i p_y = \alpha \mp i\beta - (|e| B_0/c k_0) e^{\mp i k_0 z},$$

$$p_z(\alpha, \beta, u, z) = \{u^2 - [\alpha + (|e|/c) A_{0x}(z)]^2 - [\beta + (|e|/c) A_{0y}(z)]^2\}^{1/2},$$

$$u_{0\perp} = (|e| B_0/c k_0) = m_0 \Omega_0/k_0, \qquad \Omega_0 = |e| B_0/m_0 c,$$

$$p_{\pm}(\alpha, \beta, u, z)|_{\alpha = \beta = 0} = -u_{0\perp} e^{\mp i k_0 z},$$

$$p_z(\alpha, \beta, u, z)|_{\alpha = \beta = 0} = \sqrt{u^2 - u_{0\perp}^2} = u_z(u) = \gamma(u) m_0 v_z(u),$$

$$\left(\frac{\partial}{\partial \alpha} + i \frac{\partial}{\partial \beta} \right) p_{\pm}(\alpha, \beta, z)|_{\alpha = \beta = 0} = \begin{pmatrix} 2 \\ 0 \end{pmatrix},$$

$$\left(\frac{\partial}{\partial \alpha} - i \frac{\partial}{\partial \beta}\right) p_{\pm}(\alpha, \beta, z)|_{\alpha = \beta = 0} = \binom{0}{2},$$

$$\left(\frac{\partial}{\partial \alpha} \pm i \frac{\partial}{\partial \beta}\right) p_z(\alpha, \beta, u, z)|_{\alpha = \beta = 0} = \frac{u_{0\perp}}{u_z(u)} e^{\pm ik_0 z},$$

$$\left(\frac{\partial}{\partial \alpha} \pm i \frac{\partial}{\partial \beta}\right) \tau(\alpha, \beta, u, z', z)|_{\alpha = \beta = 0} = \pm i \frac{\Omega_0/\gamma(u)}{u_z(u) k_0^2 v_z^2(u)} (e^{\pm ik_0 z'} - e^{\pm ik_0 z}),$$

$$\left(\frac{\partial}{\partial \alpha} \pm i \frac{\partial}{\partial \beta}\right) M(\alpha, \beta, u, z', z)|_{\alpha = \beta = 0}$$

$$= \pm \frac{\Omega_0/\gamma(u)}{k_0^2 v_z^2(u)} \frac{\exp\{-i[\omega/v_z(u)](z' - z)\}}{v_z(u) u_z(u)}$$

$$\times \{[\omega \mp v_z(u) k_0] e^{\pm ik_0 z'} - \omega e^{\pm ik_0 z}\},$$

$$\left(\frac{\partial}{\partial \alpha} \pm i \frac{\partial}{\partial \beta}\right) \frac{p_{\pm}(\alpha, \beta, z)}{p_z(\alpha, \beta, u, z)}\bigg|_{\alpha = \beta = 0} = \frac{1}{u_z(u)} \{2 + [u_{0\perp}/u_z(u)]^2\},$$

$$\left(\frac{\partial}{\partial \alpha} \pm i \frac{\partial}{\partial \beta}\right) \frac{p_{\mp}(\alpha, \beta, z)}{p_z(\alpha, \beta, u, z)}\bigg|_{\alpha = \beta = 0} = \frac{u_{0\perp}^2}{u_z^3(u)} e^{\pm 2ik_0 z},$$

$$\left(\frac{\partial}{\partial \alpha} \pm i \frac{\partial}{\partial \beta}\right) \tilde{G}_{\pm}(\alpha, \beta, u, z)|_{\alpha = \beta = 0}$$

$$= \mp \frac{|e|\omega}{2c} \frac{\Omega_0/\gamma(u)}{k_0^2 v_z^2(u)} \frac{(e^{\pm ik_0 z} - 1)}{u_z(u)} \tilde{A}_{\pm}(0) e^{i\omega z/v_z(u)},$$

$$\tilde{G}_z(\alpha, \beta, u, z)|_{\alpha = \beta = 0} = \frac{-\gamma(u) m_0 |e|}{2u v_z(u)} e^{i\omega z/v_z(u)}$$

$$\times \int_0^z e^{[-i\omega z'/v_z(u)]} \left\{ v_z(u) \frac{\partial \tilde{\phi}(z')}{\partial z'} + i \frac{\omega}{ck_0} \frac{\Omega_0}{\gamma(u)} \right.$$

$$\times \left. [\tilde{A}_+(z') e^{ik_0 z'} + \tilde{A}_-(z') e^{-ik_0 z'}] \right\} dz'.$$

With the above relations it becomes straightforward to derive Eqs. (24).

Note added in proof: J. M. J. Madey has communicated to the authors that the fractional energy extracted from the electron beam in the Stanford oscillator experiment has reached 0.14%, with an instantaneous peak power of 40 kW (letter dated 1/24/79); see p. 316.

ACKNOWLEDGMENTS

The authors wish to express their appreciation to R. K. Parker and I. B. Bernstein for useful discussions. Also, R. K. Parker provided Figs. 7 and 15.

REFERENCES

Bernstein, I. B., and Hirshfield, J. L. (1978). *Phys. Rev. Lett.* **40**, 761–764.
Buzzi, J. M., *et al.* (1977). *J. Phys. Lett.* **38**, L397–L399.
Coleman, P. D. (1961). *In* "Advances in Quantum Electronics." Columbia Univ. Press, New York.
Colson, W. B. (1976). *Phys. Lett.* **59A.**
Colson, W. B. (1977). *Phys. Quantum Electron.* **5**, 152–196.
Deacon, D. A. G., Elias, L. R., Madey, J. M. J., Ramian, G. J., Schwettman, H. A., and Smith, T. I. (1977). *Phys. Rev. Lett.* **38**, 892–894.
Drake, J. F., Kaw, P. F., Lee, Y. C., Schmidt, G., Liu, C. S., and Rosenbluth, M. N. (1974). *Phys. Fluids* **17**, 778.
DuBois, D. F., and Goldman, M. V. (1965). *Phys. Rev. Lett.* **14**, 544.
DuBois, D. F., and Goldman, M. V. (1967). *Phys. Rev.* **164**, 207.
Efthimion, P. C., and Schlesinger, S. P. (1977). *Phys. Rev. A* **16**, 633–639.
Elias, L. R., Fairbank, W. M., Madey, J. M. J., Schwettman, H. A., and Smith, T. I. (1976). *Phys. Rev. Lett.* **36**, 717–720.
Forslund, D., Kindel, J., and Lindman, E. (1973). *Phys. Rev. Lett.* **30**, 739.
Gilgenbach, R. M. (1978), Ph.D. Thesis, Columbia Univ.
Gilgenbach, R. M., Marshall, T. C., and Schlesinger, S. P. (1978). *Phys. Fluids* (to be published).
Gover, A., and Yariv, A. (1978). *Appl. Phys.* (in press).
Granatstein, V. L., and Sprangle, P. (1977) *IEEE Trans. Microwave Theory Tech.* **MTT-25**, 545–550.
Granatstein, V. L. Herndon, M., Parker, R. K., and Schlesinger, S. P. (1974). *IEEE Trans. Microwave Theory Tech.* **MTT-22**, 1000.
Granatstein, V. L., *et al.* (1976). *Phys. Rev. A* **14**, 1194–1201.
Granatstein, V. L., Schlesinger, S. P., Herndon, M., Parker, R. K., and Pasour, J. A. (1977). *Appl. Phys. Lett.* **30**, 384–386.
Hasegawa, A. (1978). Submitted for publication.
Hasegawa, A., Mima, K., Sprangle, P., Szu, H. H., and Granatstein, V. L. (1976). *Appl. Phys. Lett.* **29**, 542–544.
Hirshfield, J. L., and Granatstein, V. L. (1977). *IEEE Trans. Microwave Theory Tech.* **MTT-25**, 522–527.
Hopf, F. A., Meystre, P., Scully, M. O., and Louisell, W. H. (1976a). *Opt. Commun.* **18**, 413–416.
Hopf, F. A., Meystre, P., Scully, M. O., and Louisell, W. H. (1976b). *Phys. Rev. Lett.* **37**, 1342–1345.
Kapitza, P. L., and Dirac, P. A. M. (1933). *Proc. Cambridge Philos. Soc.* **29**, 297–300.
Kaw, P. K., and Dawson, J. M. (1971). *Phys. Fluids* **14**, 792.
Kroll, N. A., and McMullin, W. A. (1978). *Phys. Rev. A* **17**, 300–308.
Kwan, T., Dawson, J. M., and Lin, A. T. (1977). *Phys. Fluids* **20**, 581–588.
Lampe, M., Ott, E., and Walker, J. H. (1978). *Phys. Fluids* **21**, 42.
Lin, A. T., and Dawson, J. M. (1975). *Phys. Fluid* **18**, 201–206.
Madey, J. M. J. (1971). *J. Appl. Phys.* **42**, 1906.

Madey, J. M. J., Schwettman, H. A., and Fairbank, W. M. (1973). *IEEE Trans. Nucl. Sci.* **20**, 980.

Manheimer, W. M., and Ott, E. (1974). *Phys. Fluids* **17**, 1413.

Marshall, T. C., Talmadge, S., and Efthimion, P. (1977). *Appl. Phys. Lett.* **31**, 320–322.

McDermott, D. B., Marshall, T. C., Schlesinger, S. P. Parker, R. K., and Granatstein, V. L. (1978). *Phys. Rev. Lett.* **41**, 1368.

Miroshnichenko, V. I. (1975). *Sov. Tech. Phys. Lett.* **1**, 453–454.

Motz, H. (1951). *J. Appl. Phys.* **22**, 527.

Nishikawa, K. (1968). *J. Phys. Soc.* **24**, 916.

Pantell, R. H., Soncini, G., and Puthoff, H. E. (1968). *IEEE J. Quantum Electron.* **4**, 905–907.

Parker, R. K., and Ury, M. (1975). *IEEE Trans. Nucl. Sci.* **NS22**, 983–988.

Pasour, J. A., Granatstein, V. L., and Parker, R. K. (1977). *Phys. Rev. A* **16**, 2441–2446.

Perkins, F. W., and Kaw, P. K. (1971). *J. Geophys. Res.* **76**, 282.

Schneider, S., and Spitzer, R. (1974). *Nature (London)* **250**, 643.

Silin, V. P. (1965). *Sov. Phys.-JETP* **21**, 1127.

Silin, V. P. (1967). *Sov. Phys.-JETP* **24**, 1242.

Sprangle, P., and Drobot, A. T. (1978). NRL Memo. Rep. 3587, *J. Appl. Phys.* (to be published).

Sprangle, P., and Granatstein, V. L. (1974). *Appl. Phys. Lett.* **25**, 377–379.

Sprangle, P., and Granatstein, V. L. (1978). *Phys. Rev. A* **17**, 1792–1793.

Sprangle, P. Granatstein, V. L., and Baker, L. (1975). *Phys. Rev.* **A12**, 1697–1701.

Sukhatme, V. P., and Wolff, P. W. (1973). *J. Appl. Phys.* **44**, 2331–2334.

Walsh, J. E., Marshall, T. C., and Schlesinger, S. P. (1977). *Phys. Fluids* **20**, 709.

INDEX